Have you heard...

...what the city's
planning?

SPEED LIMIT 25
NO
ANY TIME
2
8-5 P
M-F

They want to widen Crestview to four lanes!!!
Proposed
PHPD Proj. No. 74-137
9/6/09
325xf
275xf
Proposed Street Enhancement
Mature Vegetation Reduction

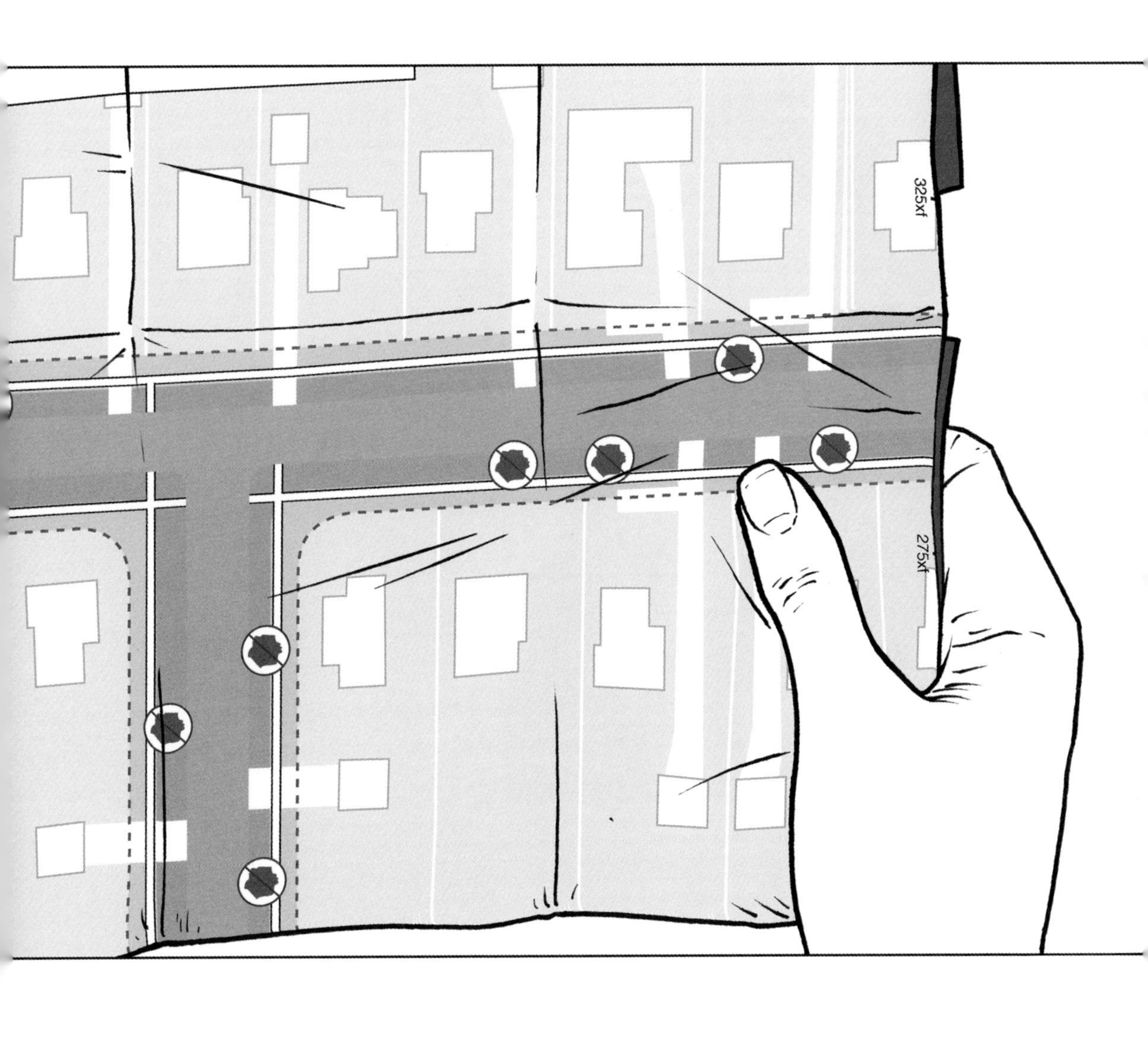
325xf
275xf

Susan, calm down. It's just a proposal.
No one's ever going to do that.
275xf

Jaki, what are you talking about!? It's on this map I downloaded from the city's website. It goes **right through my yard!!**

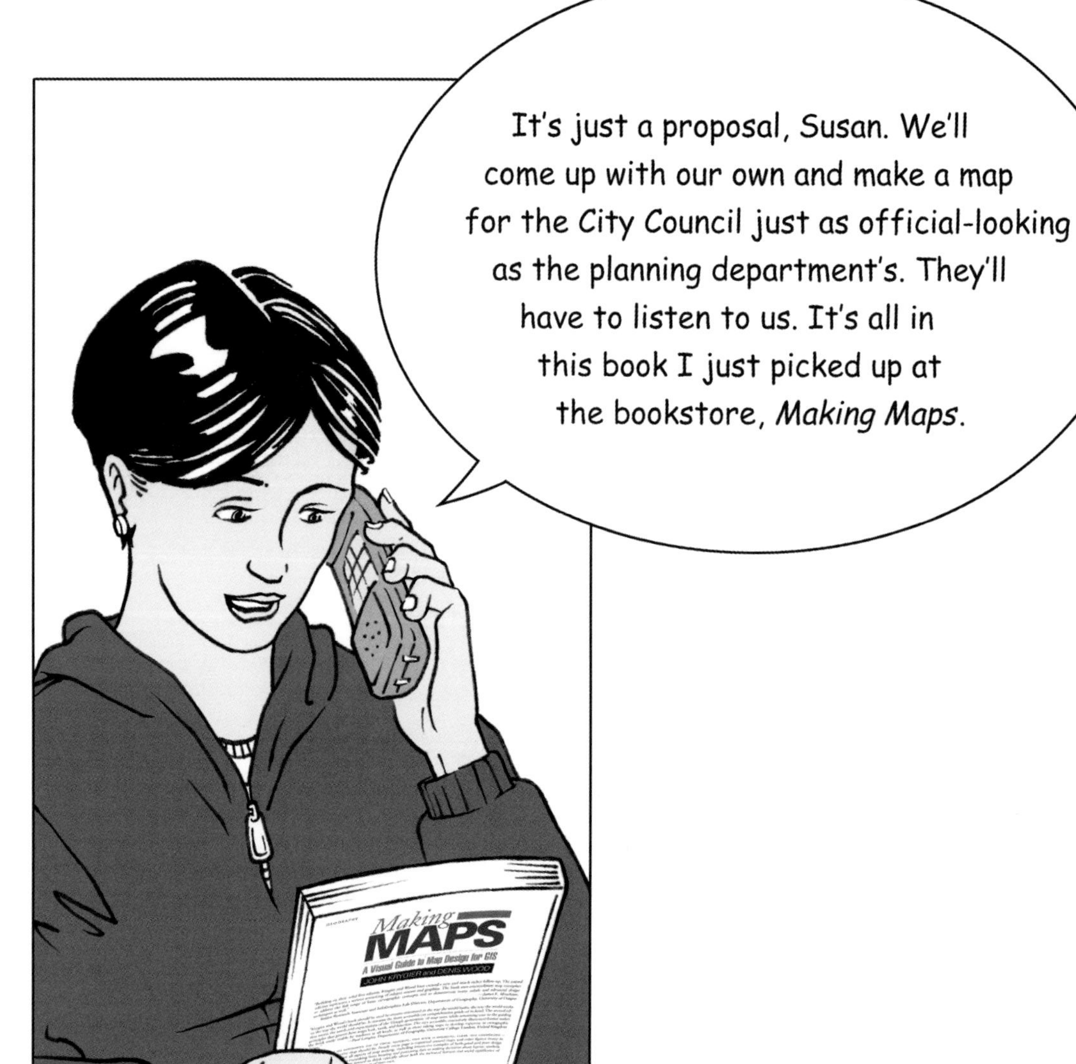
It's just a proposal, Susan. We'll come up with our own and make a map for the City Council just as official-looking as the planning department's. They'll have to listen to us. It's all in this book I just picked up at the bookstore, *Making Maps*.
Making
MAPS
A Visual Guide to Map Design for GIS
JOHN KRYGIER and DENIS WOOD

Fourth Edition

Making MAPS

A Visual Guide to Map Design for GIS

JOHN KRYGIER and DENIS WOOD

THE GUILFORD PRESS
New York London

A Division of Guilford Publications, Inc.
www.guilford.com

Printed in the United States of America.

This book is printed on acid-free paper.

Last digit is print number:

9 8 7 6 5 4 3 2 1

Library of Congress Cataloging-in-Publication Data is available from the publisher

ISBN 978-1-4625-5606-9

It's Time to Make Maps...

People communicate about their places with maps. Less common than talk or writing, maps are made when called for by social circumstances. Jaki and Susan are making maps to stop a project that seems bad for their city. Why a map? Because the city used a map. The map unambiguously expresses the city's intentions to widen Crestview Road, drawing from the maps, talk, and text of city planners. If the plan is realized, the city will also use maps to communicate its intentions to surveyors, engineers, contractors, utility companies, and others.

The maps are all of Crestview Road—all of the same place—and the maps are all different. Yet they are all equally good. Different goals call for different maps: the quality of a map is frequently a matter of perspective rather than design. Think of a map as a *kind of statement locating facts*. People will select the facts that make their case. That's what the map is for:

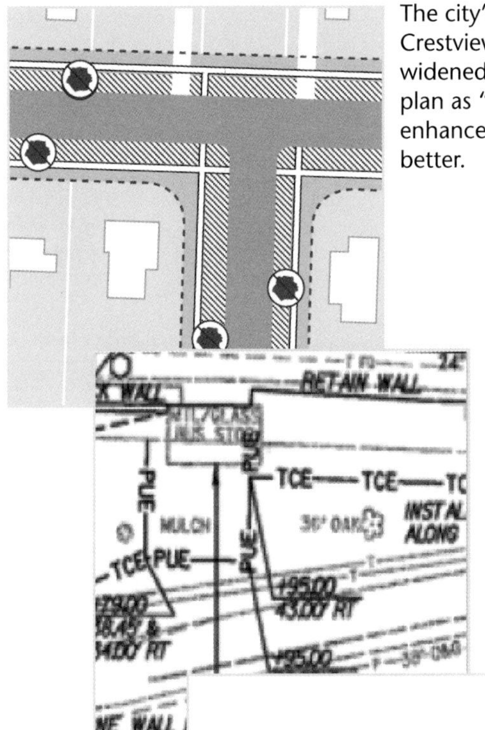

The city's case is that Crestview Road needs to be widened. They present their plan as "a new vision," an enhancement, different and better.

The city communicates to construction firms and utilities with detailed maps, making the case that the planners and engineers have done their work.

Jaki and Susan's case is that widening Crestview Road would be a terrible mistake. Time to make a map!

Making maps, making your case...

Different Goals Call for Different Maps

Jaki and Susan soon realize the plan to widen Crestview is but a piece of a larger plan to redevelop the northern and western suburbs of the city. The key feature of the plan is a connector (in solid black, below) proposed to link two major roads. The effort brings together different stakeholder groups who create equally effective maps to articulate their different perspectives on the proposed road. Though the maps may seem polemical, isolating the facts each presents is useful in focusing debate.

Goal: keeping costs low. A city map shows that its plan is the shortest and least costly route for the connector. The city's map focuses on moving traffic at the least cost to taxpayers.

Crestview Rd.
Proposed Connector

Property Values high med. low

Goal: defending neighborhood integrity. An African American community map shows how the connector rubs salt in the wound sustained by the earlier imposition of the arterial highway. The focus of their map is the further destruction of their neighborhood by the proposed connector.

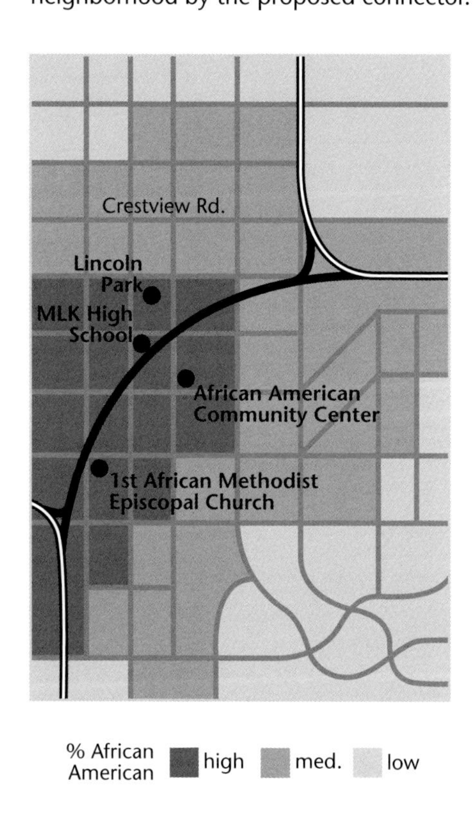

Goal: maintaining historic continuity.
The Society for Historic Preservation's map shows how the connector will affect significant properties in an existing historic district. Their map focuses on the adverse effect on significant properties and on the integrity of the historic district.

Crestview Rd.
Olmsted's Lincoln Park
Oldest Home in City
Historic City Hall
Historic "Shotgun" Houses
Oberlin Historic District

% Historic Buildings high med. low

Goal: protecting endangered wetlands.
An environmental group shows that the connector will violate the city's policy of avoiding road construction in floodplains. The Oberlin Creek watershed, already greatly impacted by over 100 years of urban growth, cannot withstand a further onslaught of development.

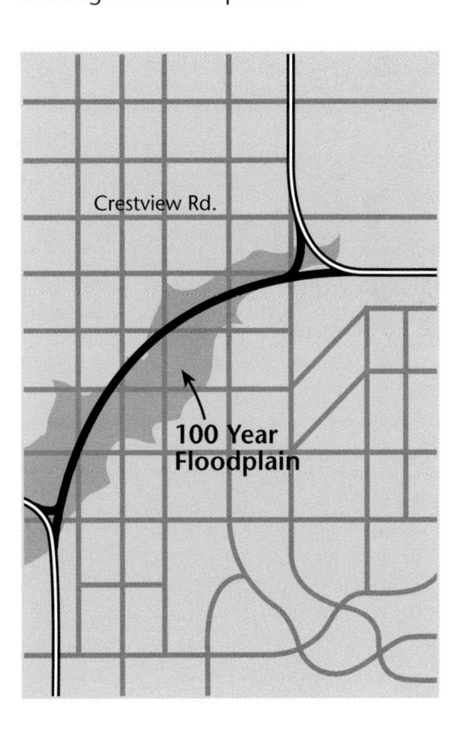

Goal: defending their street. Jaki and Susan's first map scales roads to show existing traffic counts. It suggests how much more effective it would be to widen Armitage Avenue, a street already tied into the downtown grid. Their aim is to divert attention from Crestview Road.

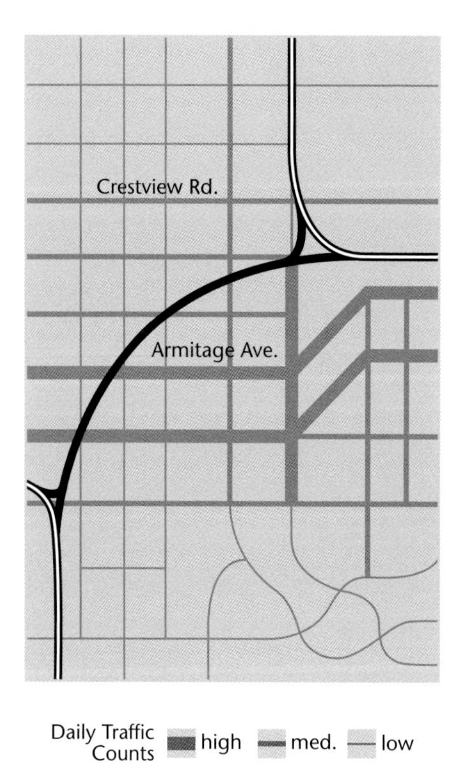

Daily Traffic Counts — high — med. — low

Goal: defeating the connector. Aware of the connector's key role in motivating the widening of Crestview, and informed by the maps produced by other groups, Jaki and Susan realize it's less that Crestview needs defending and more that the connector needs defeating: low property values correlate with historic discrimination against African Americans, with older housing, and the floodplain. The connector exploits this nexus: their new map focuses on social and environmental justice.
Jaki and Susan work out a "Social and Environmental Justice Sensitivity" metric, taking into account race, history, and environmental factors.

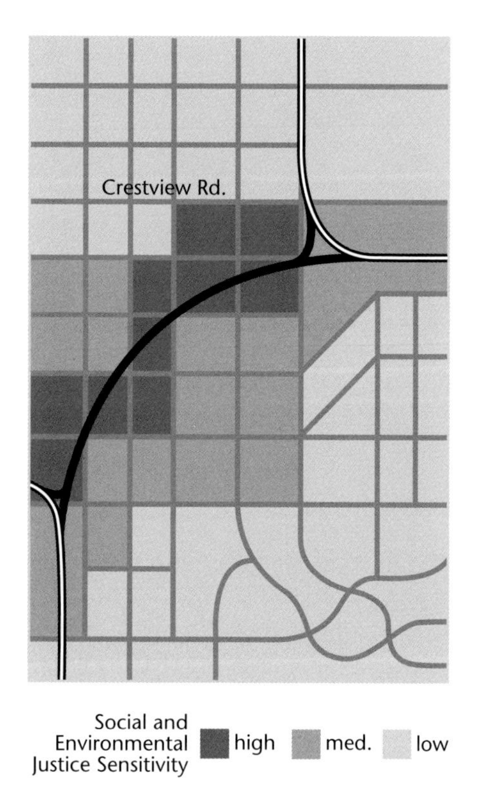

Social and Environmental Justice Sensitivity — high — med. — low

Goal: defeating the connector. When they moved from defending Crestview to defeating the connector, Jaki and Susan realized they'd shifted their attention from their neighborhood to the larger community. At first this alarmed them – maybe they were overreaching – but once they realized they could find allies in the African American community, among the historic preservationists, and in the environmental group concerned with the floodplain, they got excited.

All the stakeholder groups concerned with the city's plan got together at the African American Community Center. Jaki and Susan's Social and Environmental Justice map helped everyone see they shared a common problem. Everyone's map making had made a difference!

Contents

In December 1986 an experimental aircraft named Voyager became the first piloted aircraft to circle the earth without refueling.

	DAY 9		DAY 8		DAY 7			DAY 6		DAY 5		
Hours Aloft	216 hours	200	192 hours	184	176	168 hours	160	152	144 hours	136	128	120 hours	112	104	96 hours

Fuel on landing: 18 gallons

100° W
60° W
20° W
0°
20° E
60° E
40° N
20° N
0°
20° S
40° S
120° W
80° W
40° W
0°
40° E

United States
Atlantic Ocean
Pacific Ocean
Atlantic Ocean
Nicaragua
Costa Rica
Cameroon
Gabon
Congo
Zaire
Uganda
Kenya
Tanzania
Ethiopia
Somalia
Squall line

WNW
NNW 20
ENE 18
ESE 14
NW 10-15
E 37
E 34
E 20
E 10-20
W

Triumphant landing at Edwards AFB

Engine stalled; unable to restart for five harrowing minutes

Transition from tailwinds to headwinds

Rutan disabled by exhaustion

Oil warning light goes on

Thunderstorm forces Voyager into 90° bank

Passing between two mountains, Rutan and Yeager weep with relief at having survived Africa's storms

Flying among 'the redwoods': life and death struggle to avoid towering thunderstorms

Worried about flying through restricted airspace, Rutan and Yeager mistake the morning star for a hostile aircraft

Discovery of backwards fuel flow

Coolant seal leak

Visibility

Altitude (feet)
20,000
15,000
10,000
5,000
sea level

Distance 26,678 miles traveled
5,000 miles to go
10,000 miles to go
12,532 miles previous record

Flight data courtesy of Len Snellman and Larry Burch, Voyager meteorologists
Mapped by David DiBiase and John Krygier, Department of Geography, University of Wisconsin-Madison, 1987

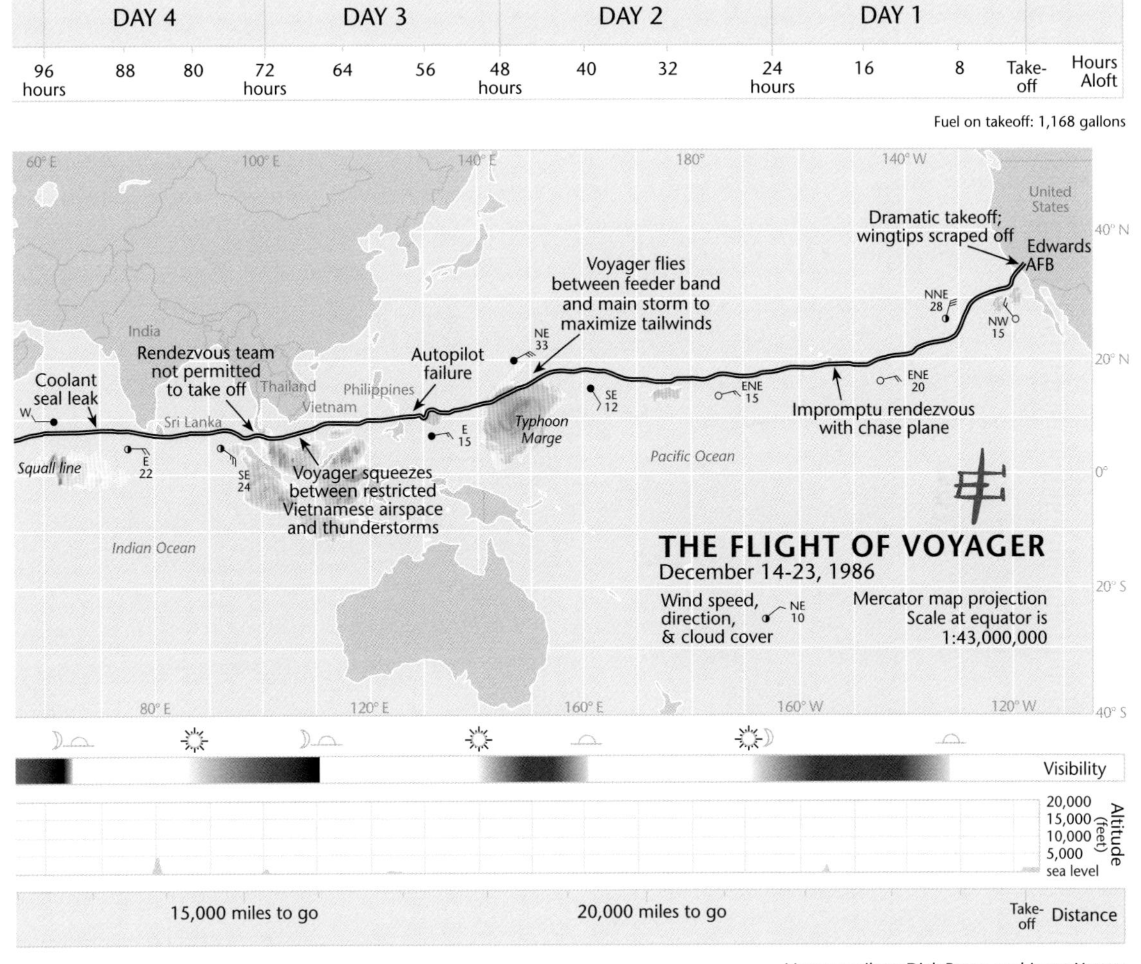

Voyager pilots: Dick Rutan and Jeana Yeager
Voyager designer: Burt Rutan

What do you need to know to make this map?

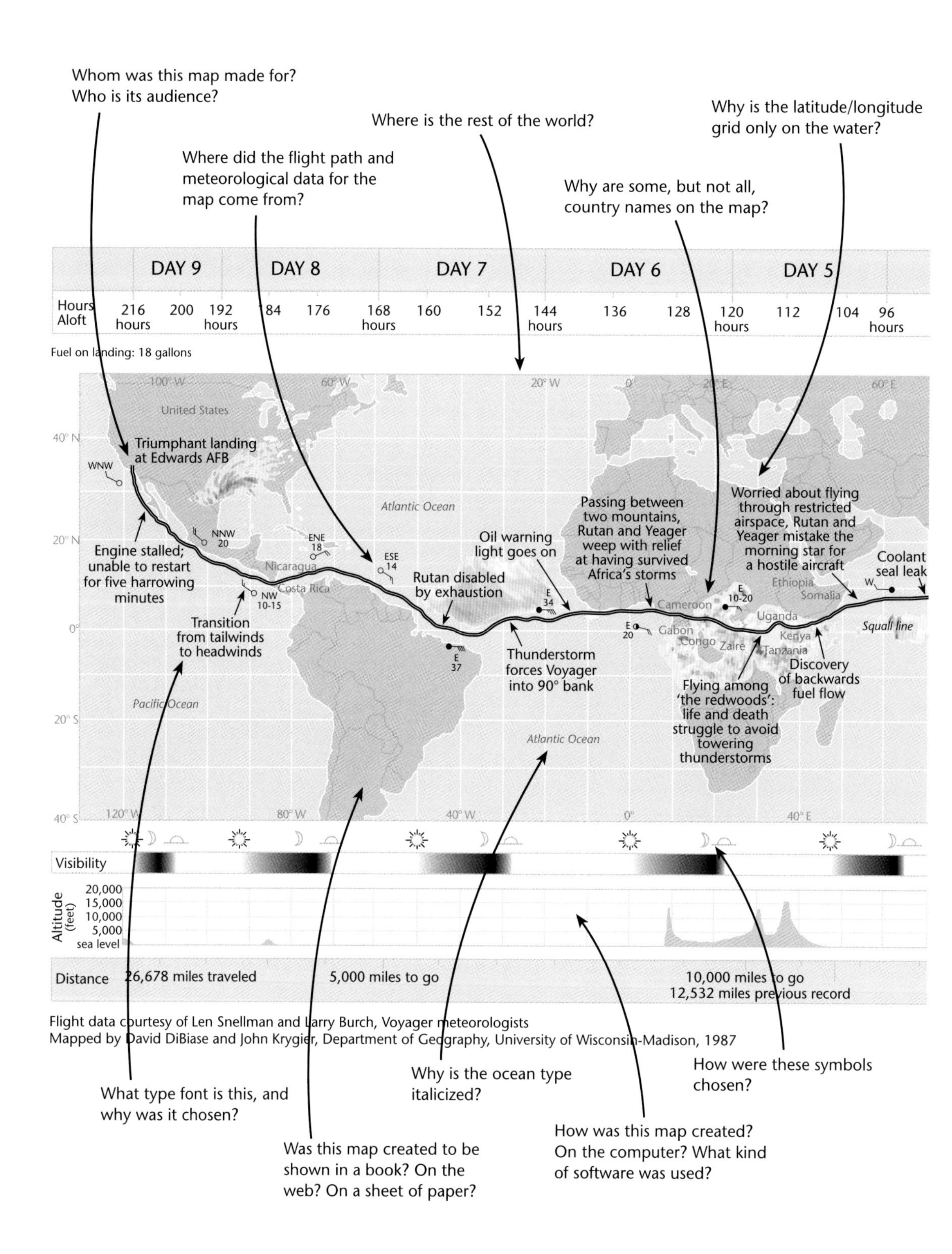

Whom was this map made for?
Who is its audience?
Where is the rest of the world?
Why is the latitude/longitude grid only on the water?
Where did the flight path and meteorological data for the map come from?
Why are some, but not all, country names on the map?
DAY 9
DAY 8
DAY 7
DAY 6
DAY 5
Hours Aloft
216 hours
200
192 hours
184
176
168 hours
160
152
144 hours
136
128
120 hours
112
104
96 hours
Fuel on landing: 18 gallons
100° W
60° W
20° W
0°
20° E
60° E
40° N
20° N
0°
20° S
40° S
120° W
80° W
40° W
0°
40° E
United States
Triumphant landing at Edwards AFB
WNW
Atlantic Ocean
NNW 20
ENE 18
ESE 14
Engine stalled; unable to restart for five harrowing minutes
Nicaragua
Costa Rica
NW 10-15
Transition from tailwinds to headwinds
Oil warning light goes on
Rutan disabled by exhaustion
E 34
E 37
Thunderstorm forces Voyager into 90° bank
Passing between two mountains, Rutan and Yeager weep with relief at having survived Africa's storms
Worried about flying through restricted airspace, Rutan and Yeager mistake the morning star for a hostile aircraft
Coolant seal leak
Ethiopia
Somalia
W
Cameroon
E 10-20
E 20
Gabon
Congo
Zaire
Uganda
Kenya
Tanzania
Squall line
Discovery of backwards fuel flow
Flying among 'the redwoods': life and death struggle to avoid towering thunderstorms
Pacific Ocean
Atlantic Ocean
Visibility
Altitude (feet)
20,000
15,000
10,000
5,000
sea level
Distance
26,678 miles traveled
5,000 miles to go
10,000 miles to go
12,532 miles previous record
Flight data courtesy of Len Snellman and Larry Burch, Voyager meteorologists
Mapped by David DiBiase and John Krygier, Department of Geography, University of Wisconsin-Madison, 1987
What type font is this, and why was it chosen?
Why is the ocean type italicized?
How were these symbols chosen?
Was this map created to be shown in a book? On the web? On a sheet of paper?
How was this map created? On the computer? What kind of software was used?

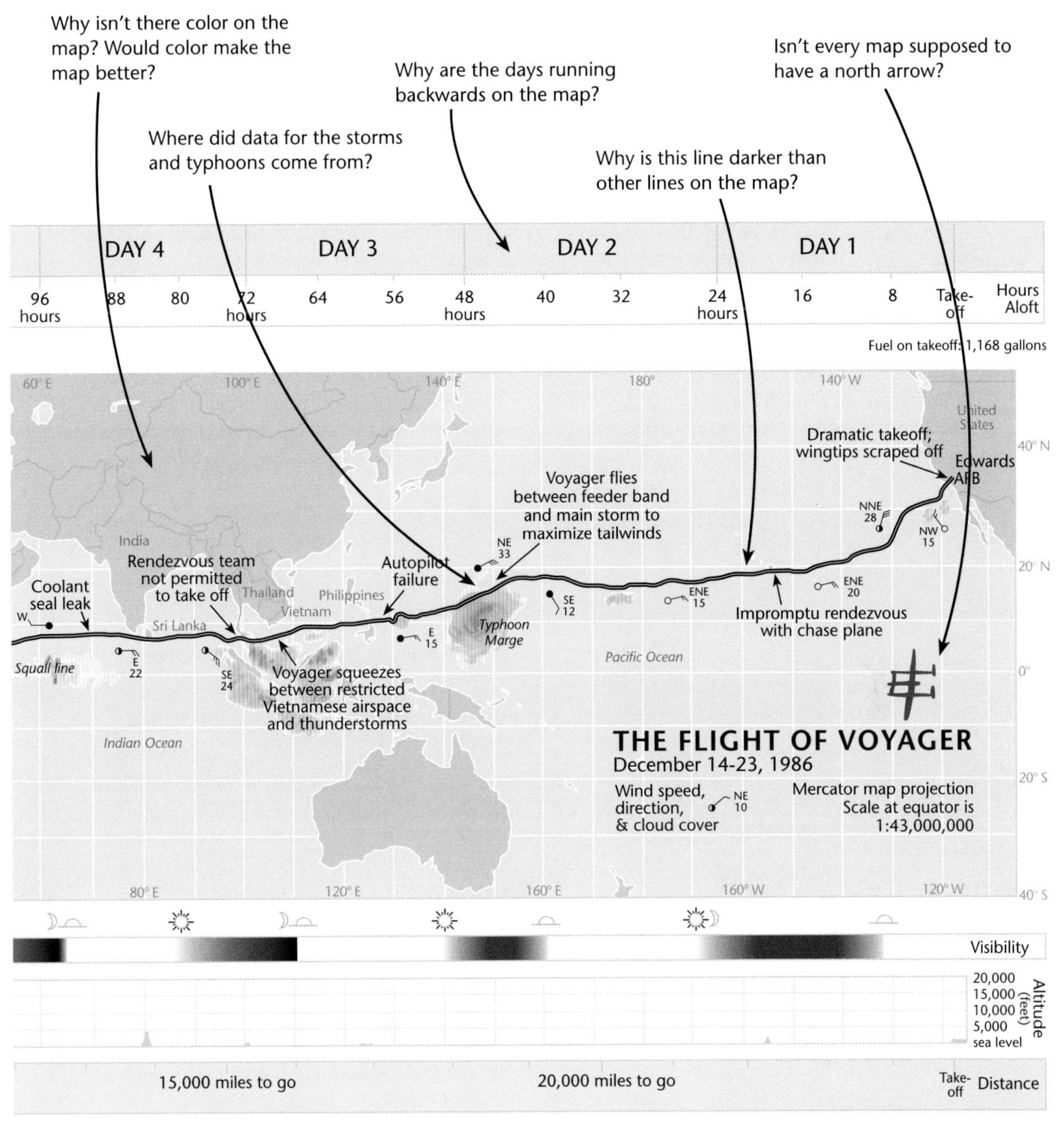

CHAPTER 1 How to Make a Map

Start by looking; what do you see? Looking at maps is easy. Not really. You can glance at the Mona Lisa in a second. But to *get* the Mona Lisa you have to look more carefully. What do you see on the Voyager map? Words, lines, continents, a grid. A story, some information with the story. What do you notice first? Black lines, gray lines, white lines ... why are they different? Making maps requires that you answer such questions, and many more. Throughout this book, in nearly every chapter, we annotate *The Flight of Voyager*. By the end of the book, you will understand how to really see – and make – a map.

Making Maps Is Hard

The bewildering array of considerations to be taken into account when making maps, shown on these two pages, should be overwhelming – at least initially. All shall be explained in subsequent chapters, in general and in relation to a series of annotated *Flight of Voyager* maps. A systematic critique of an existing map or the successful making of your own map is accomplished by considering the following issues. When making maps, think about everything before starting; then, when your map is complete, reconsider them all once again.

The Whole Map

Write out exactly what the map is supposed to accomplish: does the map meet its goals?

Are you sure a map is necessary?

Is the map suitable for the intended audience? Will the audience be confused, bored, interested, or informed?

Look at the map in its final medium: does it work?

If interactive, does your map respond to the user's choices in an intuitive way?

Does your map effectively tell a story?

Is the map, its authors, its data, and any other relevant information documented and accessible to the map reader?

Look at the map and assess what you see; is it:

confusing or clear
interesting or boring
lopsided or balanced
amorphous or structured
light or dark
neat or sloppy
fragmented or coherent
constrained or lavish
crude or elegant
random or ordered
modern or traditional
hard or soft
crowded or empty
bold or timid
tentative or finished
free or bounded
subtle or blatant
flexible or rigid
high or low contrast
authoritative or unauthoritative
complex or simple
appropriate or inappropriate

Have you evaluated your map, informally or formally?

The Map's Data

Do the data serve the goals of the map?

Is the relationship between the data and the phenomena they are based on clear?

Does the map symbolization reflect the character of the phenomena or the character of the data?

Does the origin of the data – primary, secondary, tertiary – have any implications?

Are the data too generalized or too complex, given the map's goals?

Is the map maker's interpretation of the data sound?

Are qualitative and quantitative characteristics of the data effectively symbolized?

Have the data been properly derived?

Has aggregation or districting affected the data?

Has the temporal character of the data been properly understood and symbolized?

Is the scale of the map (and inset) adequate, given the goals of the map?

What about the accuracy of the data? Are the facts complete? Are things where they should be? Does detail vary? When were the data collected? Are they from a trustworthy source?

Have you consulted metadata (data about data)?

Does the map maker document copyright issues related to the data?

Is the map copyright or copyleft licensed?

The Map's Framework

What are the characteristics of the map's projection, and is it appropriate for the data and map goals? What is distorted?

Is the coordinate system appropriate and noted on the map?

The Design of the Map

Does the title indicate what, when, and where?

Is the scale of the map appropriate for the data and the map goals? Is the scale indicated?

Does textual explanation or discussion on the map enhance its effectiveness?

Does the legend include symbols that are not self-explanatory?

If the orientation of the map is not obvious, is a directional indicator included?

Are authorship and date of map indicated?

Are inset and locator maps appropriate?

Is the goal of the map promoted by its visual arrangement, engaging path, visual center, balance, symmetry, sight-lines, and the grid?

Has the map been thoroughly edited?

Does the map contain non-data ink?

Has detail been added to clarify?

Do the data merit a map?

Do variations in design reflect variations in data?

Is the context of the map and its data clear?

Are there additional variables of data that would clarify the goals of the map?

Do visual differences on the map reflect data differences?

Do color choice and variation reflect data choice and variation on the map?

Is color necessary for the map to be successful? Does color add anything besides decoration?

Do color choices grab viewer's attention while being appropriate for your data?

Does the map's design reflect the conditions under which it will be viewed?

Are color interactions and perceptual differences among your audience accounted for?

Have symbolic and cultural color conventions been taken into account and used to enhance the goals of the map?

Do important data stand out as figure, and the less important as ground, on the map? Are there consequences of data not included on the map?

Have visual difference, detail, edges, texture, layering, shape and size, closure, proximity, simplicity, direction, familiarity, and color been used to reflect figure-ground relationships appropriate to the map's goals?

Are the level of generalization and the data classification appropriate, given the map's goals?

Do map symbols work by resemblance, relationship, convention, difference, standardization, or unconvention? Are the choices optimal for the map's goals?

How do the map symbols relate to the concepts they stand for? Is the relationship meaningful?

Have the map symbols been chosen to reflect the guidelines suggested by the visual variables?

If symbolizing data aggregated in areas, is the most appropriate method used? How will the choice affect the interpretation of the map?

What do the words on your map mean? How do they shape the meaning of the map?

Has the chosen typeface (font) and its size, weight, and form effectively shaped the overall impression of the map as well as symbolize variations in the data?

Does the arrangement of type on the map clarify, as much as possible, the data and the goals of the map?

Thinking about Making Maps

Making maps means thinking (and talking!) theoretically and philosophically about why you are making maps, their impacts, and how maps can make the world better. Catherine D'Ignazio and Lauren F. Klein draw from feminist theory for a list of principles to guide map making and data visualization. Annita Hetoevéhotohke'e Lucchesi developed principles for data sovereignty and mapping which are sensitive to Indigenous culture and science. All these principles can be thought of and applied in any map making effort.

Principles of Feminist Data Visualization

Rethink Binaries: Challenge traditional binary thinking and dualistic perspectives in data visualization. Move away from dichotomies and consider more nuanced and inclusive representations that recognize the complexity and diversity of experiences. Instead of representing data using a binary color scale (e.g., red and blue), consider a diverging color scale that incorporates multiple colors to represent a spectrum of values.

Embrace Pluralism: Encourage the inclusion of multiple voices, perspectives, and ways of knowing in the data visualization process. Recognize and embrace diversity, avoiding homogenizing narratives that may marginalize certain groups. Use a map that visualizes linguistic diversity, showcasing different languages spoken in a region, rather than focusing on the dominant language.

Examine Power and Aspire to Empowerment: Critically examine power dynamics within data collection, analysis, and visualization processes. Aim to empower marginalized communities and individuals by involving them in the decision-making processes related to data representation. Seek out and map detailed data on environmental toxins and other environmental "bads" and relate them to social and economic factors.

Consider Context: Situate data within its broader social, cultural, and historical context. Understand that data are not neutral and interpretation is influenced by context. This principle emphasizes the importance of contextual understanding for meaningful and responsible data visualization. Map the impact of historic redlining practices and interstate highway construction on current social patterns and segregation.

Legitimize Embodiment and Affect: Acknowledge the embodied experiences and emotional aspects inherent in data. Recognize that data points represent real people with feelings, experiences, and lives. Incorporate aspects of embodiment and affect into the visualization to humanize the data. Map stories from individuals affected by the overturn of *Roe v. Wade.*

Make Labor Visible: Highlight and recognize the labor involved in the data collection, analysis, and visualization processes. Acknowledge the efforts of those who contribute to the creation of data and visualizations, particularly considering the often invisible labor of marginalized groups. Include all contributors to a community mapping project, and work diligently to provide some kind of compensation for their efforts.

Principles of Indigenous Data Sovereignty and Mapping

Above all else, the protocols of the specific community you intend to map should be respected, followed, and deferred to.

Do not assume the community you are intending to map does not already have trained cartographers capable of doing the work you intend to do.

Do not solicit Indigenous people to participate in a mapping project they did not ask for.

Do not travel to an Indigenous community for mapping-based research if you are not willing to connect with the land solely in a manner that the community feels is appropriate.

Do not travel to an Indigenous community for mapping-based research without previously extensively researching the history and ongoing legacies of violent colonization as it affects that community.

Understand that as Indigenous peoples, some of our most sacred and sensitive information is the knowledge and stories we carry about our lands and significant places.

Be willing to acknowledge that open source mapping platforms could compromise the community's data sovereignty, inform them of that in the planning process, and be willing to explore alternatives and think creatively.

Understand that you may be asked to create maps that are not for public distribution, that may never be allowed to be published, because they hold sensitive information – view those moments as gifted experiences of trust, not roadblocks to publications.

Understand that the community has a fundamental right to own the data on their lands and people.

Do not assume that GIS or other Western styles of mapping are useful or desirable for a project in collaboration with an Indigenous community.

Create opportunities to help in building the capacity of the community to continue to create their own maps moving forward.

Do not expect an academic publication out of any collaboration with an Indigenous community.

This list comprises recommendations for actions that are largely at the level of the individual researcher; that said, institutions also have a role to play.

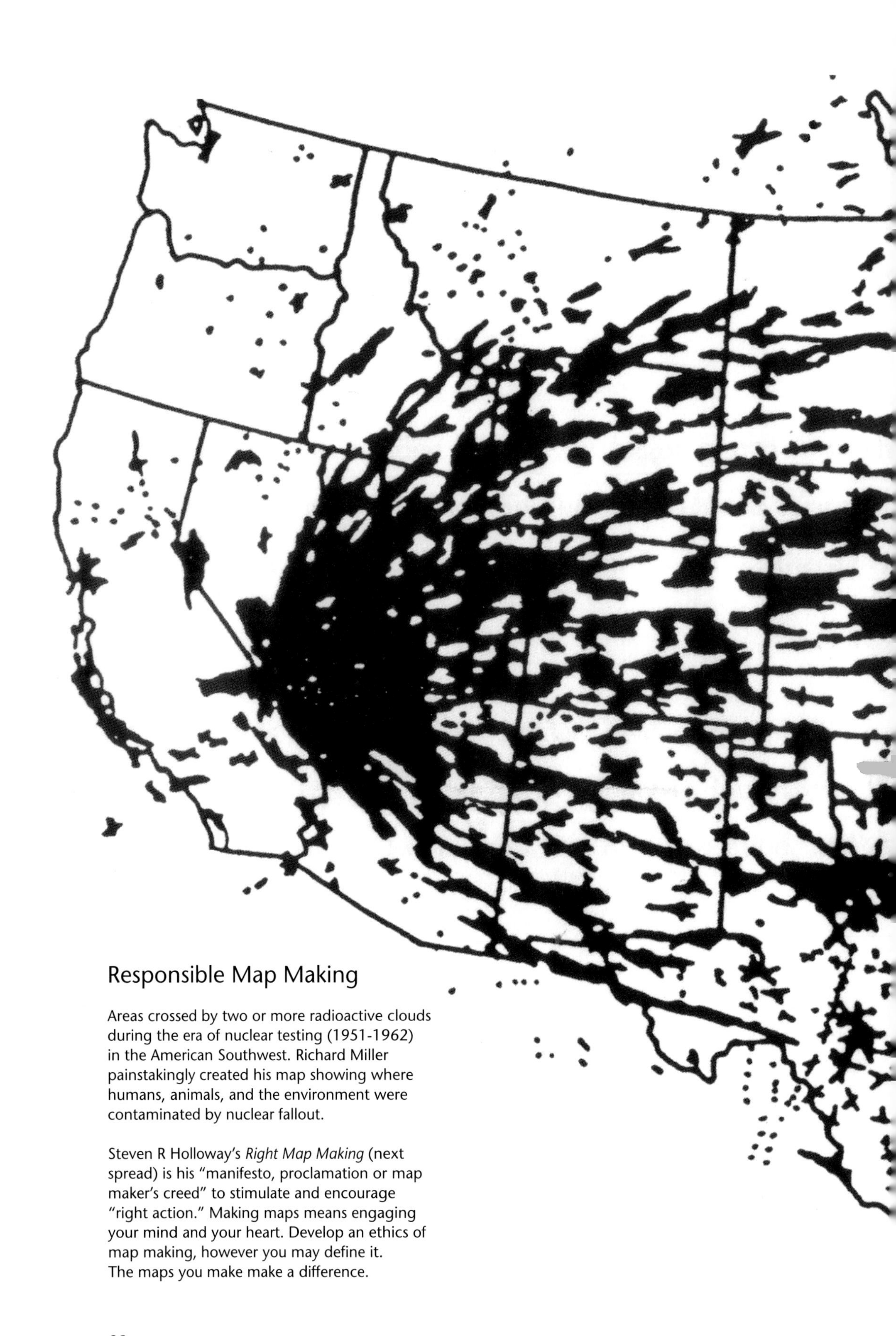

Responsible Map Making

Areas crossed by two or more radioactive clouds during the era of nuclear testing (1951-1962) in the American Southwest. Richard Miller painstakingly created his map showing where humans, animals, and the environment were contaminated by nuclear fallout.

Steven R Holloway's *Right Map Making* (next spread) is his "manifesto, proclamation or map maker's creed" to stimulate and encourage "right action." Making maps means engaging your mind and your heart. Develop an ethics of map making, however you may define it. The maps you make make a difference.

RIGHT MAP Making

"The most obvious characteristic of our age is its destructiveness." TH. MERTON

THE PROBLEM *for the maker of maps being that our maps are, in part, engaged in the active and wanton destruction of the world.*

Thus AWAKENED, *we* VOW *to take right effort & Engage in cartographic disobedience, map making "for a future to be possible."* T. N. HANH *Unacceptable it is not to* ACT.

Five Ways to MAKE MAPS for a Future to be Possible

REVERENCE; *the first precept of right map making*

From the awareness that our maps are, in part, responsible for the great and unnecessary destruction of life taking place in the world today. We vow to map and comment on spatial relationships in a manner non-harming, with reverence and with respect, and to reflect and reveal the beauty of life in a manner non-objectified, where the economic, the non-economic, and the unseen elements are given voice. We vow to recognize and incorporate story with the arguments on our maps. In agreement with M. Gandhi, *"first... non-cooperation with everything humiliating,"* we vow to refrain from economicism, the objectification of sentient beings, and cartographic pornography. Such mapping and maps reflect agreement with the first principle of right action: REVERENCE.

THE PRACTICE OF GENEROSITY; *the second precept*

From the awareness that our maps are, too often, in our self-interest, greedy consumptions of endless desire, human biased and nationalistic. We vow to engage in a mapping of that which desires to be mapped and shared, not taking that into map form that which does not belong to us; desiring to remain unmapped. We vow to be generous to all sentient beings on our maps and in our mapping. Where generosity is also the courage to leave blank on the page that which does not belong to us, not mapping to take what is not ours, and honoring the sancity of the commons. Leviticus: *"fields are not to be reaped to the border."* Such mapping and maps show agreement with the second principle of right action: GENEROSITY.

COMMITMENT TO THE RELATIONSHIP WITH THE PLACE; *the third precept*

From the awareness that our maps are, in part, reflective of a lack of relationship and commitment to the place in which we reside and map. We vow to resist the temptation to map places with which we have no intimate or committed relation. We seek to remember and honor our relationship to the place; mapping with an honesty of lines, colours and shapes, the naming of places, the unnaming as well, without gossip or intent to harm, or to divide, but rather with a clarity of intent to all sentient beings with whom we are committed to with & in the relationship. Such mapping and maps show agreement with the third principle of right action: COMMITMENT TO THE RELATIONSHIP WITH THE PLACE.

DEEP LISTENING THROUGH DIRECT ~ CONTACT & STOPPING; *the fourth precept*

From the awareness that our maps are, in part, a failure to deeply listen and have been made without stopping to directly contact and listen to the place we are mapping. We vow to refrain from mapping what we do not know to be the truth, to first stop to experience the interconnected, ever-changing and interwoven space we are privileged to map. These maps acknowledge the intimate Other, the desire for the awakened heart and mind with & in direct contact with the place itSelf. Such mapping and maps show agreement with the fourth principle of right speech: DEEP LISTENING THROUGH DIRECT-CONTACT AND STOPPING.

ON BELONGING TO ONE BODY; *the fifth precept for a future to be possible*

From the awareness that our maps are, in part, disconnected from the body of the earth. How can this be? Kabir says, *"Whose Body is it anyway?"* We vow to make our maps about the body living; our own body, the body in motion, ever-changing and interconnected, the body free from addiction and enslavement to the toxicity of drugs: ownership, objectification, disconnection, greed, capitalism, all the *isms*. We vow to map that delight in the body that serves to reduce suffering and misery. Maps, and the making of maps that respect all sentient beings; the living breathing air, the changing clouds, and the wind and the tides in motion, the soils, the interwoven rocks, the waterways and the water bodies entwined & circling, mountains rising & falling, compost building. Maps respecting and awakened to belonging to the OneBody without separation. Such mapping and maps show agreement with the fifth principle, oikos as the ecologic, economic and ecumenical whole of right livelihood: BELONGING TO ONE BODY.

My soule, saith he, is but a mappe of shoes. No substance, but a shadow for to please.

Thomas Middleton, *Wisdome of Solomon Paraphrased* (1597)

Oh no. I think we're the map police.

John, to Denis (2024)

For the execution of the voyage to the Indies, I did not make use of intelligence, mathematics or maps.

Christopher Columbus, *Book of Prophecies* (15th century)

It is not on any map; true places never are.

Herman Melville, *Moby Dick* (1851)

I presume you have reference to a map I had in my room with some X's on it. I have no automobile. I have no means of conveyance. I have to walk from where I am going most of the time. I had my applications with the Texas Employment Commission. They furnished me names and addresses of places that had openings like I might fill, and neighborhood people had furnished me information on jobs I might get.... I was seeking a job, and I would put these markings on this map so I could plan my itinerary around with less walking. Each one of these X's represented a place where I went and interviewed for a job.... You can check each one of them out if you want to.... The X on the intersection of Elm and Houston is the location of the Texas School Book Depository. I did go there and interview for a job. In fact, I got the job there. That is all the map amounts to.

Lee Harvey Oswald, Interrogation after the John F. Kennedy assassination (November 24, 1963)

More...

makingmaps.substack.com and makingmaps.net serve as an extension of this book.

Engage your thinking about maps: Matthew Edney's *Cartography: The Ideal and Its History* (2019) excavates the history of maps and cartography. Brian Harley and David Woodward's multi-volume *History of Cartography* (1987-2015) series is a marvel. Matthew Wilson's *New Lines: Critical GIS and the Trouble of the Map* (2017) critically engages digital mapping. Rasmus Grønfeldt Winther's *When Maps Become the World* (2020) tracks map thinking and spatial representations across diverse disciplines. Denis Wood's *Rethinking the Power of Maps* (2010) reinvents his classic *Power of Maps* (1992). Alan MacEachren's *How Maps Work* (2004) is a comprehensive survey of cognitive-semiotic approaches to mapping. Henk von Houtum's *Free the Map* (2024) is an engaging turn on the map. For a broad, graphically masterful survey of information design, track down Juuso Koponen and Jonatan Hildén's *Data Visualization Handbook* (2020).

This book, like all books, draws from numerous other texts, old and new, that can be consulted for more information than you'll ever want or need: J.S. Keates. *Cartographic Design and Production* (1973); R.W. Anson and F.J. Ormeling. *Basic Cartography* (1984); Arthur Robinson, Joel Morrison, Phillip Muehrcke, and A. Jon Kimerling. *Elements of Cartography* (1995); Borden Dent, Jeff Torguson, and Thomas Hodler. *Cartography: Thematic Map Design* (2008); Judith Tyner. *Principles of Map Design* (2010); A. Jon Kimerling, Aileen Buckley, Phillip Muehrcke and Juliana Muehrcke. *Map Use: Reading and Analysis* (2016); Nicolas Lambert and Christine Zanin. *Practical Handbook of Thematic Cartography* (2016); Michael De Smith, Michael Goodchild, and Paul Longley. *Geospatial Analysis: A Comprehensive Guide* (2018); Kenneth Field. *Cartography* (2018); Peter Anthamatten. *How to Make Maps* (2021); Menno-Jan Kraak and F.J. Ormeling. *Cartography: Visualization of Spatial Data* (2021); Patrick McHaffie, Sungsoon Hwang, and Cassie Follett. *GIS: An Introduction to Mapping Technology* (2023); Terry Slocum, Robert McMaster, Fritz Kessler, and Hugh Howard. *Thematic Cartography and Geovisualization* (2023); and Cindy Brewer. *Designing Better Maps* (2024). These nice folks are the "map police."

Scholarly articles and practical insights can be found in *Cartographic Perspectives* and the North American Cartographic Information Society (nacis.org), the journal *Cartographica* and the Canadian Cartographic Association (cca-acc.org), the *Cartographic Journal* and the British Cartographic Society (www.cartography.org.uk), the *Journal of Maps,* and the International Cartographic Association (icaci.org). Also see *Cartography and Geographic Information Systems, The International Journal of Geographical Information Science,* and the *Journal of Geovisualization and Spatial Analysis.*

Jonathan Crow's maproomblog.com and Matthew Edney's mappingasprocess.net do a great job of reporting on a range of intellectual and practical mapping information, including books. Daniel Huffman always is up to something cool at somethingaboutmaps.wordpress.com.

Sources: Catherine D'Ignazio and Lauren F. Klein. "Feminist Data Visualization." *Workshop on Visualization for the Digital Humanities (VIS4DH),* Baltimore. IEEE, 2016. Annita Hetoeve·hotohke'e Lucchesi. "Spatial Data and (De)colonization: Incorporating Indigenous Data Sovereignty Principles into Cartographic Research." *Cartographica* 55:3, 2020, pp. 163–169. Richard Miller, "Areas crossed by two or more radioactive clouds during the era of nuclear testing in the American Southwest, 1951-62" in *Under the Cloud: The Decades of Nuclear Testing* (Two-Sixty Press, 1999). "Right MAP Making" copyright 2007 by Steven R Holloway. Designed and produced by toMake.com Press www.tomake.co). "Right MAP Making" is intended to articulate the fundamental principles of ethical conduct in mapping and maps and to stimulate "right action." Forty letterpress prints were signed and numbered by the author.

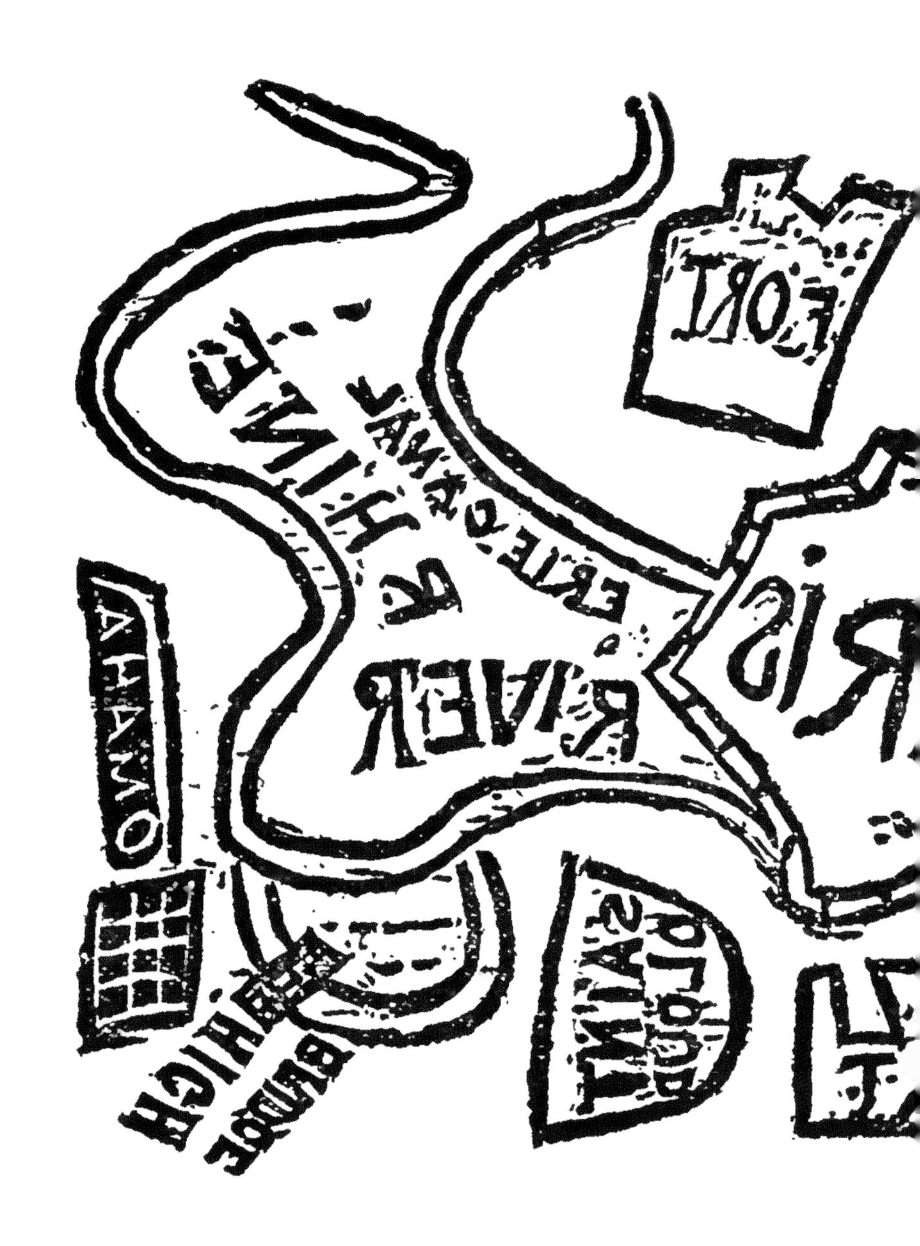
ERIE CANAL
RIVER
OMAHA
HIGH
BRIDGE
SAINT
CLOUD

What's the point?

The point, according to Mark Twain, is...

Inasmuch as this is the first time I ever tried to draft and engrave a map, or attempted anything in any line of art, the commendations the work has received and the admiration it has excited among the people, have been very grateful to my feelings. And it is touching to reflect that by far the most enthusiastic of these praises have come from people who knew nothing at all about art.

By an unimportant oversight I have engraved the map so that it reads wrong end first, except to left-handed people. I forgot that in order to make it right in print, it should be drawn and engraved upside down. However, let the student who desires to contemplate the map stand on his head or hold it before a looking-glass. That will bring it right.

The reader will comprehend at a glance that that piece of river with the "High Bridge" over it got left out to one side by reason of a slip of the graving-tool, which rendered it necessary to change the entire course of the River Rhine, or else spoil the map. After having spent two days in digging and gouging at the map, I would have changed the course of the Atlantic Ocean before I would lose so much work.

I never had so much trouble with anything in my life as I had with this map. I had heaps of little fortifications scattered all around Paris at first, but every now and then my instruments would slip and fetch away whole miles of batteries, and leave the vicinity as clean as if the Prussians had been there.

The reader will find it well to frame this map for future reference, so that it may aid in extending popular intelligence, and in dispelling the wide-spread ignorance of the day.

MARK TWAIN.

OFFICIAL COMMENDATIONS

It is the only map of the kind I ever saw.

U. S. GRANT.

It places the situation in an entirely new light.

BISMARCK.

I cannot look upon it without shedding tears.

BRIGHAM YOUNG.

It is very nice large print.

NAPOLEON.

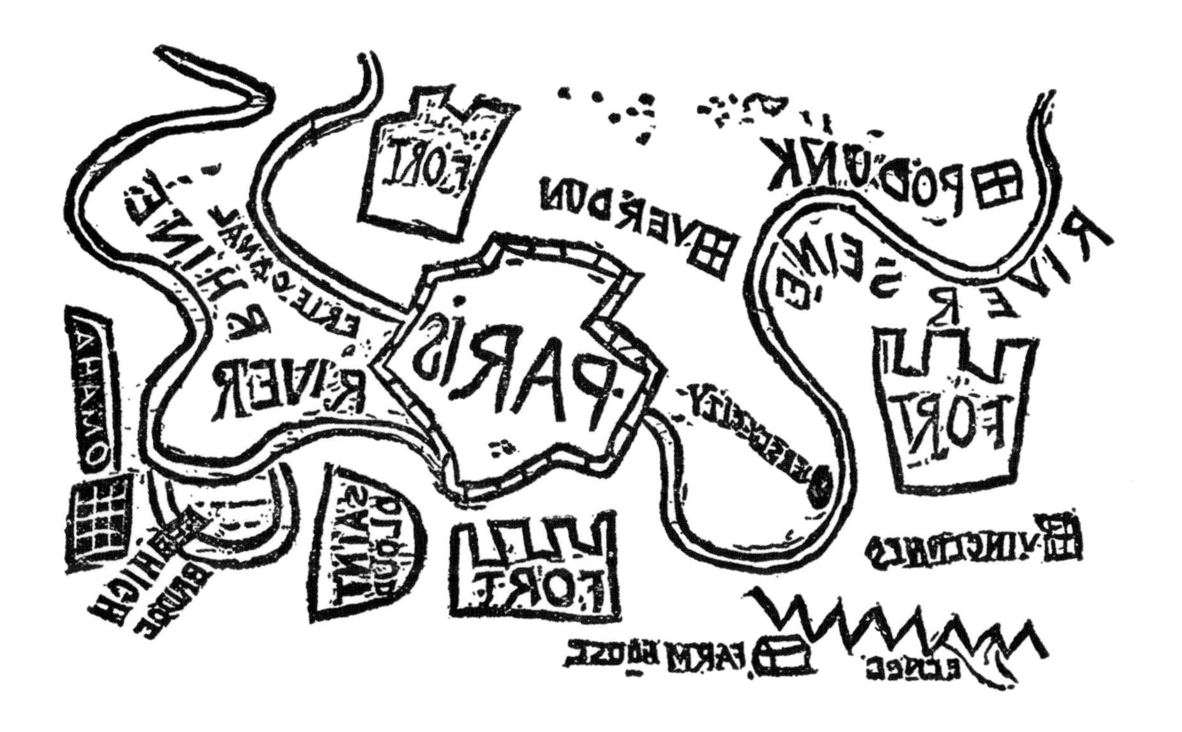

CHAPTER 2 What's Your Map For?

What was Twain's map of Paris for? To make us laugh. But first it was to make Twain laugh. It was a dark time for Twain. "He swung between deep melancholy and half insane tempests and cyclones of humor." In one of the latter moments, "He got a board and with a jackknife carved a 'crude and absurd' map of Paris under siege." The map was a parody of those found in the newspapers of the time and was wildly popular. Who's your map for? How will you show it? How will you document, evaluate, and review it? Your answers will profoundly shape your map.

But Do You Really Need a Map?

The first thing you need to decide is whether you need a map. You may not. There are secrets that don't want to be mapped. There are circumstances where maps are inappropriate. And sometimes there are more effective ways of making your point: a graph, a drawing, a photo.

The Silly

States Where Republicans Accidentally Legalized Edibles

It shouldn't be surprising that sometimes your data don't justify a map.

The Secret

Sometimes it's better not to map things you could easily map. Military sites, sacred indigenous locations, and archaeological sites are often left off maps.

Fitness company Strava released a global map of user routes in 2017. The map drew from Strava devices which collect fitness route data. The U.S. military distributed these devices to encourage fitness. The devices were worn continuously by many users, collecting jogging routes, as well as patrol routes, supply routes, and other sensitive locations. Which, in turn, were mapped by Strava for all to see.

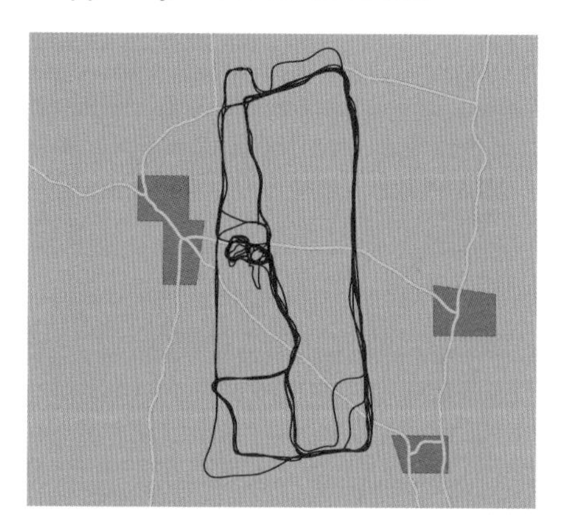

An example, above, is redrawn from the original, one of many circulated on social media. The red lines reveal routes taken by U.S. military personnel and the site of a U.S. military base.

The Inappropriate

It may be less a question of secrecy than of delicacy, privacy, or appropriateness. Not everyone is eager to have their spaces mapped.

In 1740 a surveyor working on Jacques Cassini's map of France was hacked to death in what is today the Haute-Loire by natives who saw no good coming from the work of the surveyors. In 1838 some 16 members of a surveying party in what is today Navarro County, Texas, were killed by Kickapoo, who objected to the presence of surveyors on their hunting grounds. Two years later, while surveying the Australian coast south of Brisbane, Granville Stapylton was killed by a party of Aboriginals. In 1874 a party of surveyors was massacred by Cheyenne in what is today Meade County, Kansas.

The resistance to being surveyed and mapped continues today. In 2010 a surveyor was killed and two others wounded by Maasai trying to stop the subdivision of land in Maelia Naivasha, Kenya. In 2013 a member of a surveying party at Sota, a community near Dodowa in Accra, Ghana, was butchered and three others injured while attempting a survey of disputed land.

In the past decade, mappers being paid by the U.S. Army surveyed fields collectively owned by Zapotec farmers in the highlands outside of Oaxaca, Mexico. When they learned this, the Zapotec demanded the return of the maps.

Sometimes the most effective counter-map is no map at all.

The Impossible

Not everything can be mapped.

We may be able to map the location of a roller coaster, but not what happened to us when we began to drop down that first big hill. We may be able to map where we first saw a lover, but not the way we felt. Not even the land can be mapped when the land is less topography, less dirt than ... everything: sun and moon – spirits! – included.

Land claims by native peoples are usually accompanied by maps. This is so obviously the place for a map that it seems perverse to question it, but increasingly, Indigenous peoples have been arguing that maps can't capture their relationship to the land.

In 1987 the Gitxsan and the Wet'suwet'en in British Columbia entered the Gitxsan adaawk (a collection of sacred oral traditions about their ancestors, histories, and territories) and the Wet'suwet'en kungax (a spiritual song or dance or performance tying them to the land) as evidence in their suit seeking title to their ancestral lands. In 1997 the Canadian Supreme Court found that these forms of evidence had to be accepted by Canadian courts.

Who's Your Map For?

Knowing the intended audience for your map will help you design it. Your audience may be familiar with the area being mapped or not, an expert on the mapped topic or a novice, an 8-year-old or a college student. In each case, consider how your map can function better for the people who will actually use it.

Experts

Experts know a lot about the subject of the map. Experts are highly motivated and very interested in the facts the map presents. They expect more substance and expect to engage a complex map.

Less peripheral information on map explaining content and symbols

More information, more variables of information, more detail

Follow conventions of experts: consider using a spectral (rainbow) color scheme for ordered data if the user is accustomed to using such colors to show ordered data (such schemes are usually not good for other users)

Novices

Novices know less about the map subject and may not be familiar with the way maps are symbolized. They need a map that is more explanatory. Novices may be less motivated than expert users, but they want the map to help them learn something.

More peripheral information on map explaining content and symbols

Less information, fewer variables of information, less detail

Follow map design conventions, which enhance comprehension of the map

Mike...

(1)(1)(2)(2)(3)(3)
(1)(1)(2)(2)(3)(3)
(4)(4)(5)(5)(6)(6)
(4)(4)(5)(5)(6)(6)
(7)(7)(8)(8)(9)(9)(10)(10)(11)(11)
(7)(7)(8)(8)(9)(9)(10)(10)(11)(11)
(12)(13)(14)(15)(16)(17)(17)(18)(18)(19)(19)
(205)(205) (20)(21)(22)(23)(24)(25)(17)(17)(18)(18)(19)(19)
(205)(205)(26)(27)(28)(29)(30)(31)(32)(33)(34)(35)(36)(36)(37)(37)(38)(38)
(204)(204)(39)(40)(41)(42)(43)(44)(45)(46)(47)(58)(36)(36)(37)(37)(38)(38)
(204)(204)(49)(50)(51)(52)(53)(54)(55)(56)(57)(58)(59)(59)(60)(60)(61)(61)(62)(62)
(63)(63)(64)(64)(65)(66)(67)(68)(69)(70)(71)(72)(73)(74)(59)(59)(60)(60)(61)(61)(62)(62)
(63)(63)(64)(64)(75)(76)(77)(78)(79)(80)(81)(82)(83)(84)(85)(85)(86)(86)(87)(87)(88)(88)(89)(89)
(90)(90)(91)(91)(92)(93)(94)(95)(96)(97)(98)(99)(100)(101)(85)(85)(86)(86)(87)(87)(88)(88)(89)(89)
(90)(90)(91)(91)(102)(103)(104)(105)(106)(107)(108)(109)(110)(111)(112)(112)(113)(113)(114)(114)(115)(115)
(116)(116)(117)(117)(118)(119)(120)(121)(122)(123)(124)(125)(126)(127)(112)(112)(113)(113)(114)(114)(115)(115)
(116)(116)(117)(117)(128)(129)(130)(131)(132)(133)(134)(135)(136)(137)(138)(138)(139)(139)(140)(140)
(141)(141)(142)(142)(143)(144)(145)(146)(147)(148)(149)(150)(151)(152)(138)(138)(139)(139)(140)(140)
(141)(141)(142)(142)(153)(154)(155)(156)(157)(158)(159)(160)(161)(162)(163)(163)
(164)(164)(165)(165)(166)(167)(168)(169)(170)(171)(172)(173)(174)(175)(163)(163)
(164)(164)(165)(165)(176)(177)(178)(179)(180)(181)(182)(183)(184)(185)
(186)(186)(187)(187)(188)(188)(189)(189)(190)(190)(191)(191)
(186)(186)(187)(187)(188)(188)(189)(189)(190)(190)(191)(191)
(192)(192)(193)(193)(194)(194)(195)(195)(196)(196)
(192)(192)(193)(193)(194)(194)(195)(195)(196)(196)
(197)(197)(198)(198)(199)(199)(200)(200)
(197)(197)(198)(198)(199)(199)(200)(200)
(201)(201)(202)(202)(203)(203)
(201)(201)(202)(202)(203)(203)

Social worker Mike Rakouskas's map of Wake County, North Carolina. The numbers refer to pages in the county street atlas he uses, and the shaded numbers are client sites. He uses this map to rationalize his trip planning and as an index to the atlas. It was made with a word processor. Peculiar! Clever! And perfect for Mike.

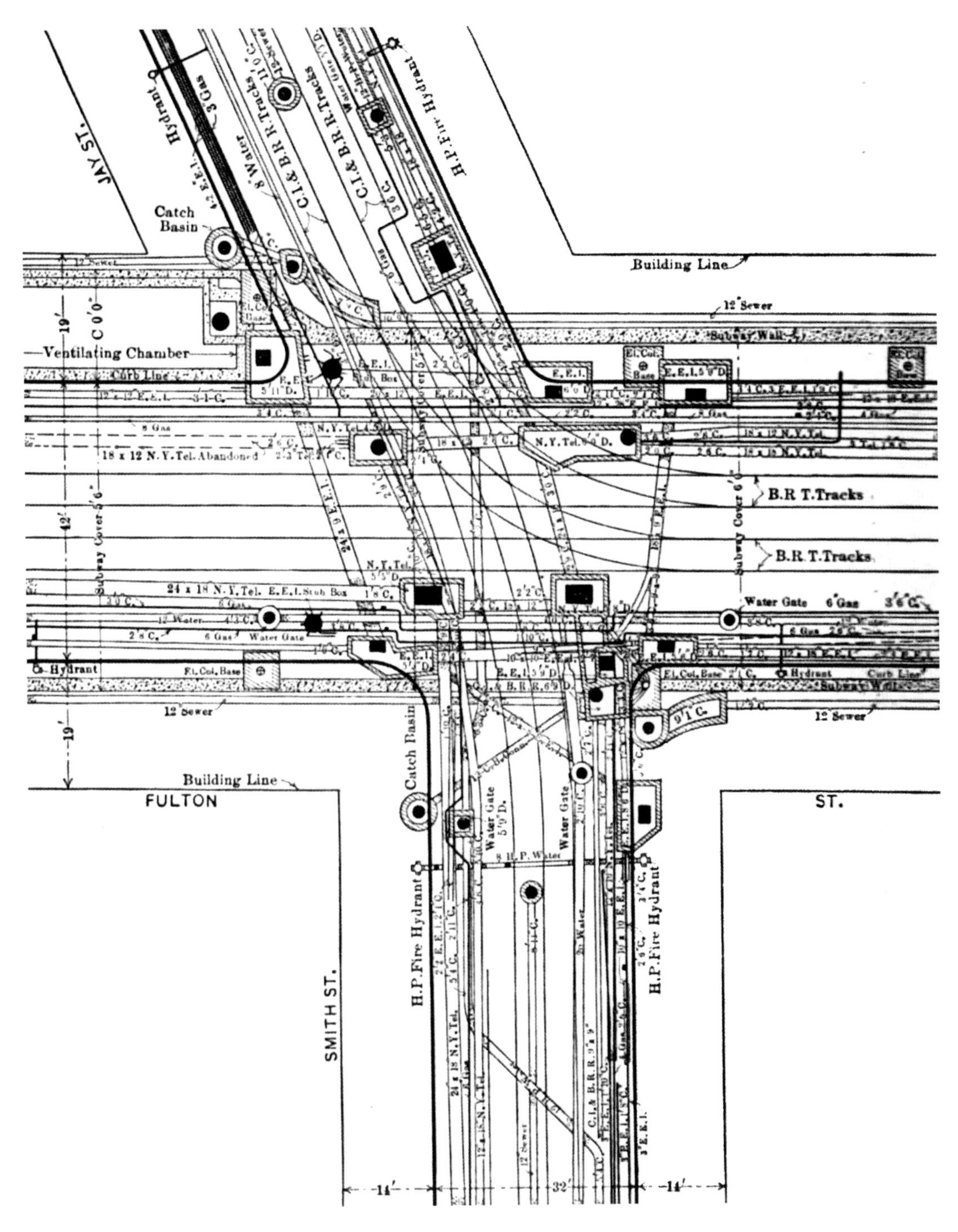

Specialized maps created by the Division of Substructures in Brooklyn, New York, around 1912 contain diverse symbols and content that make sense to few except subsurface engineers. Contemporary utility mapping builds on this tradition of specialized maps for very specialized users.

How Are You Going to Show It?

Consider the final medium of your map before making it. Most maps are made on computer monitors, but the monitor is not the final medium. Rather, it might be a cell phone screen, a piece of paper, a poster, a slide projected on a screen during a presentation, a yard sign, handbill, or protest sign. What looks great on your computer may not look so great when printed or projected or shown on a tiny phone screen.

A yard sign complete with map in opposition to a road that would really mess up a lot of stuff in Raleigh, North Carolina.

Black and White, on Paper

Most maps are created on computer monitors, with less resolution and area than is possible on a piece of paper. When paper is your final medium, design for the paper and not for the monitor. Always check design decisions by printing the map (or having your printer create a proof if your map is to be professionally printed). While all computers offer color, final printing with color is not always an option. Don't despair! Much can be done with black and white.

Map size should match final paper size, with appropriate margins

10-point type works well on a printed map, but you may have to zoom in to see it on the computer monitor

Point and line symbols can be smaller and finer on a printed map than on the computer

More subtle patterns can be used than on a computer monitor map

More data and more complex data can be included on a printed map

Substitute a range of grays and black and white for color. Remember that printers cannot always display as many grays as you can create on a monitor; subtle variations in grays may not print clearly

Black will be more intense than white; use white to designate no information or the background, dark to designate more important information

Monochrome copiers sometimes reproduce gray tones poorly

Very light gray tones may not print

Color, on Paper

Color on a computer monitor is created in a different manner than color on desktop printers or on professionally printed maps. Select colors on the computer, then print and evaluate (or ask for a proof). Always design for the final medium: adjust the colors on the monitor so they look best for the final output. The same colors will vary from printer to printer. Reproducing color is often more expensive than black and white. Finally, keep in mind that users may reproduce your color map in black and white. Will it still work?

Map size should match final paper size, with appropriate margins

10-point type works well on a printed map, but you may have to zoom to see it on the computer monitor

Point and line symbols can be smaller and finer on a printed map

More subtle patterns can be used than on a computer monitor map

More data and more complex data can be included on a printed map

Use color value (e.g., light red vs. dark red) to show differences in amount or importance. Use color hue (blue vs. red) to show differences in kind. Desktop printers cannot display as many colors as you can create on a monitor; subtle variations in colors may not print

Dark colors are more intense than light; use light colors to designate less important information and background and dark to designate more important information

Never print a color map in black and white; redesign it for black and white iPad

Amazon Fire

Computer Monitors

...another iPad

iPhone

Consider screen resolution and space limits when designing maps for final display on a computer. Desk or laptop computer monitor resolution is typically 72 dots per inch (dpi), compared to 1200 or more for printers. Computer monitors have limited area, typically 7 by 9 inches (lightest gray area on this spread) – less if the map is displayed in a web browser. All type and symbols should be visible without magnification. Avoid designs requiring the viewer to scroll to see the entire map. Use more than one map if you need more detail, or consider web tools that allow you to zoom and pan over a map.

The entire map should fit on the screen without scrolling (if pan/zoom is not possible)
Increase type size: 14-point type is the smallest you should use on a monitor
Make point and line symbols 15% larger than those on a paper map
Use more distinct patterns: avoid pattern variations that are too fine or detailed
You may have to limit the amount and complexity of data on your map, compared to a print map
Use color: but remember that some monitors cannot display billions of colors; subtle color variations may not be visible on every monitor
White is more intense than black. Take care when using white to designate the lack of information or as background color as it may stand out too much
Save static maps for the internet at 72-150 dpi. Size the map to fit in a browser window
Design your map so it works on different monitors (RGB, LCD, portables)
Interactive maps require attention to additional issues such as pan, zoom, interactivity, etc.

Samsung Galaxy Tablet

Lenovo Tablet

Samsung Galaxy Phone

Kindle

iLight

Portable Screens

Maps on smart phones, GPS units, and other portable devices pose the same design challenges as desktop monitors, with the further limitation of screen size. Typical portable monitor sizes are shown on this spread. Many portable monitors are touch-sensitive, allowing users to pan and zoom, thus overcoming some of the limitations of the small monitor size. Portable devices usually don't have cursors (or mice) and interaction is based on touch

Static maps on portable devices can follow desktop monitor design guidelines, taking into account the limited display size

Vary map design specifications with changing map scale

Generalize more as the user zooms out on the map: for example, local roads and road names disappear when zoomed out

Generalize less as the user zooms in on the map: local roads and their names appear when zoomed in

Aerial photographs may be more appropriate than maps for users with limited navigation abilities

Maps may be more appropriate than aerial photographs for users with better navigation abilities

Ground-view images may be more helpful for navigation than maps alone, but using both should increase navigation success

Map symbols should not be too complex

Colors should be more intense to account for varying lighting conditions

Serif fonts may be easier to read on portable monitors than sans serif

Interacting on Screens

According to guidelines from Axis Maps, an interactive map is digital, responds to user manipulation, and is changeable in some way. Such maps include unique map design considerations.

Map navigation: change where the map is focused. Conventional design includes:

Click and drag (or touch and drag)
Double click to zoom
Scroll to zoom
Pinch to zoom
Arrow keys to pan
Plus and minus keys to zoom

Search and filter: finding specific objects or sets of objects—where, when and what. Conventional design for *search* includes:

Set expectations of what is searchable
Provide quick feedback
Organize results in useful ways

Filtering has advantages:

Show fundamental ways data can be organized
View groups of related things, not just a specific attribute
Better explore data using numeric values
Find things that are not easily expressible as keywords

Conventional design for *filtering* includes:

Design maps and charts as filters
Make sure users know filters are active
Fewer filters for story-telling maps, more filters for exploratory maps

Information retrieval: getting more information than what the map shows. Conventional design includes:

A *floating tooltip* (popup, data probe) appears adjacent to mapped features when a cursor is over a feature

Use for maps on devices with a cursor
Use for maps with small amount of information
Include essential information only
Include click for more information

And/or a *fixed panel* which has devoted space on the screen and is activated by user interaction

Use for devices with a cursor when maps have a larger amount of information
Use for mobile devices, where expectation is to tap for information

Data manipulation: changing what the map shows. Conventional design includes:

Add or remove data layers
Change data sets
Change geographic areas
Change map symbolization, such as data classification

Visual Storytelling

Maps can be vital components in telling stories. Robert Roth delineates common visual storytelling themes that can be used in visual storytelling with maps:

Designed: A visual story has an intended purpose, with the designer making intentional decisions about composition to develop the broader lesson or moral. The goal of visual storytelling is not just to show but also to explain.

Partial: Visual stories prioritize essential information needed to follow the narrative, often emphasizing a small set of key characters, places, and events while ignoring others. Visual stories privilege brevity over completeness.

Intuitive: Visual stories are partial abstractions making them easier to understand due to reduced information complexity and the manner in which they reflect human experience.

Compelling: Visual stories offer a deep, contextualized account versus a superficial, sanitized overview. Visual stories capture our attention through a vivid array of graphics, images, and other multimedia, that, when combined with maps, develop a rich sense of place.

Relatable: Visual stories promote empathy; the audience puts themselves into the story setting, assumes the roles of the characters, and draws from personal experience to add context. Visual storytelling humanizes maps.

Memorable: Visual stories tie seemingly unrelated information together in a memorable way through logical continuity. Visual stories do not just capture memories, they also shape them.

Situated: Visual stories present meaning from a grounded perspective from somewhere and someone in contrast to an objective view from nowhere. Stories that include the designer's voice in the telling can be more compelling and relatable, open to multiple interpretations, and inspire empowerment.

Persuasive: Visual stories argue from their situated positions and invite counter stories. Persuasive visual stories mobilize their audience to act in a specific way.

Political: Visual stories exercise power by promoting particular voices and interpretations while obscuring others. Visual storytelling stakes new claims, empowering people and communities to reassert control over their surroundings.

Fluid: The meanings of visual stories are not static, and change across cultures and through time. Visual stories are living, with interactive technologies enabling continuous curation of visual stories and story maps as they evolve.

Projections

For presentations, maps may be shown on a large screen with a computer projector. When projected, white and lighter colors will be more intense, black and darker colors subdued. Computer projectors vary in the amount of light they can project. Some projectors wash out colors. Consider previewing your projected map and adjusting the projector. Projected maps must be designed with the viewing distance in mind (find out the size of the room). A map projected to an audience in a small room can have smaller type and symbols than a map projected in an auditorium. Always check that the map is legible from the back of the room in which the map will be displayed.

Greater map size is offset by the increased viewing distance

Increase type size so that smallest type is legible from the back of the room

Increase point and line symbol size to be legible from the back of the room

More distinct patterns: avoid pattern variations that are too fine or detailed

You may have to limit the amount and complexity of data on your map, compared to print maps

Older or lower-output projectors may wash out colors, so intensify your colors for projection

If your map will be projected in a dark room, use black as background, darker colors for less important information, and lighter colors for more important information

If your map will be projected in a well-lighted room, use white as background, lighter colors for less important information, and darker colors for more important information

Posters

Posters are similar to projected maps, although usually viewed in well-lighted conditions. Viewers should be able to see key components of the map (such as the title) from afar, then walk up to the map and get more detail. Design the poster, then, so information can be seen both close and at a distance. The size of poster maps is limited by the largest printer you can use; always check color and resolution of the printer used to reproduce your poster. You may want to request a test print of the colors you plan to use to evaluate your color choices.

Design the map title and mapped area so they are legible from across the room

The majority of type, point, and line symbols should be slightly larger than on a typical printed map, but not as large as on a monitor or projected map. Design this part of the map so it is legible from arm's length

More complex information can be included on a poster map than on a computer monitor or projected map

Follow color conventions for color printed maps

Most posters are viewed in a well-lighted room, so use white as background, lighter colors for less important information, and darker colors for more important information

Viewing Distance

The relationship between viewing distance and the size of graphics and text on projections or posters can be approached by using a rough rule of thumb.

First, determine what you expect your viewers to *do* with your projection or poster.

Analytical Decision Making: Viewers will analyze details and make critical decisions; full engagement with content. For example, a projected map showing a correlation between drinking water contaminated by perfluorooctanoic acid (C8) and cancer in southern Ohio in a lawsuit against DuPont, the chemical's manufacturer.

Basic Decision Making: Viewers will assess details and make basic decisions but are not dependent on seeing all details; active engagement with content. For example, a poster at a student research symposium showing the relationship between urban heat islands and park space in Chicago, Illinois.

Passive Viewing: Viewers recognize content (graphics and text) but need not analyze or make decisions based on the display; passive engagement with content. For example, a map on a slide showing the location of a study area in Mexico City focused on gentrification and protest.

Second, how *close* should your viewers be to your projection or poster, given what they need to do?

Analytical Decision Making: Your viewers should be within a distance no more than *4 times* the diagonal size of the screen or poster.

Basic Decision Making: Your viewers should be within a distance no more than *6 times* the diagonal size of the screen or poster.

Passive Viewing: Your viewers should be within a distance no more than *8 times* the diagonal size of the screen or poster.

If these rough criteria are not met, you can increase both text and graphic size to compensate (using suggestions on the previous page) or attempt to rearrange your viewers.

Keep in mind that many other factors come into play: the age (and eyesight) of your viewers, the resolution of the projector, room lighting, etc. Always try to evaluate your projection or poster ahead of time in actual viewing conditions.

Document, Evaluate, Review

Constantly cast a critical eye on your work. Document what you do and continually evaluate whether the map is serving its intended goal, meeting the needs of its intended audience, and working well in its final medium.

Documentation

What were those six great shades of red I used on that map I made last month? What font did I use on the last poster map? How long did it take me to make that map for the annual report last year? Where did we get that great data set? Was it licensed? Who printed that large format map for us last year? How much did it cost to print and fold those color maps?

Documentation of the details involved in making a map may seem tedious but can save time and effort in future map making, both for yourself and others who may need to make similar maps. Working toward a few general styles that are effective for specific types of commonly produced maps is useful. Documentation of mapped data is vital if the map is to be published.

Document what you do and update it when appropriate. Share your documentation with the world.

Editing

How editable should your map be?

Do you anticipate updates?

How does an update cycle work, and what can you do now in building the map structure to make updating easier?

How editable do you need the map to be by others?

Have you written up details of the map and what tools you'll need to make updates?

Do you need others to be able to fix typos or make minor changes? Check to see what tools they have, and be prepared to adjust their expectations.

Documenting General Issues

Document your goal for the map and

...the intended audience, and what you know about them

...the final medium, and details about the medium that will affect map design and reproduction

...the amount of time it takes to create the map, and any major problems and how you solved them

Keep copies of the map as well as information on where it was published or presented

Documenting Data

Document the source of the data, including contact information and copyright information

...the age, quality, and any limitations of the data

...how the data were processed into a form appropriate for mapping

...map projection and coordinate system information

Documenting Design

Document specifics of map size, scale, and sketches of layouts

...a list of information on the map, arranged in terms of importance, and associated symbols

...data classification and generalization information

...sources and details of map symbols

...details of type size, font, etc.

...color specifications for all colors used

...design problems encountered and solutions

...software problems encountered and solutions

Formative Evaluation

Ongoing formative evaluation is as simple as asking yourself whether the map is achieving its goals throughout the process of making the map. Formative evaluation implies that you will "re-form" the map so it works better, or maybe even dump it! It is never too late to bail if the map is not serving your needs. It is a good idea to ask others to evaluate your map as well: What do you think of those colors? Can you read that type from the back of the room? Does what is most important on the map actually stand out? What is the boss going to think? Simply engaging your mind as you make your map, and being open to criticism and change, will lead to a better map.

Mapmaking pros almost always seek formative feedback on their work and make appropriate changes.

Who is your target audience? Let some of them take a look before you finalize your work.

Stakeholders in a collective and participatory project should always be part of your formative evaluation process.

Ask...

Is this map doing what I want it to do?

Will this map make sense to the audience I envision for it?

How does the map look when printed, projected, or viewed in the final medium, and what changes will make it better?

Are the chosen scale, coordinate system, and map projection appropriate?

Does the layout of the map with its legend look good? Could layout be adjusted to help make the map look better and easier to interpret?

Does the most important information on the map stand out visually? Does less important information fall into the background?

Are the data on the map too generalized or too detailed, given the intent of the map?

Does the way I classified my facts help to make sense out of them? Would a different classification change the patterns much?

Do the chosen symbols make sense, and are they legible?

Is the type appropriate, legible, and is its size appropriate, given the final medium?

Is color use logical (e.g., value for ordered data, hue for qualitative data) and appropriate, and will the chosen colors work in the final medium?

Do I want a series of simpler maps, or one more complicated one?

Is a handout map needed, if presenting a map on a poster or projected?

...then re-*form* your map.

Impact Evaluation

Impact evaluation is a range of informal and formal methods for evaluating the finished map. It may be your boss or a publisher reviewing the map, or public feedback on the map's efficacy. You should begin any map making with a clear sense of who may have the final say on the acceptability of your map, and factor in their wants, needs, and requirements at the beginning of the process.

Caribou Calving Areas
Arctic National Wildlife Refuge (ANWR)

Percent Likelihood of High Density Calving Area

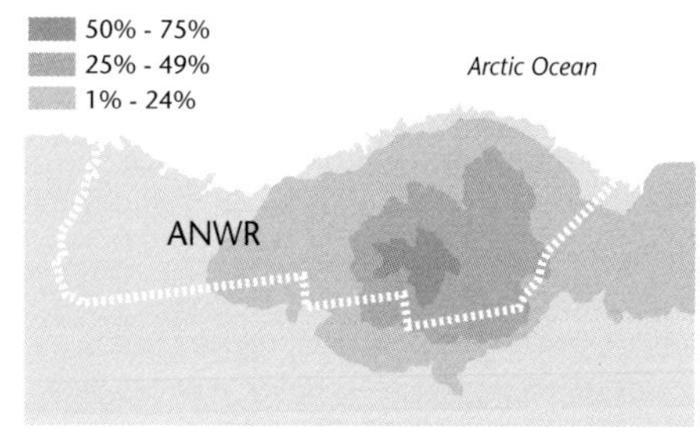

Ian Thomas, a contractor for the U.S. Geological Survey, was fired, allegedly for making maps of caribou calving areas in the ecologically and politically sensitive Arctic National Wildlife Refuge. Thomas argues he was fired for publicizing facts that would undermine the push for oil exploration in the refuge. Others claim the maps were based on out-of-date information beyond Thomas's area of expertise and had nothing to do with his firing. In either case, it is obvious that making maps can piss off your boss.

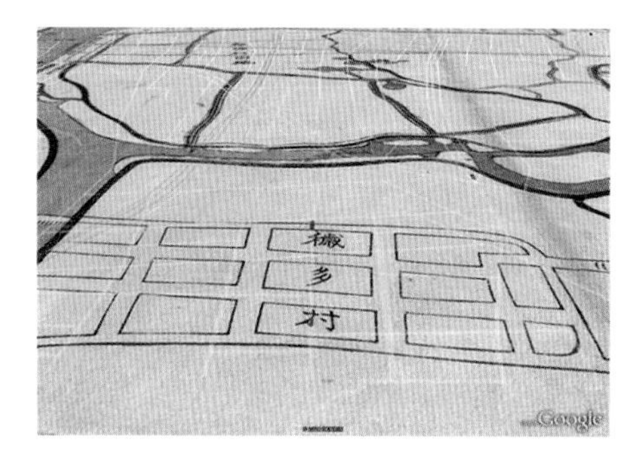

An old Japanese map from the David Rumsey digital map collection was added to Google Earth in early 2009. A label on the map described a village as populated by "eta," the untouchable caste of burakumin (translation, "filthy mass"). Because some jerks in Japan discriminate against the burakumin, it is common practice to remove such references. Rumsey initially decided not to censor the map, but after an uproar, the offending nomenclature was removed.

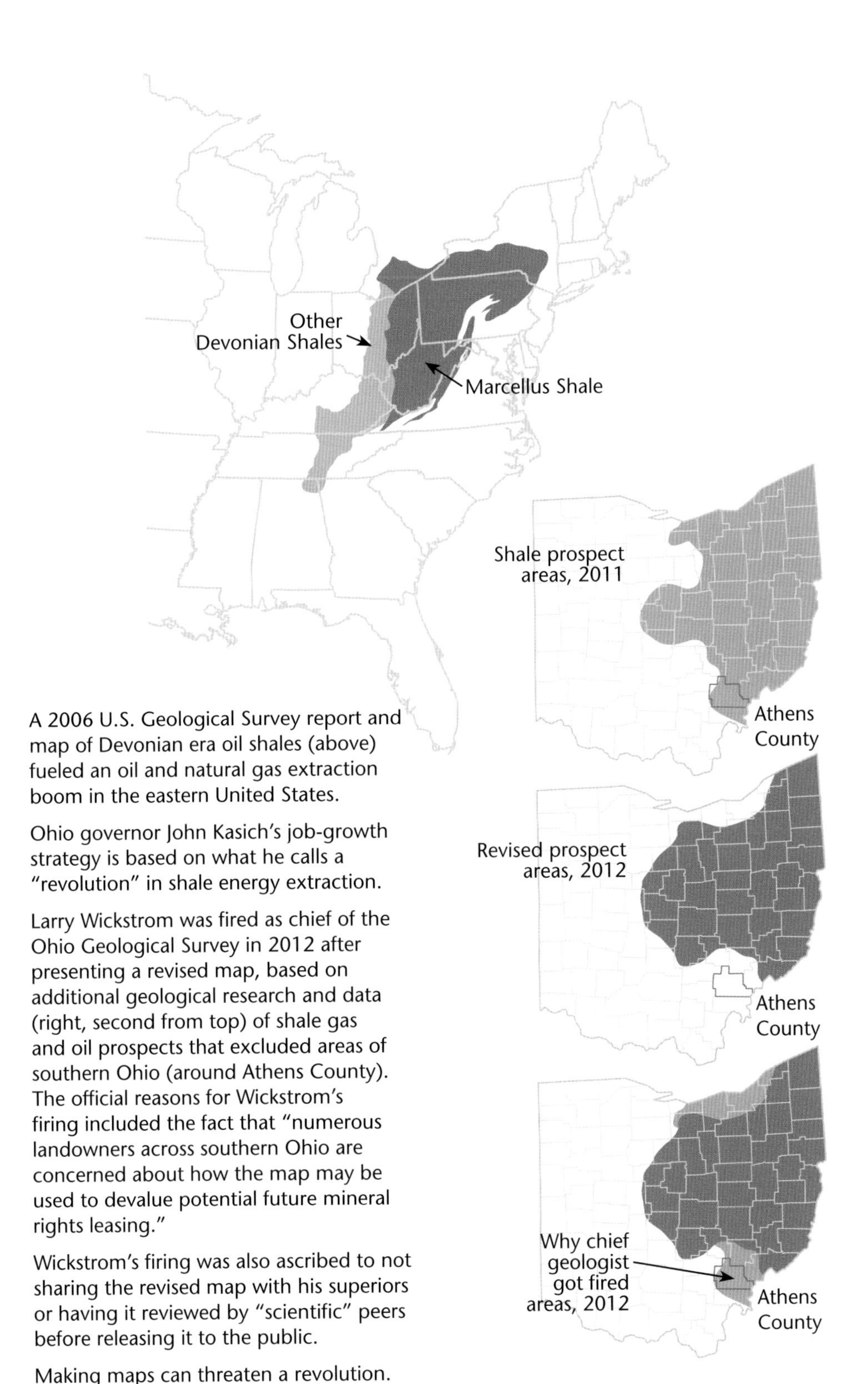

A 2006 U.S. Geological Survey report and map of Devonian era oil shales (above) fueled an oil and natural gas extraction boom in the eastern United States.

Ohio governor John Kasich's job-growth strategy is based on what he calls a "revolution" in shale energy extraction.

Larry Wickstrom was fired as chief of the Ohio Geological Survey in 2012 after presenting a revised map, based on additional geological research and data (right, second from top) of shale gas and oil prospects that excluded areas of southern Ohio (around Athens County). The official reasons for Wickstrom's firing included the fact that "numerous landowners across southern Ohio are concerned about how the map may be used to devalue potential future mineral rights leasing."

Wickstrom's firing was also ascribed to not sharing the revised map with his superiors or having it reviewed by "scientific" peers before releasing it to the public.

Making maps can threaten a revolution.

The Flight of Voyager map was published in 1987 in the book *Voyager* by Jeana Yeager, Dick Rutan, and Phil Patton

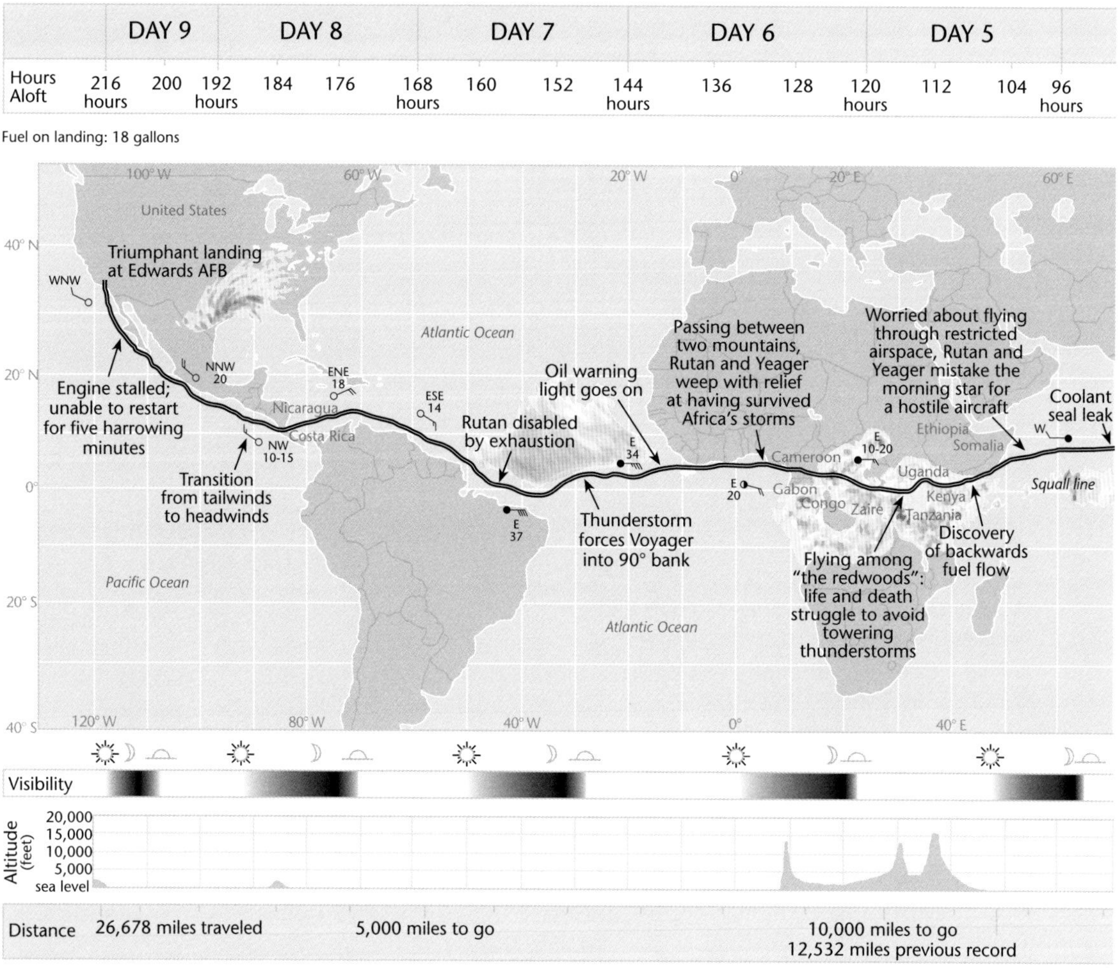

Flight data courtesy of Len Snellman and Larry Burch, Voyager meteorologists
Mapped by David DiBiase and John Krygier, Department of Geography, University of Wisconsin-Madison, 1987

David DiBiase and John Krygier designed and made a map to tell the story of Voyager and its pilots. The map was created for a map design course at the University of Wisconsin-Madison taught by David Woodward.

The map was made for readers of the book *Voyager* (1987), with its general, educated audience, including those with a specialist interest in flight and aerospace. Given the audience, the map was designed to contain a significant amount of information including detailed data, sure to resonate with pilots.

The map was split between the front and back book endpapers, half in the front and half in the back. Each endpaper was 9″ high and 12″ wide.

The map was designed to be viewed at (a book-holding) arm's length.

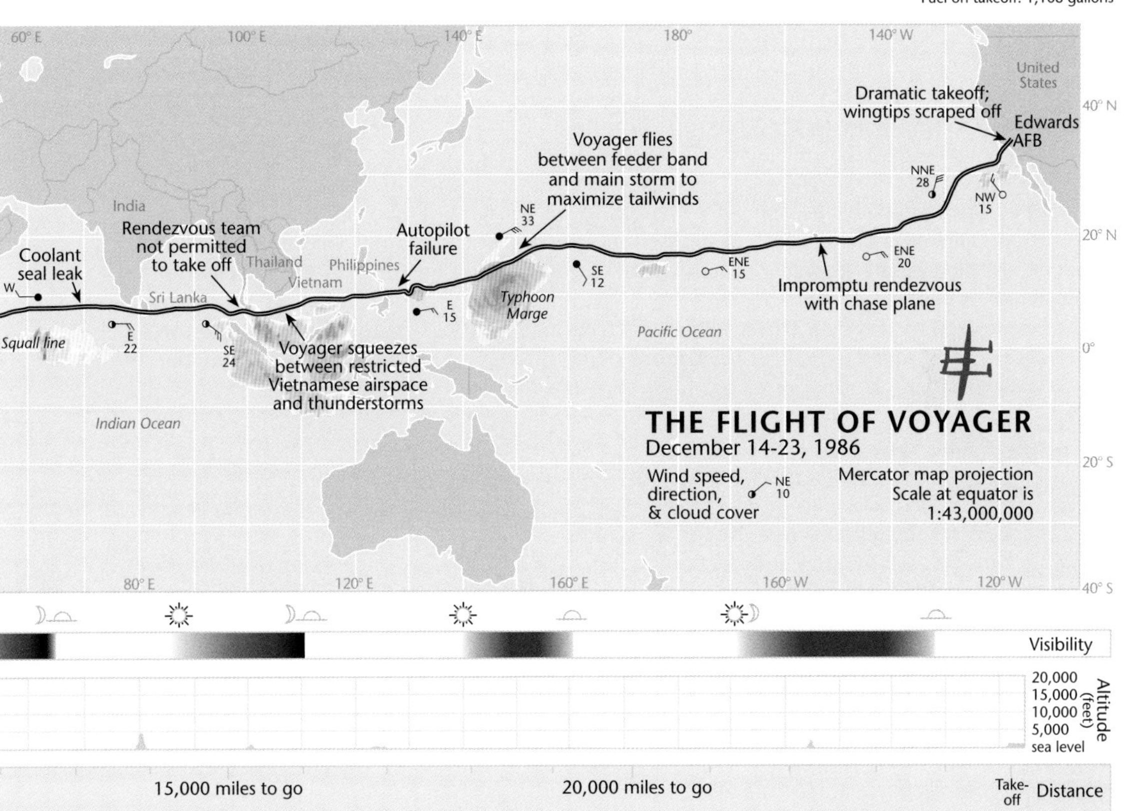

Voyager pilots: Dick Rutan and Jeana Yeager
Voyager designer: Burt Rutan

The publisher of *Voyager*, Knopf, allowed us black and one color for the map. We chose deep red for the most important information (such as the flight path and related text). The map was redesigned in monochrome for *Making Maps.* The map still works!

Details of the map's design – line weights, type size, percent gray of different areas on the map, etc. – were documented, as we were taught in David Woodward's course and at the University of Wisconsin-Madison Cartographic Lab. Formative evaluation was ongoing throughout the process, and the editors at Knopf provided the final edit and evaluation of the map.

Maps are full of the impertinence of the arbitrary.

Brian O'Doherty, *American Masters* (1973)

The most remarkable escape story of all concerns Havildar Manbahadur Rai.... He escaped from a Japanese prison camp in southern Burma and in five months walked 600 miles until at last he reached the safety of his own lines. Interrogated by British intelligence about his remarkable feat, Manbahadur told them that ... he had a map, which before his capture had been given to him by a British soldier in exchange for his cap badge. He produced the much creased and soiled map. The intelligence officers stared at it in awe. It was a street map of London.

Byron Farwell, *The Gurkhas* (1984)

Borders are scratched across the hearts of men
By strangers with a calm, judicial pen,
And when the borders bleed we watch with dread
The lines of ink across the map turn red.

Marya Mannes, *Subverse: Rhymes for Our Times* (1959)

More...

If you're making your map for any of the usual reasons, there's more than enough to read in the books listed at the end of Chapter 1. If you've got something else in mind, they're unlikely to be of much help.

But there are a lot of other things to look at. Eden Kinkaid and Cassidy Schoenfelder edited a special issue of *You Are Here: A Journal of Creative Geography* on Counter/Cartographies (2023). *This Is Not an Atlas: A Global Collection of Counter-Cartographies* (2019) is a big ole' tome of map experiments. Wesley Jones' *Fantasy Mapping: Drawing Worlds* (2020) and *Fantasy Mapping: Drawing Realms and Kingdoms* (2023) takes on the burgeoning world of mapping fantasy worlds. Hans Ulrich Obrist's *Mapping It Out: An Alternative Atlas of Contemporary Cartographies* (2014) is crammed with examples of novel maps and map-fusions intended to blow your complacency if not your mind. Laura Kurgan's *Close Up at a Distance: Mapping, Technology and Politics* (2013) maps and theorizes mass graves, incarceration patterns, and disappearing forests in nine beautiful case studies. Denis Wood's *Everything Sings: Maps for a Narrative Atlas* (2013) is a sequence of maps intended to be read as poems or as chapters in a story. His introduction describes the difficulty he had in breaking out of the map mold. Frank Jacobs's *Strange Maps: An Atlas of Cartographic Curiosities* (2009) is a collection less strange than the title implies, but rich with suggestive directions.

Community and land use occupancy mapping is a whole other kettle of salmagundi. Two great textbooks are Terry Tobias's *Living Proof: The Essential Data-Collection Guide for Indigenous Use-and-Occupancy Map Surveys* (2000) and Alix Flavelle's *Mapping Our Land: A Guide to Making Maps of Our Own Communities and Traditional Lands* (2002). Joe Bryan and Denis Wood's book *Weaponizing Maps: Indigenous Peoples and Counterinsurgency in the Americas* (2015) finds indigenous mapping in the sites of (largely) U.S. counterinsurgency efforts.

In the U.K., Common Ground's Parish Maps Project heads to the same place along a wholly different path. Common Ground (www.commonground.org.uk) offers Sue Clifford and Angela King's useful pamphlet *From Place to PLACE: Maps and Parish Maps* (1996). Kim Leslie's extraordinary atlas *A Sense of Place: West Sussex Parish Maps* (2006) not only has gloriously reproduced examples of nearly a hundred parish maps made for the millennium observance, but also many descriptions by residents of how they made them. This too is a furiously ongoing activity.

If you're interested in making maps as art, the place to start is Katherine Harmon's *The Map as Art: Contemporary Artists Explore Cartography* (2009). You might also want to take a look at Huw Lewis-Jones's *The Writer's Map: An Atlas of Imaginary Lands* (2018).

Sources: The Mark Twain map and text were published in the *Buffalo Express* newspaper on September 17, 1870. The edibles map was redrawn from one posted to r/dataisbeautiful by mrwiseman (no date). The Strava Global Heatmap is available at strava.com. The 1912 Brooklyn subsurface utility map is from George Tillson *Street Pavements and Paving Materials* (Wiley & Sons, 1912). The *NO to Century Boulevard* yard sign is used courtesy of Ron Rozzelle. Interactive map guidelines from www.axismaps.com. Visual storytelling guidelines from Robert E. Roth. "Cartographic Design as Visual Storytelling." *The Cartographic Journal,* 58:1, 83-114, 2021. Viewing distance rules from Alan Brawn and Jonathan Brawn's "Image Size vs. Viewing Distance" (PDF, 2013). The calving map is recreated from data and maps originally found at Ian Thomas's website. The Google eta/burakumin map is used courtesy of Google and David Rumsey. The Marcellus shale thickness map was adapted from *Assessment of Appalachian Basin Oil and Gas Resources* by Robert Milici and Christopher Swezey (United States Geological Survey, Open-File Report 2006-1237). The Ohio maps were redrawn from "With Shale, There's Lots at 'Play'" by Spencer Hunt in *The Columbus Dispatch,* April 1, 2012. *The Flight of Voyager* map is redrawn from the original map created by David DiBiase and John Krygier (1987). The story of Voyager is documented in the book of the same name by Jeana Yeager, Dick Rutan, and Phil Patton (Knopf, 1987).

FOR
YOUR
CONVENIENCE
BY
Paul Pry

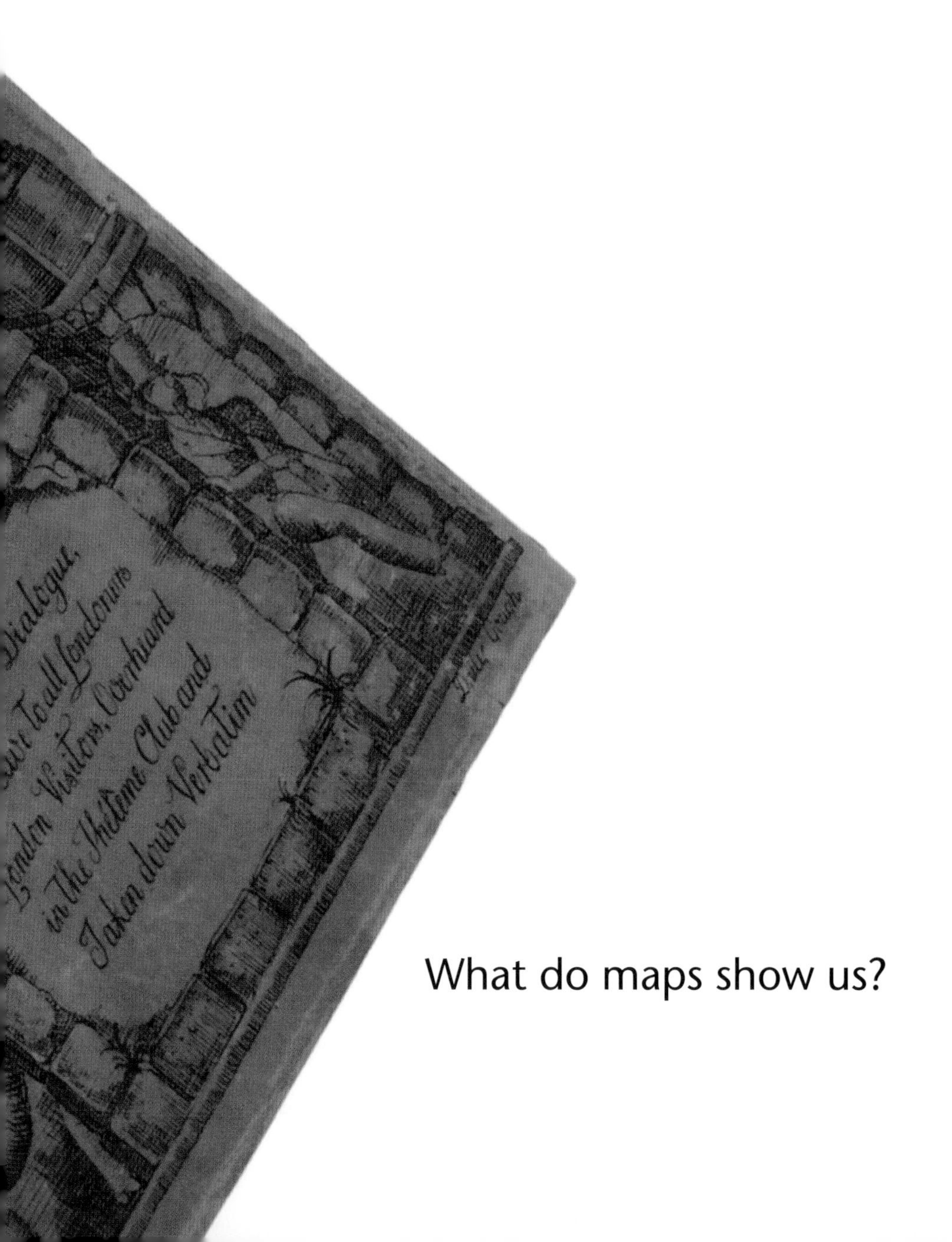

What do maps show us?

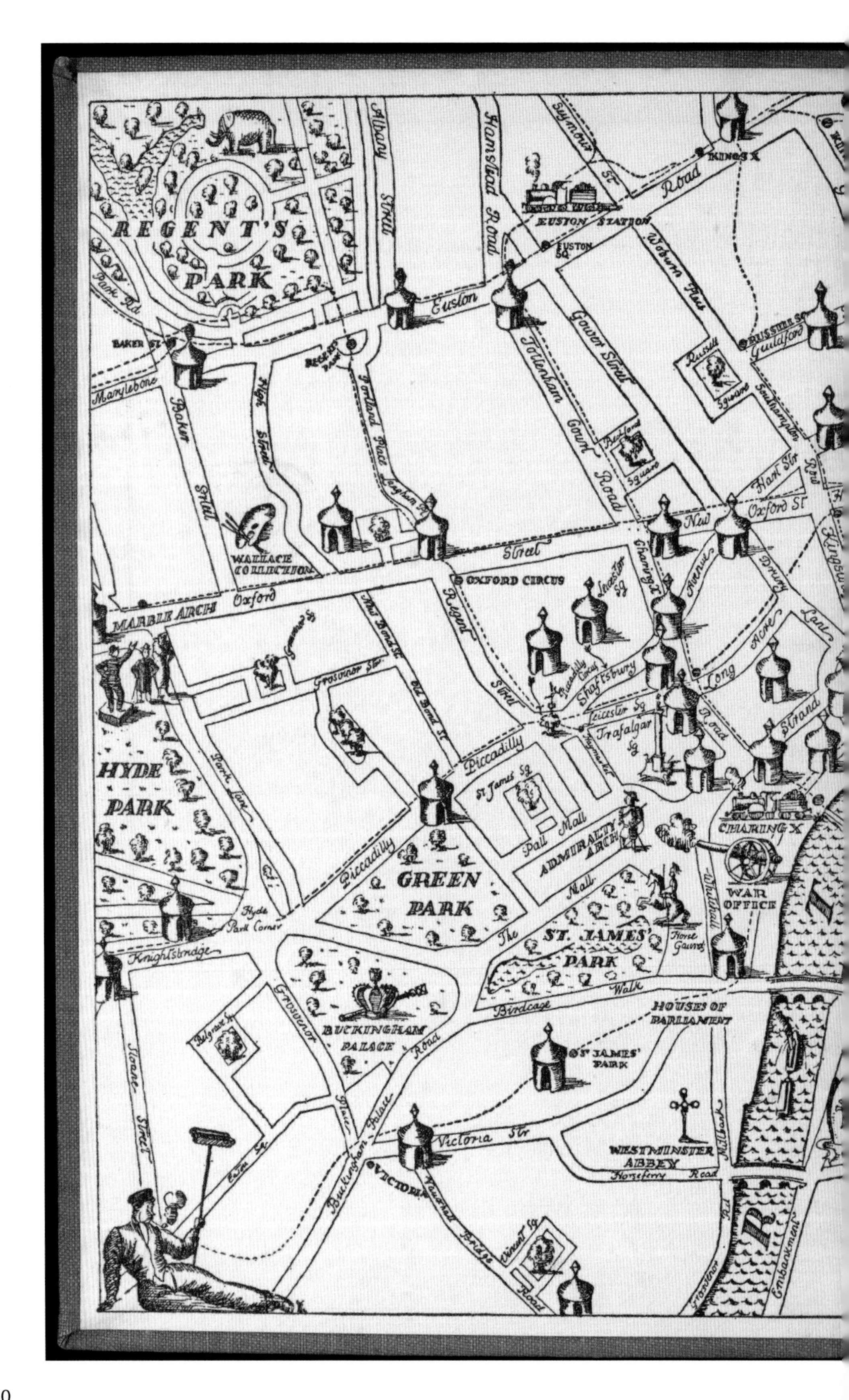
REGENT'S PARK
Park Rd
BAKER ST
Marylebone
Baker Street
High Street
Portland Place
Langham Pl
Albany Street
Hamstead Road
Seymour St
KINGS X
Road
EUSTON STATION
EUSTON SQ
Euston
Tottenham Court Road
Gower Street
Woburn Place
RUSSELL SQ
Guildford
Russell Square
Southampton Row
Bedford Square
Hart Str
New Oxford St
Charing X
Avenue
Drury Lane
Oxford Street
OXFORD CIRCUS
WALLACE COLLECTION
MARBLE ARCH
Oxford
New Bond St
Grosvenor Sq
Grosvenor St
Old Bond St
Regent Street
Leicester Sq
Piccadilly Circus
Shaftsbury
Long Acre
Leicester Sq
Trafalgar Sq
Haymarket
Strand
Road
Piccadilly
St James Sq
Pall Mall
ADMIRALTY ARCH
CHARING X
HYDE PARK
Park Lane
Piccadilly
GREEN PARK
Hyde Park Corner
WAR OFFICE
Whitehall
The Mall
ST. JAMES' PARK
Horse Guards
Knightsbridge
Grosvenor
Birdcage Walk
HOUSES OF PARLIAMENT
BUCKINGHAM PALACE
Road
Belgrave Sq
ST JAMES' PARK
Sloane Street
Eaton Sq
Buckingham Palace Road
Place
Victoria Str
VICTORIA
Vauxhall Bridge Road
Vincent Sq
WESTMINSTER ABBEY
Horseferry Road
Millbank
Grosvenor Rd
Embankment

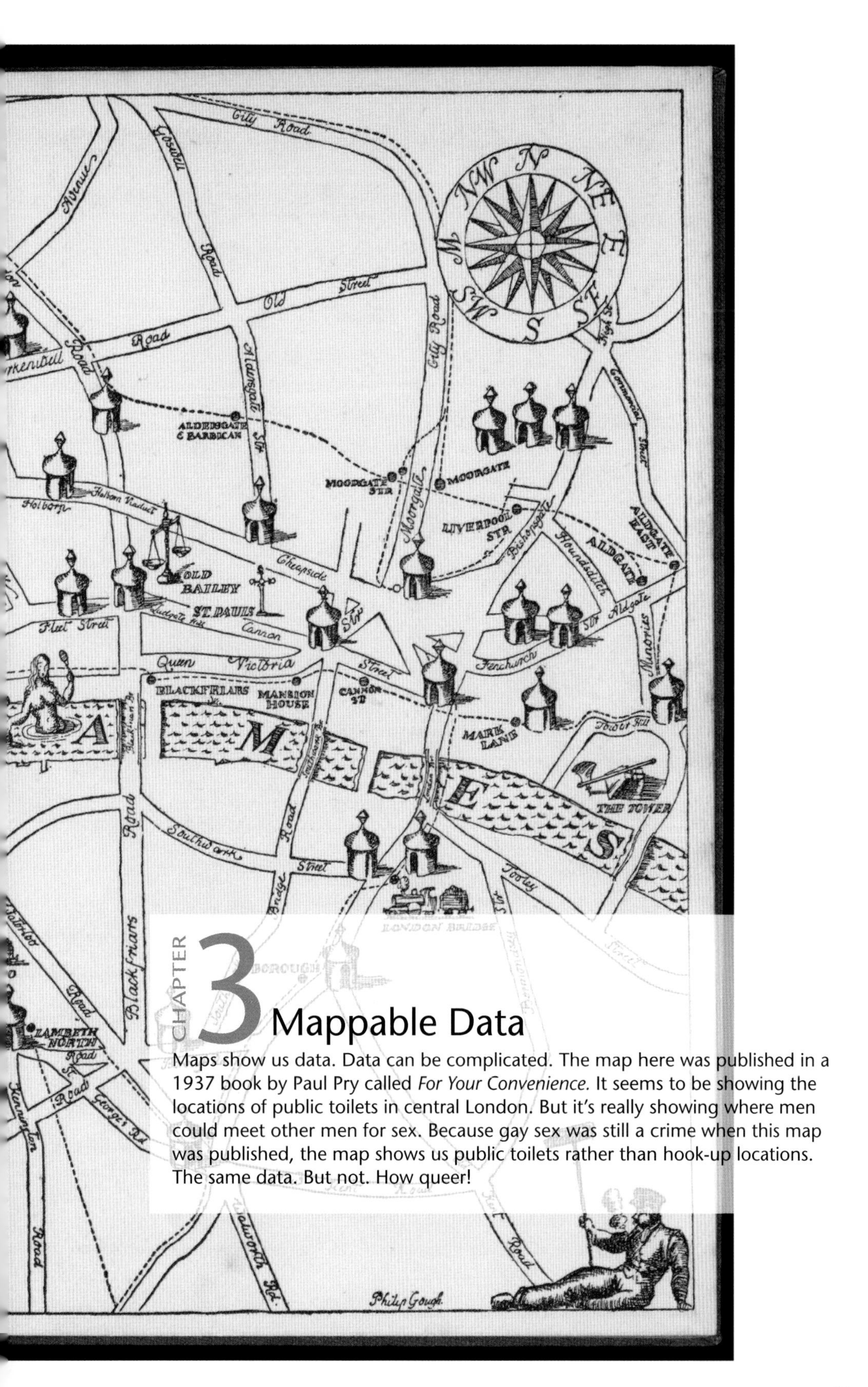

CHAPTER 3 Mappable Data

Maps show us data. Data can be complicated. The map here was published in a 1937 book by Paul Pry called *For Your Convenience.* It seems to be showing the locations of public toilets in central London. But it's really showing where men could meet other men for sex. Because gay sex was still a crime when this map was published, the map shows us public toilets rather than hook-up locations. The same data. But not. How queer!

Turning Phenomena into Data

Phenomena are all the stuff in the real world. Data are records of observations of phenomena. Maps show us data, not phenomena. Carefully consider the data you are mapping, how they relate to the stuff in the world, how they are similar, how they are different, and how that may affect our understanding of the phenomena. Differentiate, for example, between individual and aggregated data:

A map of individual pin oak trees in the southern part of the Clintonville neighborhood in Columbus, Ohio. The phenomena are trees, and the data, individually mapped by members of the local "urban arboretum," retain this individuality.

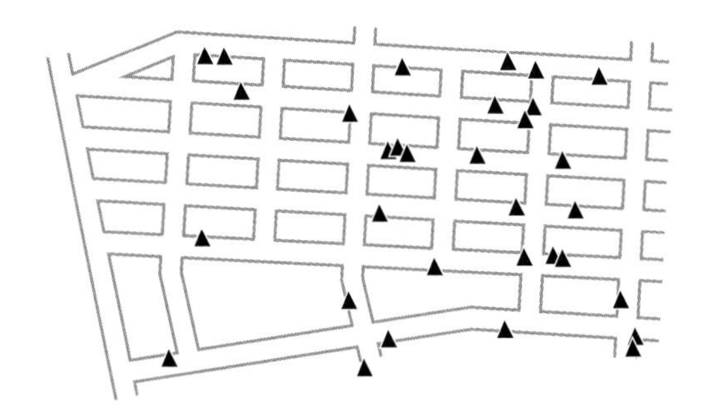

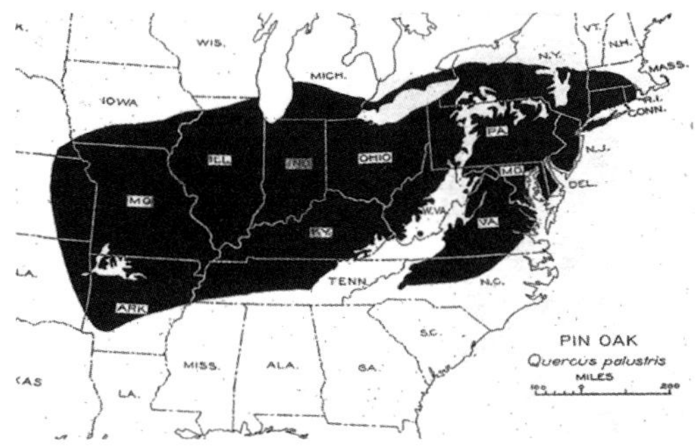

An old map shows the range of the pin oak tree in the U.S. The phenomena are not trees but the potential range of a species, and the data – aggregated from other maps – embody this abstraction.

A map of major vegetation zones in the eastern U.S. The area labeled "D" is broadleaf deciduous forest. The adjacent "M" is mixed broadleaf and needleleaf forest. The phenomena here are neither trees nor species, but forest, and the data – aggregated from many different maps – embody this still greater abstraction.

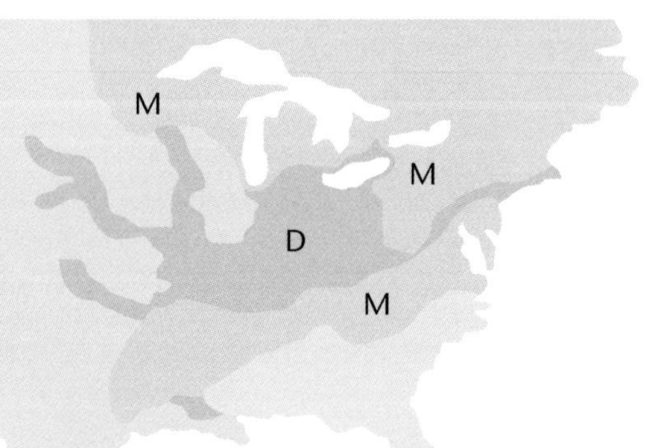

Aggregating data *changes phenomena*, as here, from individual trees, to species, to forest.

Another issue is variation: how phenomena, and so data, change over space and time. Continuous phenomena vary more or less smoothly. Good examples include atmospheric pressure or temperature. Discrete phenomena exist at some places and not others (although they may move around). Good examples are people, pin oak trees, and Studebaker Champ pickup trucks. There is no necessary relationship between phenomena and data, as it's possible to have different kinds of data, as below left, for a single phenomenon.

Air temperature varies *continuously* and is everywhere, but thermometers in weather stations can only record it at points. We can map temperature as collected revealing the structure of the *data*...

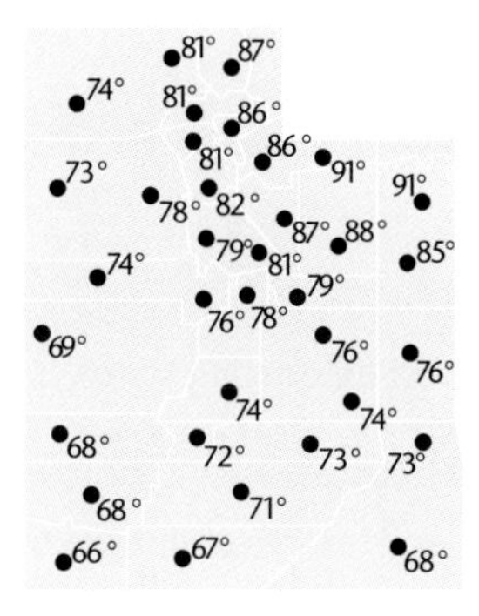

...or we can transform the point data into continuous data by interpolating data likely to exist between the readings. We then create a continuous surface on the map with this interpolated data to reveal the structure of the *phenomena:*

Humans are *discrete* phenomena. The U.S. Census counts the number of people at individual addresses during each decennial census. We could map people as collected at addresses, except we can't: the Census Bureau keeps this resolution of data secret to protect privacy.

Four people live at
100 East 3rd Avenue
Mexican Hat, UT 84531

We map detailed census data into single-value areas as small as a city block, but usually counties or states. The map below implies that people are continuous throughout each county, but we know that not to be the case. Here the map shows us more about the structure of the *data:*

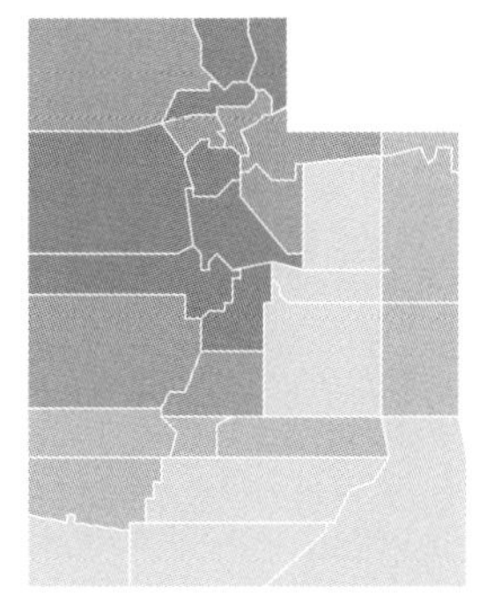

Creating and Getting Data

Data are records of observations of phenomena. These records may be made by machines (like those made by a recording thermometer) or by the map makers themselves. All these are primary data, that is, records of observations made in the environment itself. Maps made from primary data can be considered evidence. Most map makers use secondary or tertiary data sources created and published by others, but it's surprisingly easy to create mappable data yourself. It is common, and often necessary, to combine primary, secondary, and tertiary data sources on a map.

Primary Data Sources

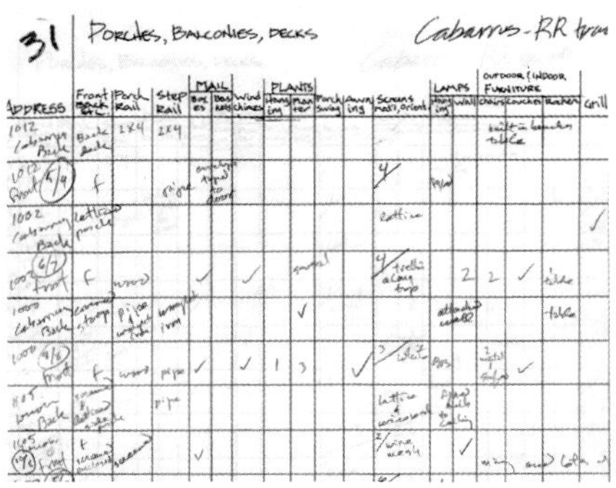

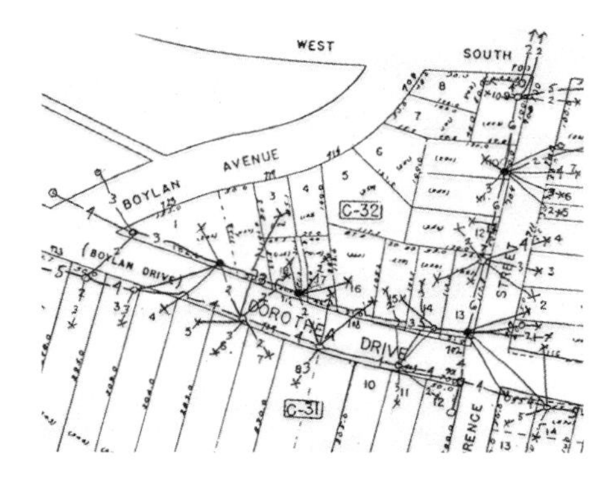

Collecting data at addresses. Researchers recorded 20 different facts about the exterior of houses (above: data collection sheet). Such data can be geocoded (address-matched) using geographic information systems (GIS) or web geocoding tools. Geocoding provides coordinates for your addresses, allowing them to be mapped with GIS software.

Global positioning systems (GPS). Satellites relays signals used by a GPS receiver to determine the location of the device. Inexpensive GPS receivers provide location and elevation. More sophisticated devices allow you to append attribute information (data at the location) and export the data so that it can be mapped in GIS.

Smart phones. Smart phones generate approximate locations using your position relative to cell towers when you have location services on. Cell phones may supplement these data using wi-fi and GPS signals. Smart phone applications allow you to use your phone to collect locations and attributes at those locations. Companies and governments collect these locations too... from you. Be afraid.

Data collection on existing maps. A city property map was used to record the location of electrical poles and power lines. Such data can be digitized, scanned, or added (by hand) to, and used with, existing layers of data in GIS and other mapping software.

Remote sensing imagery. Images of the earth, taken from airplanes or satellites. Imagery is available at different resolutions and can include non-visible energy such as infrared. DIY imagery can be collected with balloons or quadcopter drones, stitched together in graphics software, and georectified and georeferenced using GIS. Imagery can be used to generate mappable data: roads can be traced or vegetation types delineated. Remotely sensed imagery can be combined with map data in GIS.

Crowdsourcing. Websites can collect mappable data from anyone who can access the site and enter information. Thousands of users from around the world are constructing the OpenStreetMap.org open source map of the world. In essence, the "crowd" is the data source.

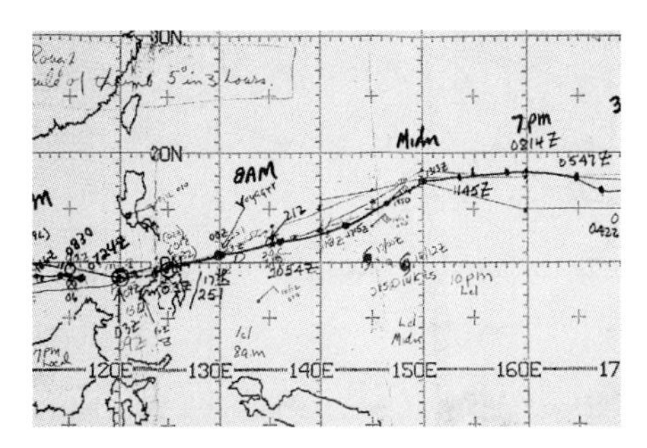

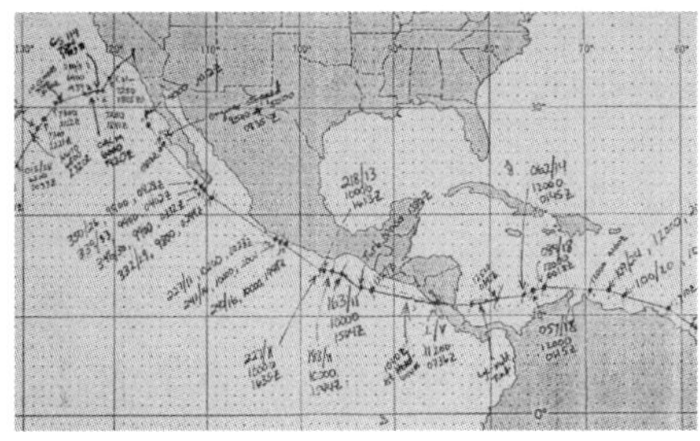

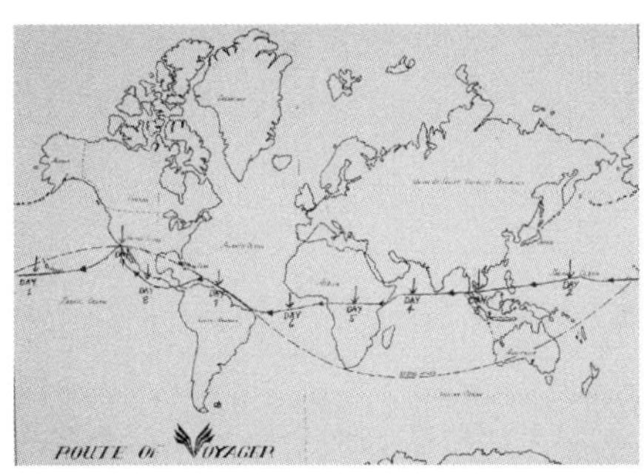

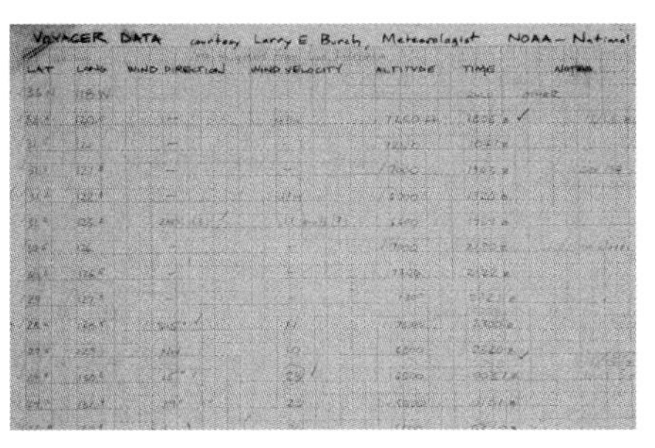

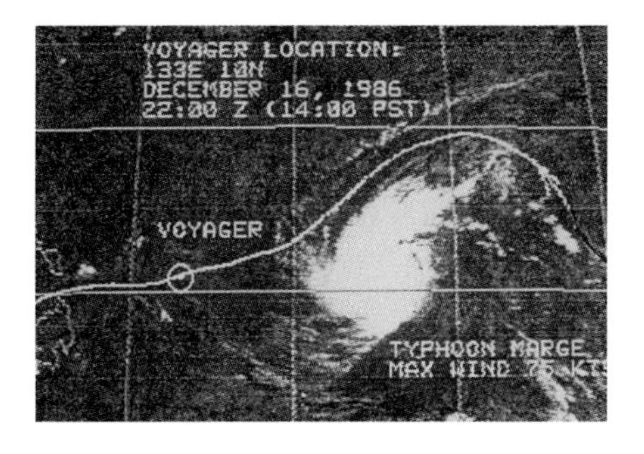

***Primary data for* The Flight of Voyager.** Primary data for the Voyager map were provided by Len Snellman and Larry Burch, the two meteorologists responsible for monitoring weather conditions for the flight. The data consisted of annotated maps, data tables, and satellite imagery.

Snellman and Burch hand-compiled, on maps (left, top three maps), detailed data including a series of flight-path locations (latitude and longitude) for Voyager with associated wind direction, wind velocity, airplane altitude, and time.

DiBiase and Krygier created a table of relevant data (left, bottom) to guide the mapping of the Voyager path and associated flight data.

Weather information was documented in satellite images (above) that served as the basis for the depiction of storms on the final Voyager map.

Gathering data for your map can take a lot of time.

Secondary Data Sources

Secondary data are derived from primary data: aggregations of traffic counts, generalizations of vegetation types.

Scanned and digitized paper maps
Federal government data, including census, economic, environmental data
Regional and local government data, including detailed road, property, and zoning data
For-profit data providers
Public domain data providers
Non-governmental organizations such as the UN and World Bank
Data "scraped" or copied from the internet and placed in a spreadsheet or other file

Secondary data sources for the Voyager map consisted of a basic Mercator map projection (below), place names from an atlas, and sunrise and sunset information from an ephemeris (below, right).

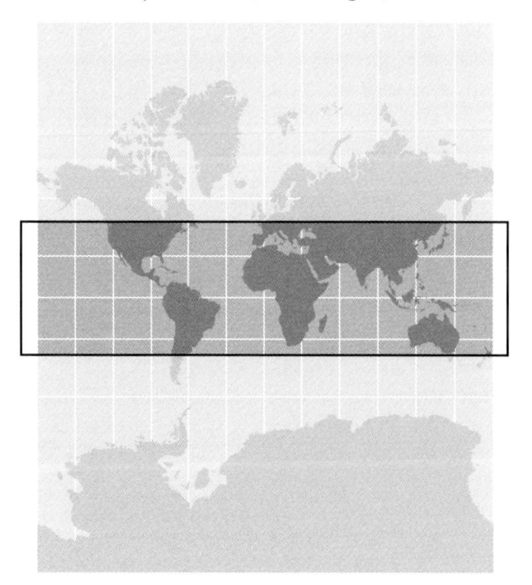

Tertiary Data Sources

Secondary data can be assembled in turn, thus resulting in tertiary data. If maps made from primary data are evidence, and maps made from secondary data are like reports of evidence, maps made from tertiary data would be akin to indexes of law cases.

Maps are often made from other maps. Map makers don't think about all this assemblage as generalization, but it is. Each step away from the phenomena makes the map less and less about the phenomena and more and more about the data.

Secondary and tertiary data sources are increasingly available as *open data layers,* free of cost and easily accessible with common mapping and GIS software.

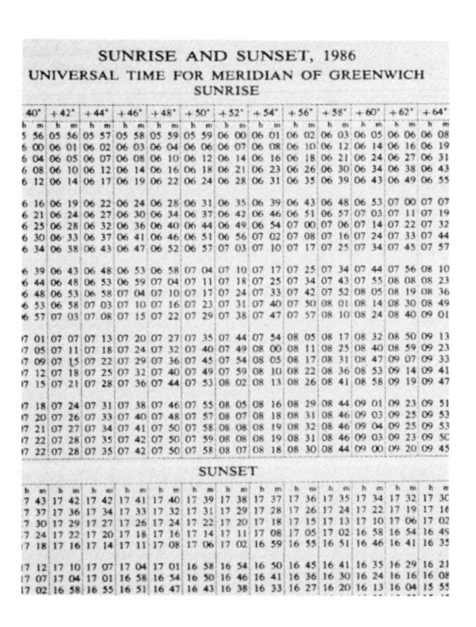

SUNRISE AND SUNSET, 1986
UNIVERSAL TIME FOR MERIDIAN OF GREENWICH

SUNRISE

40°	+42°	+44°	+46°	+48°	+50°	+52°	+54°	+56°	+58°	+60°	+62°	+64°
h m	h m	h m	h m	h m	h m	h m	h m	h m	h m	h m	h m	h m
5 56	05 56	05 57	05 58	05 59	05 59	06 00	06 01	06 02	06 03	06 05	06 06	06 08
6 00	06 01	06 02	06 03	06 04	06 06	06 07	06 08	06 10	06 12	06 14	06 16	06 19
6 04	06 05	06 07	06 08	06 10	06 12	06 14	06 16	06 18	06 21	06 24	06 27	06 31
6 08	06 10	06 12	06 14	06 16	06 18	06 21	06 23	06 26	06 30	06 34	06 38	06 43
6 12	06 14	06 17	06 19	06 22	06 24	06 28	06 31	06 35	06 39	06 43	06 49	06 55
6 16	06 19	06 22	06 24	06 28	06 31	06 35	06 39	06 43	06 48	06 53	07 00	07 07
6 21	06 24	06 27	06 30	06 34	06 37	06 42	06 46	06 51	06 57	07 03	07 11	07 19
6 25	06 28	06 32	06 36	06 40	06 44	06 49	06 54	07 00	07 06	07 14	07 22	07 32
6 30	06 33	06 37	06 41	06 46	06 51	06 56	07 02	07 08	07 16	07 24	07 33	07 44
6 34	06 38	06 43	06 47	06 52	06 57	07 03	07 10	07 17	07 25	07 34	07 45	07 57
6 39	06 43	06 48	06 53	06 58	07 04	07 10	07 17	07 25	07 34	07 44	07 56	08 10
6 44	06 48	06 53	06 59	07 04	07 11	07 18	07 25	07 34	07 43	07 55	08 08	08 23
6 48	06 53	06 58	07 04	07 10	07 17	07 24	07 33	07 42	07 52	08 05	08 19	08 36
6 53	06 58	07 03	07 10	07 16	07 23	07 31	07 40	07 50	08 01	08 14	08 30	08 49
6 57	07 03	07 08	07 15	07 22	07 29	07 38	07 47	07 57	08 10	08 24	08 40	09 01
7 01	07 07	07 13	07 20	07 27	07 35	07 44	07 54	08 05	08 17	08 32	08 50	09 13
7 05	07 11	07 18	07 24	07 32	07 40	07 49	08 00	08 11	08 25	08 40	08 59	09 23
7 09	07 15	07 22	07 29	07 36	07 45	07 54	08 05	08 17	08 31	08 47	09 07	09 33
7 12	07 18	07 25	07 32	07 40	07 49	07 59	08 10	08 22	08 36	08 53	09 14	09 41
7 15	07 21	07 28	07 36	07 44	07 53	08 02	08 13	08 26	08 41	08 58	09 19	09 47
7 18	07 24	07 31	07 38	07 46	07 55	08 05	08 16	08 29	08 44	09 01	09 23	09 51
7 20	07 26	07 33	07 40	07 48	07 57	08 07	08 18	08 31	08 46	09 03	09 25	09 53
7 21	07 27	07 34	07 41	07 50	07 58	08 08	08 19	08 32	08 46	09 04	09 25	09 53
7 22	07 28	07 35	07 42	07 50	07 59	08 08	08 19	08 31	08 46	09 03	09 23	09 5
7 22	07 28	07 35	07 42	07 50	07 58	08 07	08 18	08 30	08 44	09 00	09 20	09 45

SUNSET

h m	h m	h m	h m	h m	h m	h m	h m	h m	h m	h m	h m	h m
7 43	17 42	17 42	17 41	17 40	17 39	17 38	17 37	17 36	17 35	17 34	17 32	17 3
7 37	17 36	17 34	17 33	17 32	17 31	17 29	17 28	17 26	17 24	17 22	17 19	17 1
7 30	17 29	17 27	17 26	17 24	17 22	17 20	17 18	17 15	17 13	17 10	17 06	17 02
7 24	17 22	17 20	17 18	17 16	17 14	17 11	17 08	17 05	17 02	16 58	16 54	16 4
7 18	17 16	17 14	17 11	17 08	17 06	17 02	16 59	16 55	16 51	16 46	16 41	16 35
17 12	17 10	17 07	17 04	17 01	16 58	16 54	16 50	16 45	16 41	16 35	16 29	16 21
17 07	17 04	17 01	16 58	16 54	16 50	16 46	16 41	16 36	16 30	16 24	16 16	16 08
17 02	16 58	16 55	16 51	16 47	16 43	16 38	16 33	16 27	16 20	16 13	16 04	15 55

Interpreting Data

Making maps requires other maps, thinking, and interpretation. Erwin Raisz's fabulous landform map of Mexico (excerpt, bottom) began with aerial photos and topographic maps, like the map with interpretive annotations by Raisz (immediately below). Raisz reviewed and interpreted the data he had – contours and images – and from these created caricatures of the key landforms in the region. Raisz's interpretations resulted in a map that is more than the sum of its original data sources.

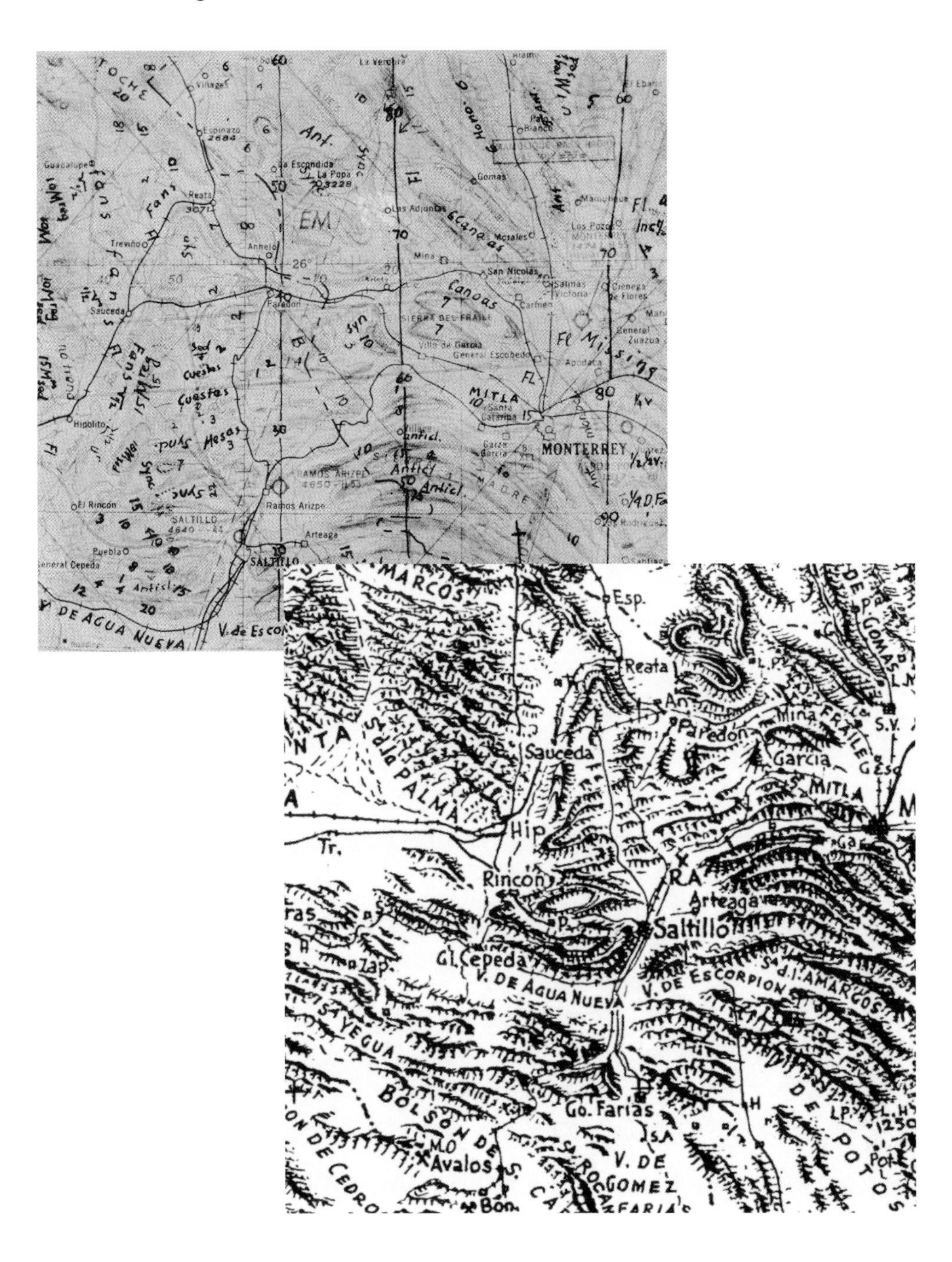

Data Characteristics

At a basic level, mappable data can be categorized as either qualitative (differences in kind) or quantitative (differences in amount). Such data distinctions guide analysis and map symbolization. Digital mappable data, qualitative or quantitative, are organized and stored in two ways: vector or raster. Vector data consist of points that can be connected into lines or areas. Raster data consist of a grid of cells, each with a particular value or values.

Qualitative

Differences in kind. Also called nominal data.

Cities as named locations
Dominant race in provinces
Location of different shrike species on a map of the world

Symbolization: shown with symbols, pictographs, or icons; or with differences in color hue (red, green, blue), as such colors are different in kind, like the data.

Quantitative

Differences in amount: includes ordinal, interval, and ratio data.

Cities grouped by total population
Total number of Hispanics by provinces
Number of loggerhead shrikes counted in a nature preserve

Symbolization: shown with differences in color value (dark red, red, light red), as such colors suggest more and less, like the data.

Levels of quantification

Ordinal: order with no measurable difference between values.
Low-, medium-, and high-risk zones

Interval: measurable difference between values, but no absolute zero.
Temperature Fahrenheit (30° is not half as warm as 60°)

Ratio: measurable difference between values, with an absolute zero value.
Total population in countries

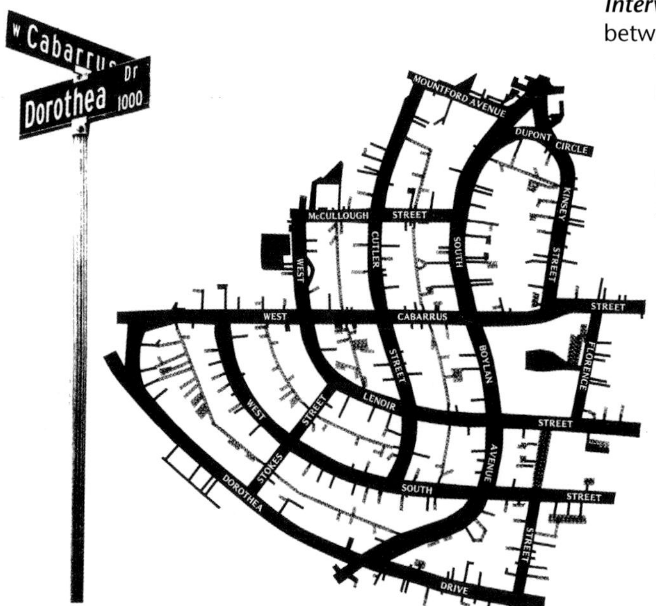

Things are complicated!

And unduly so!

The difference between streets and alleys in the Boylan Heights neighborhood (left) is *qualitative* (alleys serve a different purpose than streets) and *quantitative* (streets are larger and carry more traffic). The map symbols reflect this complexity.

Vector Data

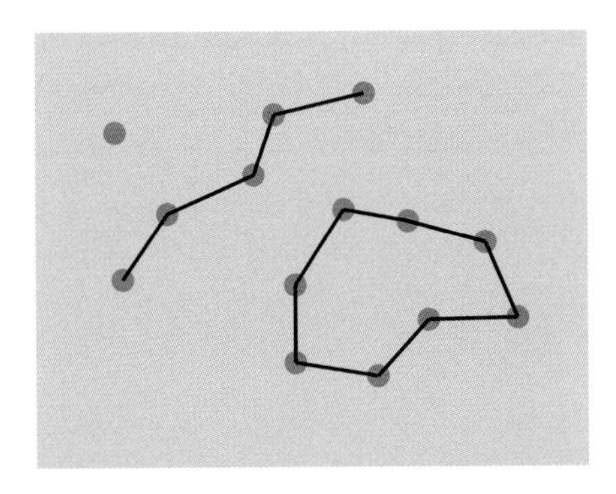

Vector data consist of located points (nodes), lines (a connected series of points), and areas (a closed, connected series of points, also called polygons). Collectively, these are often called feature geometry. Attribute information can be appended (joined) to a point, line, or area (using a common column of data) and stored in a related database. A line standing for a road includes attributes such as name, width, surface, etc. Design characteristics can be appended to points, lines, and areas.

Sources and use: GPS collects vector data; public and private sources of mappable data (USGS, Census TIGER, KML/KMZ) provide data in vector format. Common GIS software uses vector format data. Graphic design software, such as Illustrator, also use vector data, making the conversion of GIS output into graphic design software relatively easy.

Raster Data

Raster data consist of a grid with values associated with each grid cell. Higher-resolution raster files have smaller cells. Remotely sensed imagery is raster: each cell contains a level of energy reflected or radiated from the earth in the area covered by the cell. Raster data can have points (one cell), lines (a series of adjacent cells), or areas (a closed series of adjacent cells). Raster data can also include attributes.

Sources and use: Common raster data include satellite and aerial imagery available from public and private sources. Most GIS software allows you to use raster and vector data together. GIS software, such as the open-source GRASS, works with raster data. Image editing software, such as Photoshop or the open-source GIMP, use raster data and can import raster GIS output.

Geographic data are displayed on many web maps as **tiles** (tiled web maps or "slippy maps" – they slip around as you interact with them). Set at a particular size (usually 256 x 256 pixels), tiles are typically made from vector data transformed into squares of raster data, rendered ahead of time for quick display on computer monitors and handheld devices.

Transforming Data

Raw data, whether primary or secondary, may need to be transformed, or standardized, in order to make a map maker's point. It may be more useful to use totals instead of individual instances; it could make more sense to report phenomena as so many per unit area; an average temperature might be more meaningful than a bunch of daily highs and lows; or if your point has to do with change, rates might be helpful. There is always a *motivation* behind data transformations; choose wisely for an effective map.

Total Numbers

The total number of some phenomenon associated with a point, line, or area.

Amount of pesticide in a well
Pounds of road kill collected in a county

Symbolization: Variation in point size or line width. Represent whole numbers in areas with a scaled symbol for each area.

Densities

The number of some phenomenon per unit area. Divide the population in a country by its area.

Doctors per square km in a country
Adult bookstores per square mile in U.S. cities

Symbolization: Variation in color lightness and darkness in the areas.

Averages

Add all values together and divide by the number of values in the data set. Can be associated with points, lines, and areas.

Average monthly rainfall at a weather station
Average age of murder victims in U.S. cities

Symbolization: Variation in point size or line width. Variation in color lightness and darkness in areas.

Rates

The number of some phenomenon per unit time. May be associated with points, lines, or areas.

Cars per hour along a road
Murders per day in major cities

Symbolization: Variation in point size or line width. Variation in color lightness or darkness in areas.

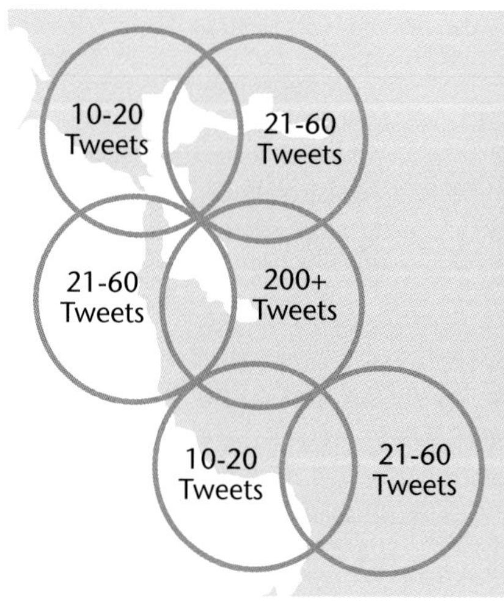

Tweeting "earthquake": The U.S. Geological Survey collected the number of Twitter messages with the word "earthquake" to assess the location of earthquakes around San Francisco, California. Total numbers can be misleading, as many more people live in the area with the highest number of tweets. Instead, transform the data into the percent of tweets per total population.

Aggregating

Data can also be put into different areas, typically at a different scale.

Average Credit Score (FICO)
By County, 2019

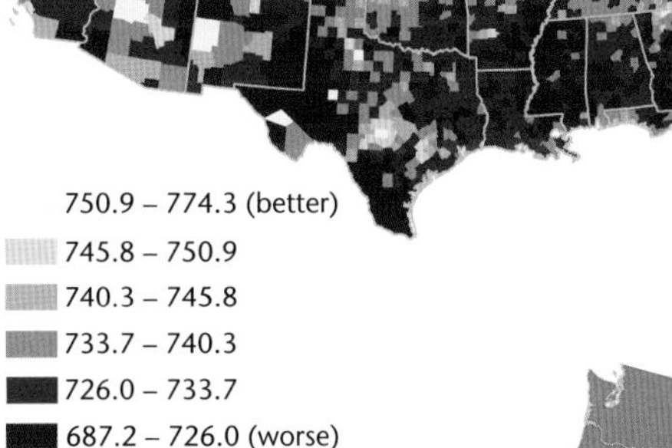

The modifiable areal unit problem (MAUP) describes a *scale effect:* how different aggregation schemes using the same data produce different results. Most spatial data is not distributed uniformly: aggregating the data into different scale areas can thus obscure important data characteristics.

A state like Texas has wide variation in credit score from county to county with some in the highest (better) range and some in the lowest.

When the county level data are aggregated to states and the classification scheme modified, Texas becomes part of the low credit score group, despite the actual county by county details.

Average Credit Score (FICO)
By State, 2019

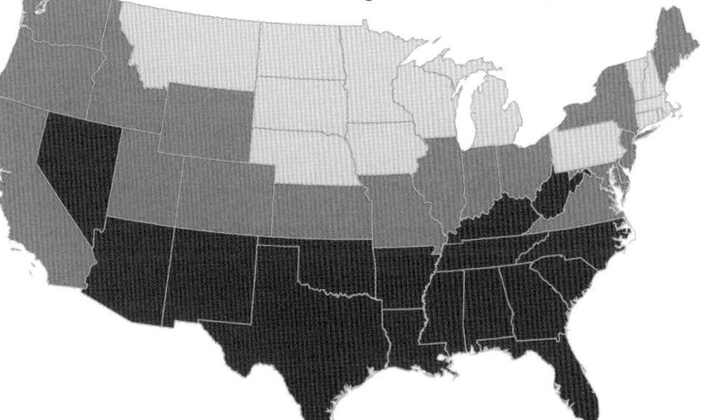

Statistical methods can assess the effects of the MAUP when you aggregate data to a different scale. You can also visually interrogate the effects of aggregating your data to different scale maps.

Districting and Redistricting

The modifiable areal unit problem (MAUP) also describes a *zonal effect:* how different-shaped areas imposed on the same spatial data produce different results.

Durham

When politicians draw voting districts, they adjust the shape of the districts to benefit their political party in elections. Gerrymandering is named after Massachusetts Governor Elbridge Gerry who, in 1812, configured an electoral district, which resembled a salamander, to benefit his political party. Gerrymandering is the creative use of the MAUP for political purposes. It's also unfair and anti-democratic.

North Carolina's 12th Congressional District was drawn, in 1992, to pack a large number of Black voters into a single, 150 mile long area from Charlotte to Durham.

Charlotte

Gerrymandering illustrates the zonal effect of the MAUP. Fifty people in an area need to be divided up into five districts. You can draw the districts in different ways to get different outcomes:

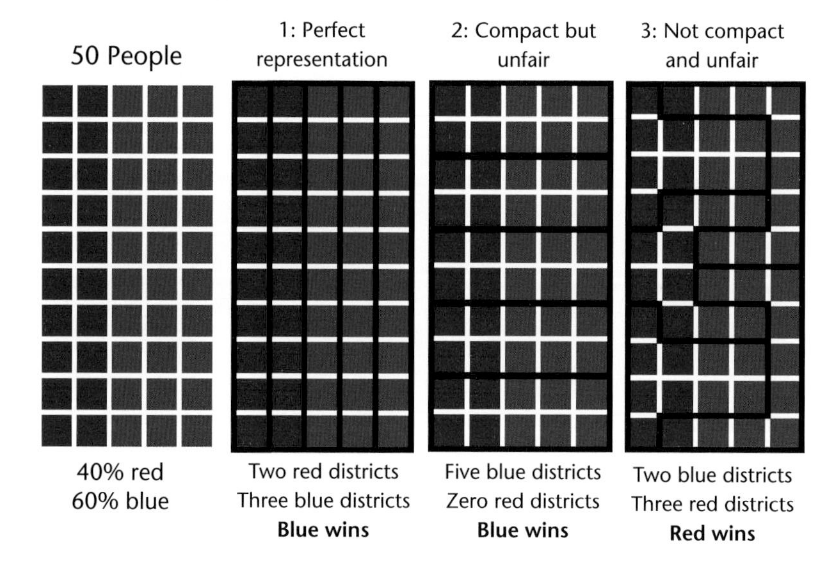

Time and Data

Map makers talk about space as though the spatial location of a phenomenon were key to its understanding. Location is important, but no more so than *when* it occurred. Every phenomenon happens at some place and at some time. Many phenomena change over time, and single maps or sequences of maps at regular intervals can effectively reveal spatial and temporal patterns. Such a sequence naturally suggests animation. But not all sequences require animation.

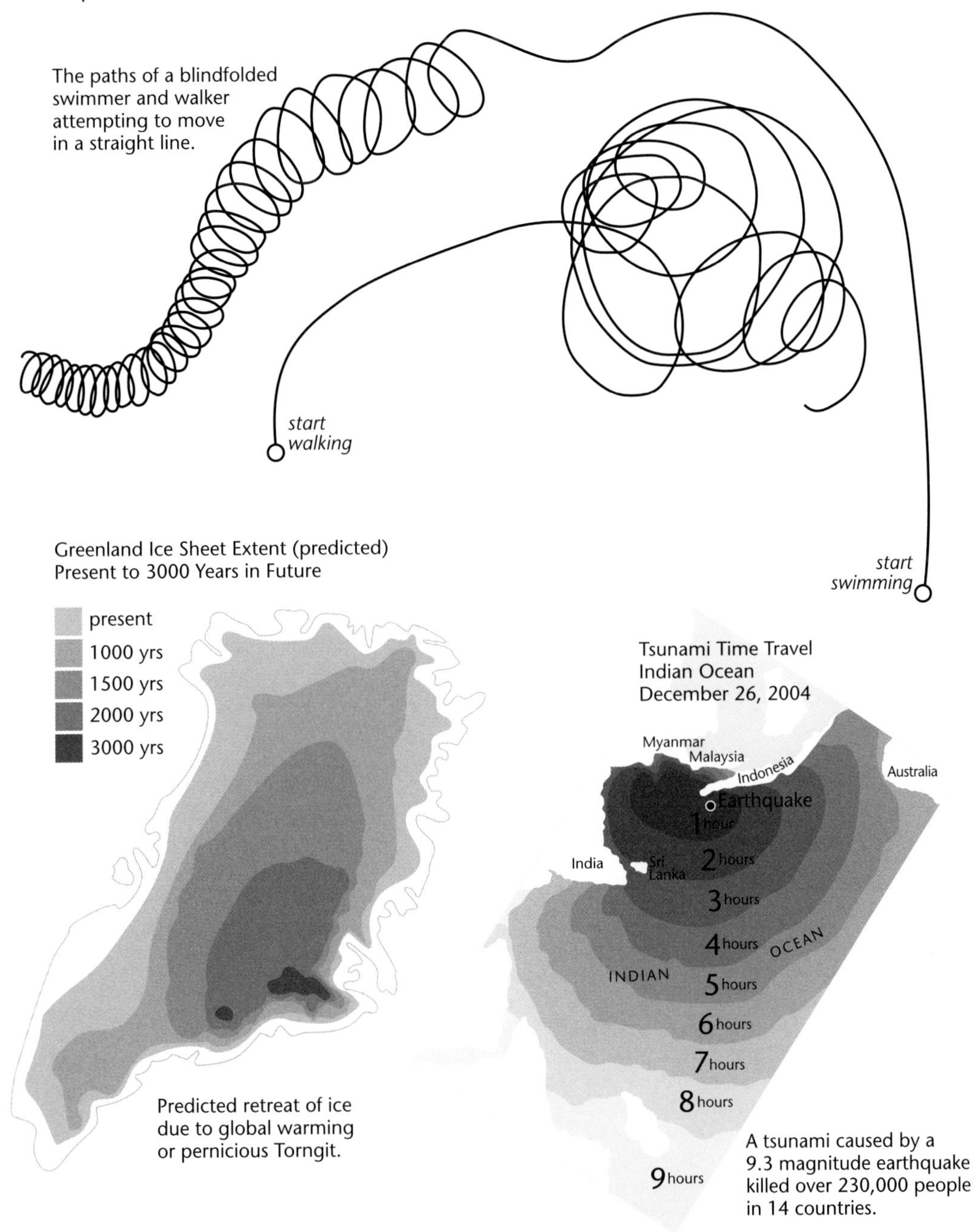

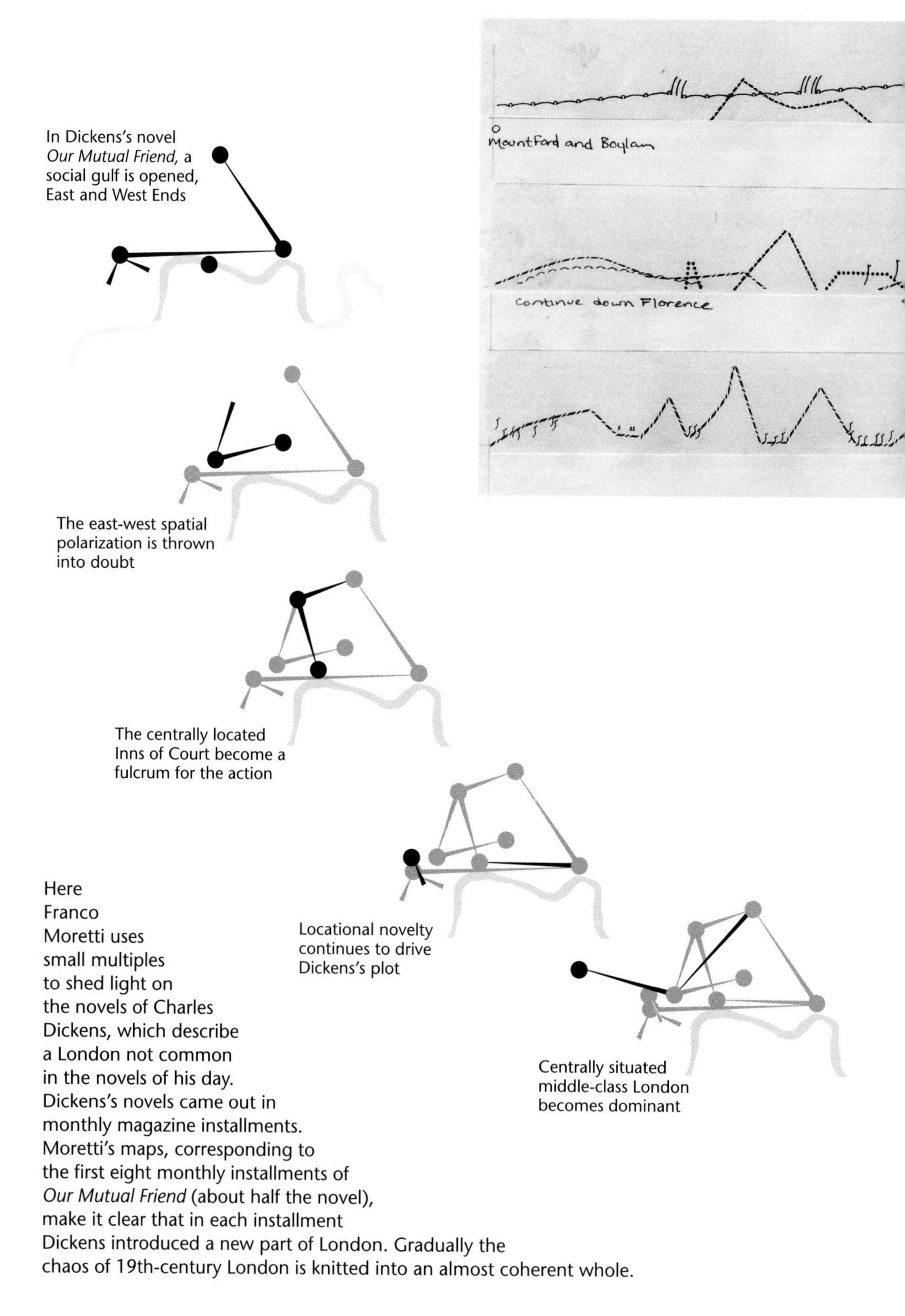

Mountford and Boylan
continue down Florence
In Dickens's novel *Our Mutual Friend,* a social gulf is opened, East and West Ends
The east-west spatial polarization is thrown into doubt
The centrally located Inns of Court become a fulcrum for the action
Locational novelty continues to drive Dickens's plot
Centrally situated middle-class London becomes dominant

Here Franco Moretti uses small multiples to shed light on the novels of Charles Dickens, which describe a London not common in the novels of his day. Dickens's novels came out in monthly magazine installments. Moretti's maps, corresponding to the first eight monthly installments of *Our Mutual Friend* (about half the novel), make it clear that in each installment Dickens introduced a new part of London. Gradually the chaos of 19th-century London is knitted into an almost coherent whole.

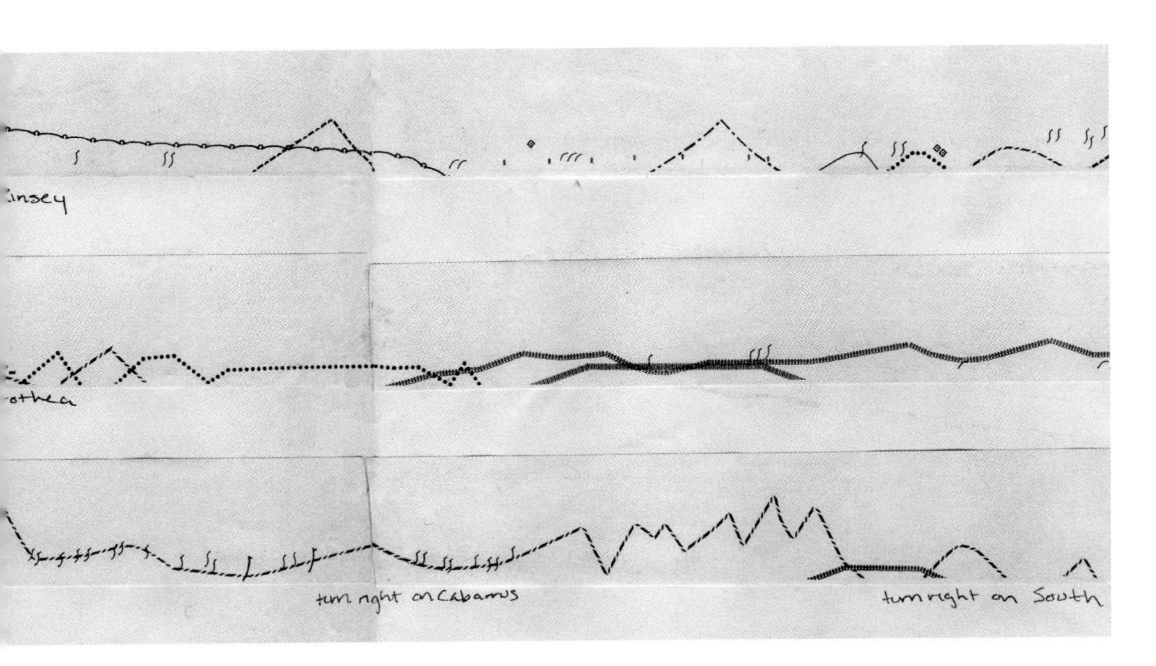

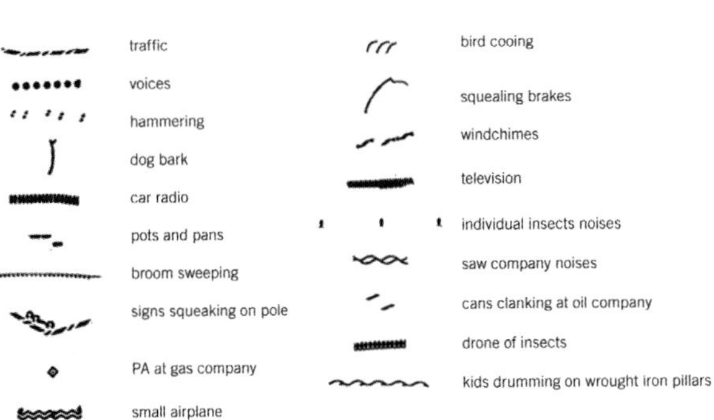

Sounds (above) heard during neighborhood walks (below) in Raleigh, NC, in 1982. Three of the dozen walks are shown above. Different sounds are indicated by different symbols, volume by height above the baseline, and time by horizontal distance.

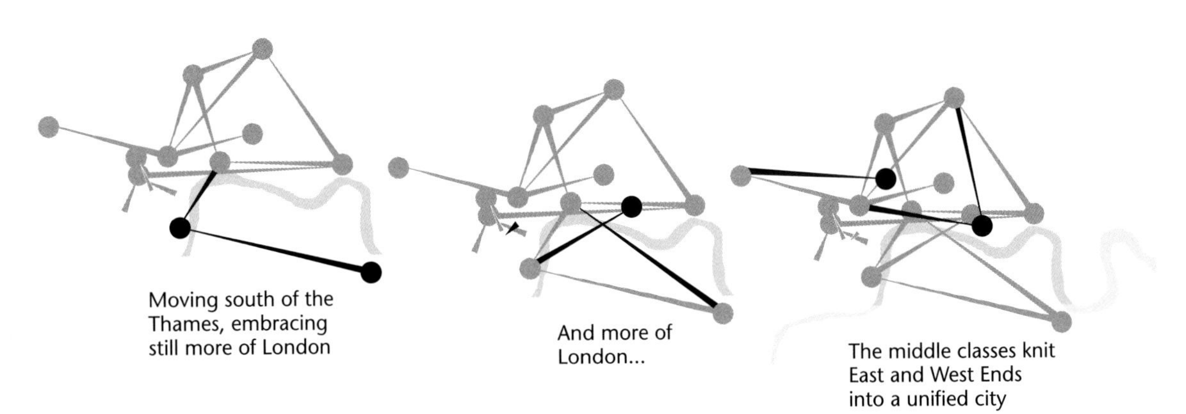

Moving south of the Thames, embracing still more of London

And more of London...

The middle classes knit East and West Ends into a unified city

Data Accuracy

There are many types of accuracy associated with data and maps. Accuracy is not precision, which is how specific or detailed the data is. "It's hot today" may be accurate, but is not precise. "It reached 89.443° fahrenheit today" is both accurate and precise. One approach to assessing accuracy is to ask a series of questions about your data.

Ways to Think about Data Accuracy

Are the facts accurate? An Israeli tourism ad on the London Underground included a map showing the occupied West Bank, the Gaza Strip, and the Golan Heights as parts of Israel. They're not.

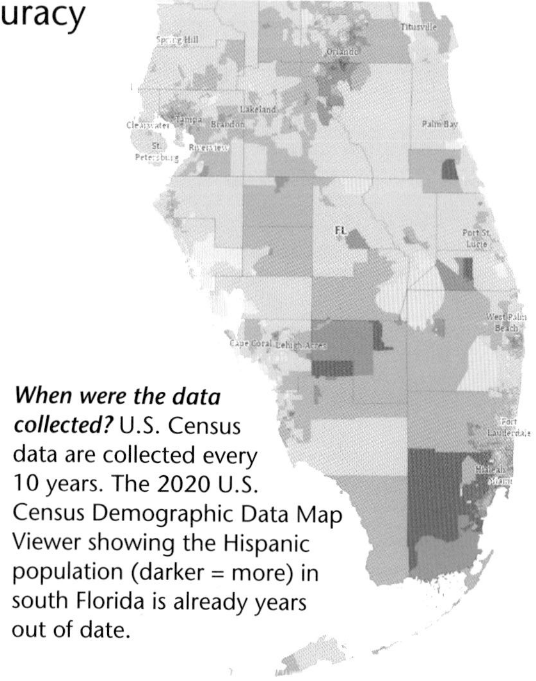

When were the data collected? U.S. Census data are collected every 10 years. The 2020 U.S. Census Demographic Data Map Viewer showing the Hispanic population (darker = more) in south Florida is already years out of date.

What are the assumptions behind the data? U.S. Federal Emergency Management Agency flood maps fail to take into account recent, significant climate change. Studies have shown that 14.6 million properties, rather than the 8.7 million shown on the maps, are susceptible to 100-year floods.

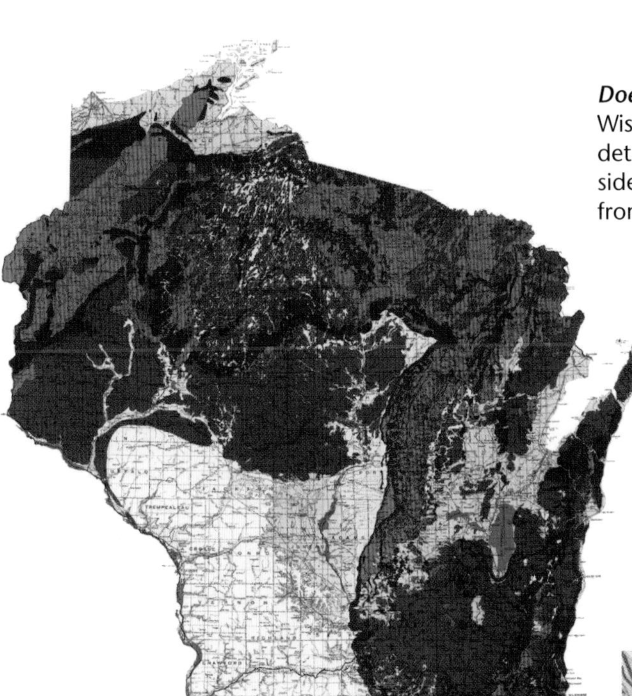

Does detail vary across the data set? A Wisconsin sand and gravel map includes detailed county data from some areas (east side of map) with crude, state scale data from other counties (west side of map).

Are the data from a trustworthy source? Some guy on Reddit says where you shouldn't live in Columbus.

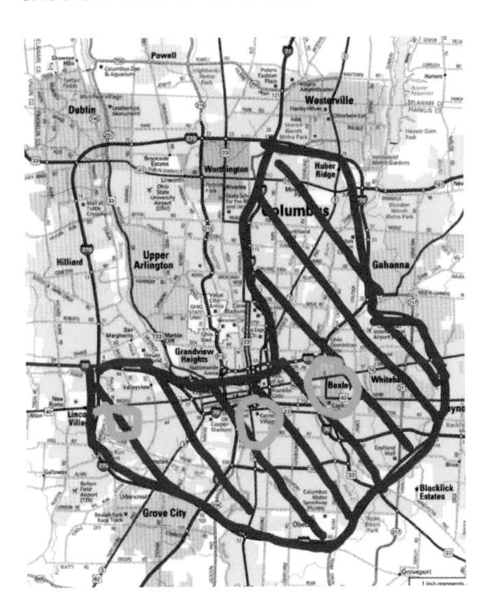

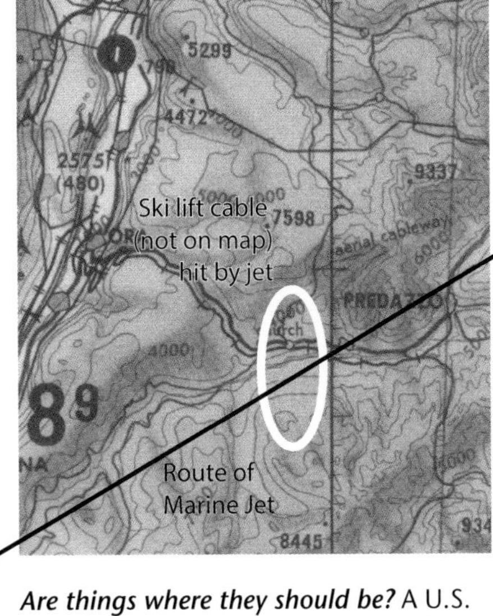

Are things where they should be? A U.S. Marine Corps jet struck and severed a ski lift cable spanning an Alpine valley in Cavalese, Italy, in 1998, sending the ski lift gondola crashing to the earth, killing 20 people. The jet crew did have a map of the area, but not one that showed the ski lift cable. The cable was shown on Italian maps, but the Pentagon prohibits the use of maps made by foreign nations.

Digital Data

Widespread use of geographic information systems and the development of extensive databases of digital GIS data require an understanding of metadata and copyright. The rapid growth of digital geographic data generated by your cell phone and other devices portends a human future with ever-decreasing locational privacy.

Metadata and Standards

Metadata are data about data. Dependable digital geographic data include such detailed information as

Identification information: general description of the data
Quality information: which defines the data quality standards
Spatial data organization information: how spatial information in the data is represented
Spatial reference information: coordinate and projection information
Entity and attribute information: map data and associated attributes
Distribution information: data creator, distributor, and use policy
Metadata reference information: metadata creator
Citation information: how to cite information when used
Temporal information: when data were collected, updated
Contact information: how to contact data creator

Geographic data standards in the U.S. have been set by the Federal Geographic Data Committee (www.fgdc.gov). Geographic data providers often follow such guidelines.

A fundamental benefit of geographic data and software standards is *interoperability*, defined by the IEEE (Institute of Electrical and Electronics Engineers) as "the ability of two or more systems or components to exchange information and to use the information that has been exchanged." Strive for such standards when creating your own geographic data.

Copyright

Copyright is a form of protection provided by the laws of a country to the creators of original works. Intended to reasonably compensate them for their efforts, it was originally of limited duration and included exceptions such as the fair use doctrine. The exceptions have been weakened over time, making current copyright laws more burdensome. In the U.S., copyrights include:

Reproduction of copies of the original copyrighted work
Preparation of derivative works based on the original copyrighted work
Distribution/sale/transfer of ownership of the original copyrighted work

Maps, globes, and charts are covered under U.S. copyright law. This copyright does not extend to the data, the facts themselves. What is copyrighted is the representation of the facts – the line weights, the colors, the particular symbols – as long as this representation includes an "appreciable" amount of original material. So, you can make a map based on the data included on a copyrighted map, but you can't photographically reproduce it.

Works created since 1978 are automatically copyrighted, and there's no way to tell if something is copyrighted or not by looking at it (unless it has a notation to that effect). Given this opacity, it's best to assume that works are copyrighted until you learn otherwise.

Copyleft

Copyleft refers to an array of licensing options encouraging reuse, reproduction, distribution of, and modifications to creative works within certain parameters. Copyleft counters the restrictions and prohibitions of copyright in that only "some" rather than "all" rights are reserved by the creator of a work. Copyleft strategies are integral to the philosophy behind open-source software and collectively created, crowdsourced data (such as Wikipedia).

The GNU General Public License is used for open-source software. For example, the raster GIS software GRASS is available under a GNU license.

Creative Commons Licenses are used for other creative works, including maps, geographic data, and reproductions of historical maps. Six licenses are offered under Creative Commons, including the *Attribution-ShareAlike License:*

You are free to: share – copy, distribute, and transmit the work in any medium; adapt – remix, transform, and build upon the material for any purpose, even commercially, under the following conditions:

Attribution: You must give appropriate credit, provide a link to the license, and indicate if changes were made. You may do so in any reasonable manner, but not in any way that suggests the licensor endorses you or your use.

Share-Alike: If you remix, transform, or build upon the material, you must distribute your contributions under the same license as the original.

Public Domain

Creative works and content neither owned nor controlled by anyone are said to be in the public domain. Public domain works may be used by anyone for any purpose without restriction.

Every January 1, literature, movies, music scores, and other works released 95 years earlier enter the public domain.

Everything copyrighted is subject to the fair use doctrine. Fair use is a legal doctrine that promotes freedom of expression by permitting the unlicensed use of copyright-protected works in certain circumstances.

As a general rule of thumb, federal government maps and data in the U.S. are in the public domain.

Public domain and copyleft licensed maps and geographic data are a great idea when generating data and making maps. Consider using copyleft licensing on maps and geographic data you create. Creativecommons.org provides easy methods for licensing your work.

You and Your Things Are the Data

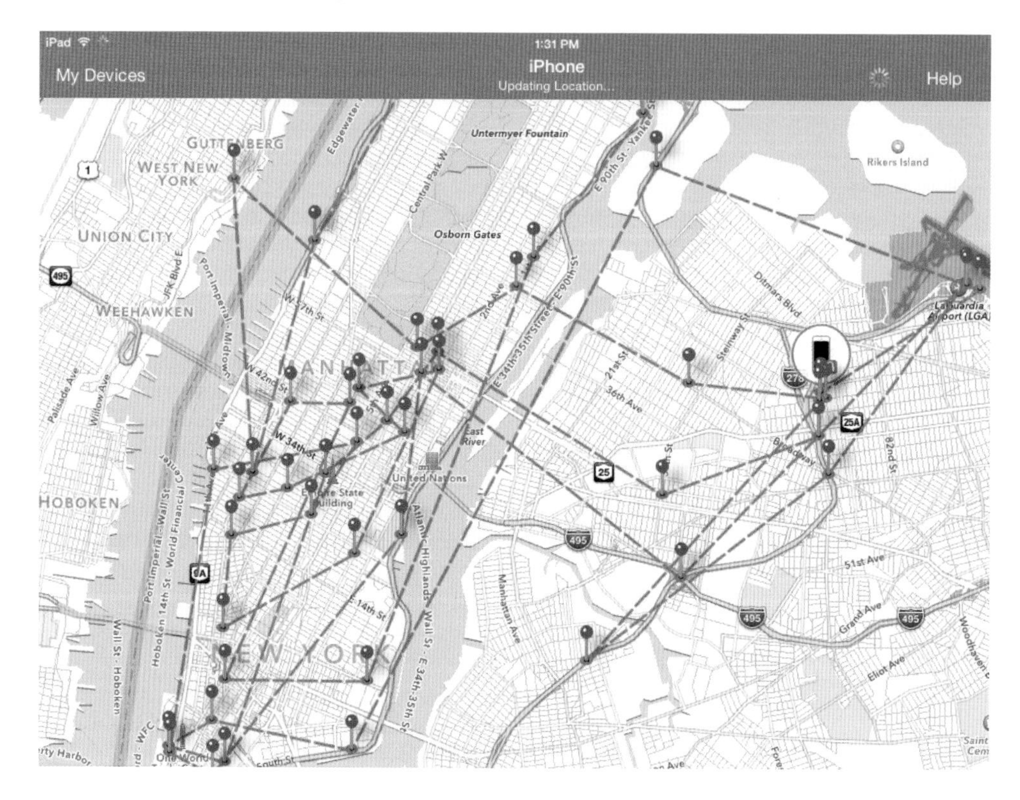

Maile Meloy left her phone in a cab in New York City, and the phone kept track of its adventures, documented on the map above. Location Services on your phone are great for navigating and finding out what's going on around you. Alas, the digital geographic data your phone creates is not just yours: Google, Apple, your service provider, and others have access to the same information.

Location Services

Your phone (or any cellular device) is in constant communication with cell phone towers and can generate your location by using cell tower triangulation:

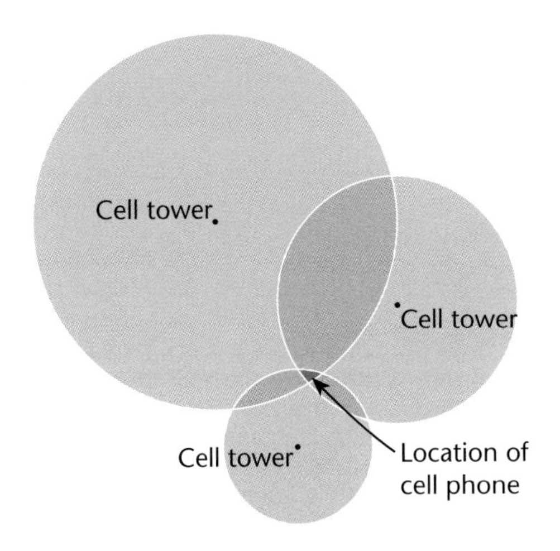

Global positioning systems (GPS) and wi-fi triangulation supplement cell tower data in providing location services for your device. But this comes at a cost – a loss of your locational privacy.

Apple tracks your frequent locations "to provide you with personalized services, such as predictive traffic routing." You can also find your phone if you lose it. Google tracks your location, and those of other people, to generate current traffic conditions on its maps.

Unlike some parts of the government, corporations don't have to follow any laws protecting locational privacy, although they claim to have internal guidelines.

Locational Privacy

Locational privacy (according to the Electronic Frontier Foundation) is "the ability of an individual to move in public space with the expectation that under normal circumstances their location will not be systematically and secretly recorded for later use."

Our smartphones, apps, watches, cars and other devices are saturated with geographic data: locating – geotagging – our photographs, tweets, proximity to food and shopping, traffic, and our friends. All these data can be mapped, by you or whomever else has access to the data. The creation of these data is not secret – you turned on location services, or allowed your social media to include a location with each post. But most of us are not aware how systematic and extensive a source of geographic data we are. Watch out when you protest or storm the U.S. Capitol!

Locational Specificity

A "Location Intelligence" industry (worth nearly $15 billion a year) collects, trades, and sells user location data, much of it collected through smartphone apps. The U.S. National Security Agency (NSA), military, and law enforcement are clients of these locational data sellers.

I wanna hang a map of the world in my house. Then I'm gonna put pins into all the locations that I've traveled to. But first I'm gonna have to travel to the top two corners of the map so it won't fall down.

Mitch Hedberg (no date)

The *Atlas* maps, writing, and illustrations were done by people who live in thatch-roof, wooden houses they made themselves and who eat food they grew themselves. They got up early in the dark morning hours to make wood fires to cook tortillas and warm coffee before walking to their milpas to cultivate corn and beans, and then mapped their fields, rain-forest hunting grounds, traditional medicine places, and ancient ruins.

Maya Atlas (1997)

Ignorance exists in the map, not in the territory.

Eliezer Yudkowsky, *Mysterious Answers to Mysterious Questions* (2007)

More...

Data are a topic of much interest among philosophers of science. What are data? How are they created? How are they used? The literature's vast, and an interesting place to start is Bruno Latour's *Pandora's Hope: Essays on the Reality of Science Studies* (1999). For a review of GIS, mapping, and data, try David DiBiase's open-source web textbook *The Nature of Geographic Information* (www.e-education.psu.edu/natureofgeoinfo). Books about specific application areas of mapping and GIS and the data and analysis associated with them have proliferated. For very different perspectives on mappable data, see some of the books listed at the end of the preceding chapter. For literary mapping see Franco Moretti, *Atlas of the European Novel* (1998) and *Graphs, Maps, Trees* (2005). For other humanities mapping see *Geographic Information Systems in the Humanities: An Introduction* by Ian Gregory and Alistair Geddes (2020).

Data are the subject of a number of specialized journals, including *IEEE Transactions on Big Data,* the *Data Science Journal, Journal of Data Mining and Knowledge Discovery, Journal of Data Mining and Digital Humanities. Data and Society* is concerned with data's social implications (datasociety.net).

On gerrymandering, see Mark Salling's *Redistricting: A Guide for the GIS Community* (2021). It's succinct, but it's all here.

There are lots of sites on the web for free mappable data – just search around. The Creative Commons web pages have super resources on copyright and copyleft licensing. Search for "map copyright" on the web for a diversity of materials on the subject.

Sources: Book dust jacket and map from Paul Pry. *For Your Convenience.* Routledge, 1937. Courtesy of Routledge. Pin oak map from E.N. Munns, *Distribution of Important Forest Trees of the United States* (U.S. Department of Agriculture, Washington, 1938). Vegetation zones map redrawn from *Goode's World Atlas* (Chicago, 1990). Erwin Raisz's Mexico maps from his "A New Landform Map of Mexico" (*International Yearbook of Cartography* 1, 1961). Boylan Heights street and alley map from Denis Wood, *Everything Sings: Maps for a Narrative Atlas* (Siglio Press, 2010). Delaware County, Ohio, air imagery courtesy of the Delaware County, Ohio (auditor.delco-gis.org). Earthquake tweet data from the U.S. Geological Survey's *Twitter Earthquake Detector.* FICO Score data from Sumit Agarwal, Andrea Presbitero, Andre F. Silva, and Carlo Wix. "Who Pays For Your Rewards? Redistribution in the Credit Card Market," *Finance and Economics Discussion Series 2023-007.* Washington: Board of Governors of the Federal Reserve System, 2023. Gerrymandering diagram adapted from "This is the best explanation of gerrymandering you will ever see." Washington Post, March 1, 2015. Blind walker and swimmer data from Emily Davis, 1928, "Why Lost People Go in Circles" (*The Science News-Letter* 14:378). Greenland ice sheet change data from NASA's *Earth Observatory* website (earthobservatory.nasa.gov). Indian Ocean tsunami data from the National Oceanic and Atmospheric Administration (noaa.gov). Charles Dickens's *Our Mutual Friend* data from Franco Moretti, *Atlas of the European Novel* (Verso,1998). Boylan Heights neighborhood sound diagram from Denis Wood, *Everything Sings: Maps for a Narrative Atlas* (Siglio Press, 2010). Wisconsin sand and gravel map based on *Glacial Deposits of Wisconsin* map (Wisconsin Geological and Natural History Survey, 1976). The Cavalese map is courtesy of the Harvard Map Library, Harvard University.

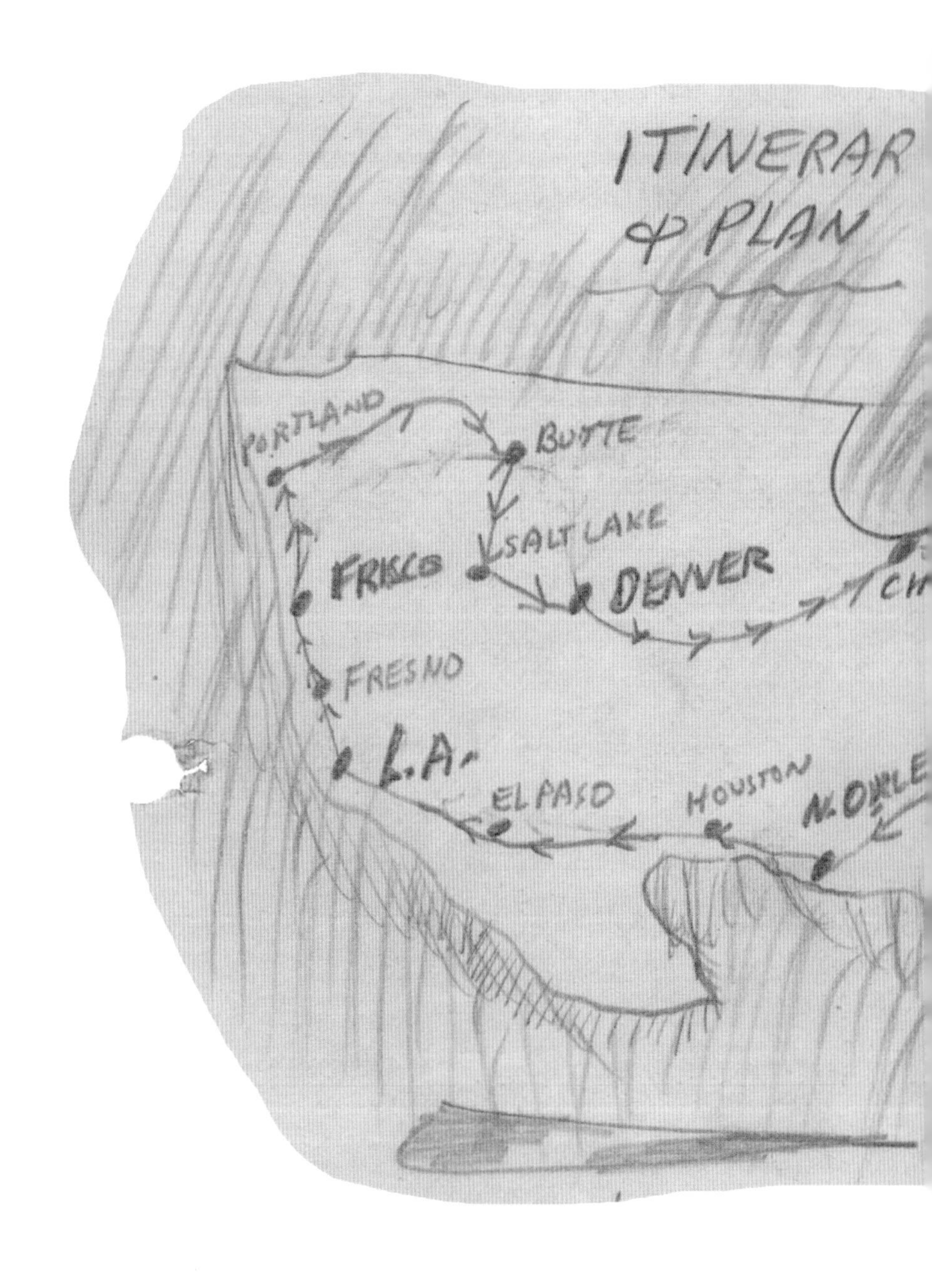
ITINERAR
& PLAN
PORTLAND
BUTTE
SALT LAKE
FRISCO
DENVER
FRESNO
L.A.
EL PASO
HOUSTON
N. ORLE

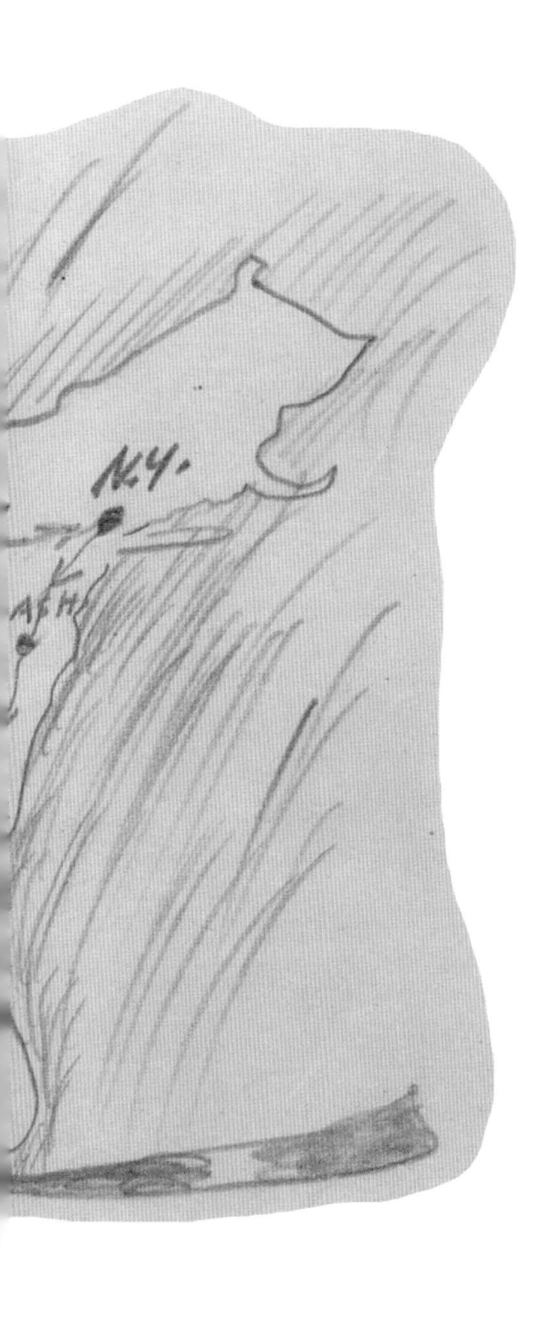

How is it made?

ON THE ROAD
Reverting to a Simpler style —
Further draft & beginnings — Nov. 1949
ITINERARY & PLAN
PORTLAND
BUTTE
SALT LAKE
FRISCO
DENVER
CHI
DET.
N.Y.
WASH
FRESNO
L.A.
EL PASO
HOUSTON
N. ORLEANS
① New York Jail
② Times Square I
③ Road to New O'leans
④ New Orleans
⑤ Road to Frisco
⑥ Frisco (+ Valley)
⑦ Road to Butte
⑧ Butte
⑨ Road to Denver
⑩ Denver
⑪ Road to New York
⑫ Times Square Again
From May to May
CHARACTERS
Red Moultrie
Clem Lemke
Slim Jackson
Old Bull
Dean Pomeray
Marylou
Evelyn Johnson
Mrs. Moultrie
Elena
Old Moultrie
Laura Moultrie
Laurette
And Various Shades

CHAPTER 4

Map Making Tools

The reason a map is being made suggests appropriate tools. Even if Jack Kerouac had access to GIS software, it would not have served him as well as the pencil he used for the maps he made while working on his Beat Generation bible, *On the Road.* Notoriously typed on a 120-foot scroll during a three-week creative outburst in 1951, this work was preceded by years of travel, note taking, draft making, and maps. The map here, from Kerouac's 1949 notebook "Night Notes," reimagines the second road trip described in the book. The pencil, with its ease of use, erasability, and low cost is a great map making and thinking tool. But GIS software can be, too. Choose appropriate tools based on what you need to do.

Making Maps without Computers

You certainly don't need a computer to make maps. Indeed, map making with pencils and paper is appropriate, inexpensive, and effective in many instances. A sketch map made with pencils and paper may be your final map, or it may be a vital step in the process of producing a map with other tools. Jack Kerouac's hand-drawn map helped him envision *On the Road*. Visual thinking and discovery are not limited to any specific map making tools.

Old-school mapping tools. Back when the Voyager map was made (1987), the map making tools of choice were lighted drafting tables, scribers, technical pens, peel-coat film, and stick-up type (right). Within five years such tools would be replaced by computers, software, digitizing tables, scanners, and mice. Map making tools change quickly, but map design principles should transcend these changes.

Empower without power. Computer mapping doesn't work when you don't have a computer. Sketch mapping (left), in this case in the Philippines, engages community members in compiling cultural and economic resources. In parts of the world, computers, computer skills, and electricity may not exist. Sketch maps are as useful as their computer counterparts and certainly may be digitized for use on the computer.

Win the election. Volunteers for the Democratic presidential candidate go door to door in a key city in a swing state. Using a map printed from an internet site and a pen, they mark their opponent's supporters as an X and their supporters as an O. A filled O means a supporter who may not vote (such as an elderly person with no transportation). Addresses and phone numbers of these supporters are entered into a spreadsheet. These people will be called on election day, urging them to vote and offering them transportation to the polls.

Sketch and discover. Abraham Verghese used maps to help think about his HIV-infected patients. Dr. Verghese practiced medicine in rural Tennessee. He and his colleagues were stunned when HIV-infected patients began to dominate their practices. What was this urban problem doing in rural Tennessee? "There was a pattern in my HIV practice. I kept feeling if I could concentrate hard enough, step back and look carefully, I could draw a kind of blueprint that explained what was happening here..." Dr. Verghese borrowed a map of the U.S. from his son. With the map spread on his living room floor, he marked where his HIV patients lived. He labeled the map Domicile, but he could have called it "Birthplace," for most of his patients were men who had come home to die.

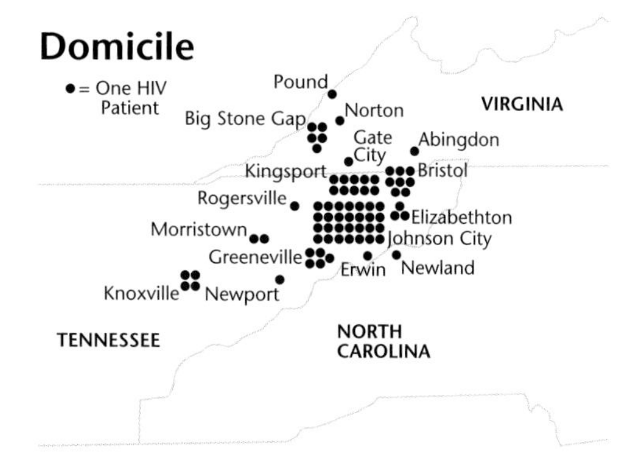

Dr. Verghese next mapped where his HIV patients lived between 1979 and 1985. The places on the Acquisition map "seemed to circle the periphery of the United States" and were mostly large cities. "As I neared the end, I could see a distinct pattern of dots emerging on this larger map of the USA. All evening I had been on the threshold of seeing. Now I understood." Dr. Verghese learned of a circuitous voyage, a migration from home and a return, ending in death. It was "the story of how a generation of young men, raised to self-hatred, had risen above the definitions that their society and upbringing had used to define them."

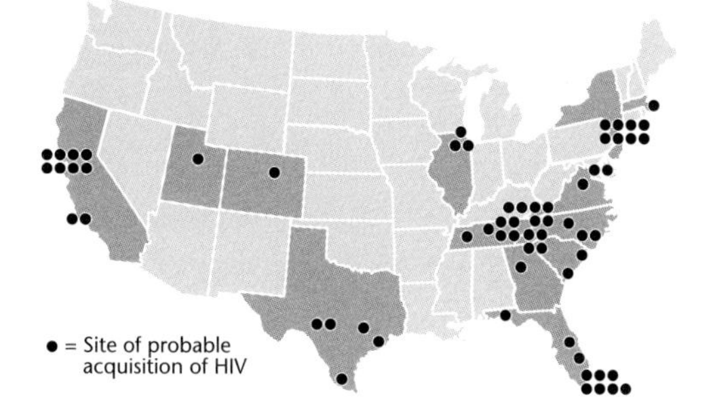

The maps Verghese made on his living room floor with pencils and paper (and redrawn here) might not be much to look at, but the thinking they inspired was invaluable.

Making Maps with Computers

Most maps are made with computers and software. Geospatial software imports or creates data and maps referenced to a coordinate system (such as latitude and longitude). Making maps means making diverse software work together. Mapping software changes constantly. Online discussion forums and conferences are vital for serious mapmakers.

Mappable Data

Geospatial data sources provide data with geographic coordinates, data that can be matched (such as addresses) or joined to geospatial data (such as FIPS codes).

Open Source data: OpenStreetMap, Natural Earth, USGS Earth Explorer, Copernicus Open Access Hub, NASA Earthdata Search, United States Census Bureau

Commercial data: ESRI ArcGIS Living Atlas of the World, DigitalGlobe, Airbus OneAtlas, Google Earth Engine, Policymap

Mapping Software

Geographic Information Systems (GIS) capture, store, analyze, manage, and present spatial or geographic data. Such software typically runs on a desktop or laptop computer or server.
Open Source: QGIS, GRASS
Commercial: ArcGIS, MapInfo

Remote sensing software analyzes raster data collected by satellites, aerial photography, drones, and other remote sensing technology. Such image analysis is computationally demanding.
Open Source: Orfeo Toolbox (OTB), SNAP
Commercial: ENVI, Erdas Imagine

Global Positioning Systems (GPS) generate geographic data from coordinated earth/satellite system. GPS software processes GPS data for use in other software.
Open Source: GPSBabel
Commercial: Garmin Base Camp

Programming languages allow customization of geospatial software and the development of mapping applications.
Open Source: Python, JavaScript, SQL (databases)
Commercial: C++, C#,

Online mapping sites provide online geographic data for navigation and general reference with a few basic GIS, GPS and mapmaking functions.
Open Source: OpenStreetMap
Commercial: Google Maps, Google MyMaps, Google Earth, Apple Maps

Web mapping software help create, display, and interact with web maps,typically with scripting languages and libraries of map app elements.
Open Source: Leaflet, Open Layers
Commercial: Map Box, ArcGIS Online

Bridge Software transfers GIS (spatial) data to graphic design software.
Commercial: ESRI Adobe Interface, MAPublisher, Geographic Imager

Graphic design software imports maps created with other geospatial software for further design, layout, and preparation for printing, the web, and other media. Maps typically lose their coordinates when imported into such software.
Open Source: GIMP, Inkscape, Krita
Commercial: Adobe Illustrator, Photoshop, InDesign, CorelDraw

Discussion Forums and Conferences

GeoNet - Esri Community
Geospatial World Forum
GIS Lounge Forum
GIS Stack Exchange
OpenStreetMap Forum
Reddit - r/gis
The Spatial Community

ESRI User Conference
International Cartographic Association
International Conference on Cartography and GIS
NACIS: North American Cartographic Information Society
State of the Map Conference

Workflow

Workflow is a specific map's creation process: initial conception, planning, data gathering, appropriate tools, design, and completion in the intended medium. Workflow is related to the broader map making process, reflected in the order and content of chapters in this book. Any mapping project requires drawing ideas from the general map making process and applying them strategically to a specific workflow. Professional project managers coordinate all these aspects of workflow and the people involved in the process. Nat Case helped to develop these workflow guidelines.

Workflow requires strategically refining the general map making process to fit your specific map.

Workflow requires thinking about the end of the process at the beginning. When does the map have to be finished? What is your deliverable? A digital file to a printer for a book? Publication on the web? Printed on laser printer? Are there intermediate milestones with their own deadlines, and how are they structured?

Workflow is often (but now always) collaborative. Who asked you to make the map? Are you clear on all their expectations? Who can help with data? Who can help with software? Who helps with the final map (say, a professional printer, or webmaster if the map is online). If you are part of a group working on a larger project, yours is one of multiple workflows: how does your workflow contribute to the broader group workflow?

Workflow is an interactive process. For each project, engage in trial and error and be prepared to make adjustments to reach your final goal. Seek feedback and constructive criticism along the way.

Workflow is a learning process. Throughout a project (and career) you'll learn about new kinds of data, software, and techniques. Check online videos, discussion forums and presentations by pros at conferences. *Workflow requires knowing when to depart your comfort zone. Be flexible.*

Workflow subdivides your efforts. Workflow requires splitting your project into sections (possibly layers in software), usually linked to specific data sources, each of which will require different kinds of acquisition strategies, processing via software, organization, design, and finalizing as part of the final map. Consider which subdivided section can run simultaneously, and which sequentially.

Workflow should be documented, for the future. Workflow documentation can be as simple as sources documentation, software documentation, process outline, and linking of final layers to specific sources. Basic question: who is this documentation for, and what do you need to include for them to understand details of your workflow?

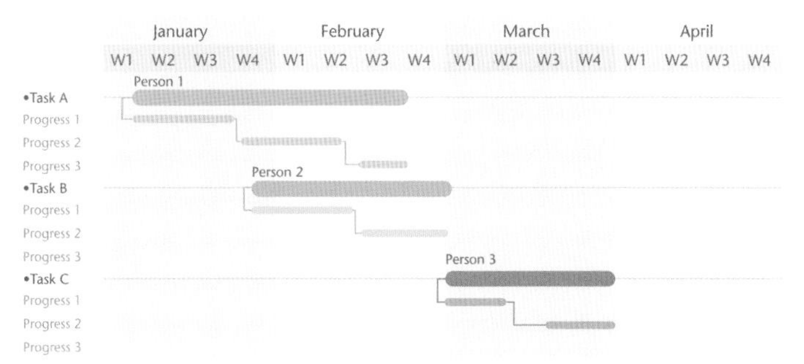

Serious project managers use a Gantt Chart to map out workflow and guide people and projects

The Washington Post
DEPARTMENT OF DATA
Why the South has such low credit scores
Average credit score

That's a cool map. As a client of (and author for) Guilford Publications, Denis and I are responsible for new editions of this book, *Making Maps.* That means keeping an eye out for cool mappable data: like this county level map of US credit scores. I'll need to remake this map for the book: the next step is to find the data.

Where's the data? "Source: Sumit Agarwal, Andrea Presbitero, André Silva and Carlo Wix" led me to a paper with some maps, but no data. Hmmm. Might be easier to contact the author of the *Washington Post* article, Andrew van Dam.

Analysis by Andrew Van Dam
Staff writer | + Follow
February 17, 2023 at 6:00 a.m. EST

Help arrives. Andrew responded. He got the data from the U.S. Federal Reserve Board, who were responsible for the data used in the study. The FRB forwarded the data to me, as a spreadsheet.

A2

	A	B
1	county_fips	fico_bucket
2	1001	2
3	1003	3
4	1005	2
5	1007	1
6	1009	2
7	1011	1
8	1013	2
9	1015	2
10	1017	1
11	1019	3
12	1021	2
13	1023	1
14	1025	2
15	1027	2
16	1029	1
17	1031	2
18	1033	2

Editing data for ArcGIS Pro. In Excel, I converted the FIPS column to text and added a 0 before the four digit county FIPS codes to help the join in ArcGIS Pro.

H15

	A	B
1	FICO Bucket	Range
2	1	687.2-726.0
3	2	726.0-733.7
4	3	733.7-740.3
5	4	740.3-745.8
6	5	745.8-750.9
7	6	750.9-774.3

Joining spreadsheet, classifying, and color in ArcGIS Pro. The data were already classified into "buckets" so I used the FRB's classification. Colors came from an orange range provided in the software.

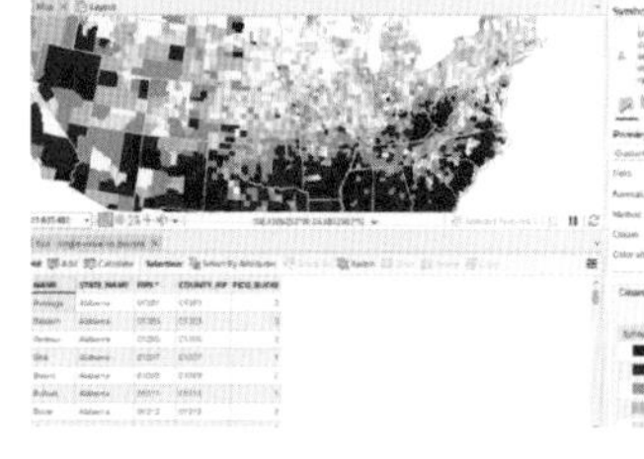

Export to Adobe Illustrator. The ArcGIS Pro Adobe export option did not work, so I saved as a vector PDF and emailed it to my Mac. That worked! Once in Illustrator, I adjusted the colors and line weights based on my documented specs.

InDesign to the Printer. The map was placed into InDesign where this book has been designed and constructed. I'll send all the files to William at Guilford and he'll prep the files and export them for the printer of the book.

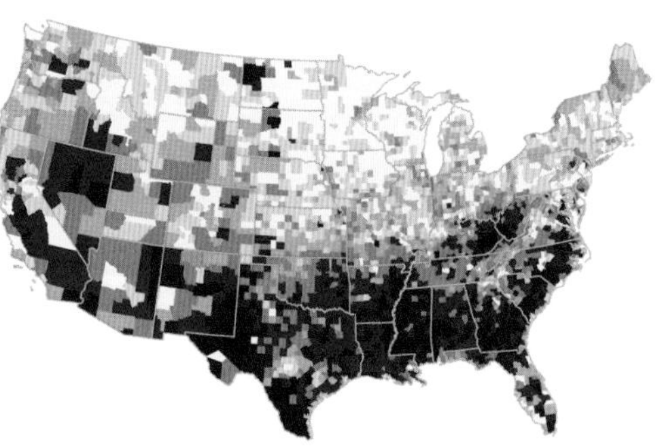

Artificial Intelligence

ChatGPT: Artificial intelligence (AI) refers to the development of computer systems that can perform tasks that typically require human intelligence. These tasks include learning, reasoning, problem-solving, perception, language understanding, and decision-making.

JBK: How will AI impact map design?

ChatGPT: AI will have a profound impact on map design, influencing how maps are created, customized, and presented.

Automated Map Layout: AI algorithms can automatically arrange map elements such as labels, symbols, and legends in a visually pleasing and organized manner. This can save time for cartographers and improve the overall aesthetics of the map.

Dynamic Styling: AI can adapt map styles based on user preferences or the specific context of use. For example, an AI-powered map might adjust its color scheme and visual representation to match a specific theme or cater to individuals with color vision deficiencies.

Personalized Maps: AI can create maps tailored to individual users or specific target audiences. By analyzing user data and preferences, AI can generate maps that highlight relevant points of interest or provide information based on the user's interests and needs.

Style Transfer: AI can transfer map styles from one map to another. This means a map designer could take the aesthetic elements from a well-designed map and apply them to a new map, maintaining a consistent visual identity across different projects.

Data-Driven Design Decisions: AI can analyze large datasets to identify patterns and correlations that can inform map design decisions. For instance, AI might determine the most effective way to display a particular dataset to enhance understanding and insights.

Real-Time Map Updates: AI can dynamically update map information in real-time, ensuring that the map remains current and relevant. This is especially useful for digital maps that are frequently updated with live data, such as traffic conditions or weather information.

Context-Aware Maps: AI can analyze user context, such as location and time of day, to provide maps with contextually relevant information. For instance, a map app could display nearby restaurants during lunchtime.

Automated Data Integration: AI can assist in integrating diverse datasets into maps seamlessly. By understanding the relationships between different data sources, AI can help create more comprehensive and informative maps.

Natural Language Interaction: AI-powered map design tools may allow designers to interact with the software using natural language, making it easier to specify design preferences and customize map elements.

Enhanced 3D Mapping: AI can improve the design of 3D maps by automatically generating realistic terrain models and rendering textures, leading to more immersive and visually appealing representations.

Despite these advantages, it's crucial for map designers to retain creative control and ensure that the AI-generated maps align with the intended purpose and user experience.

Whereas maps, like guns, must be accurate, they have the additional advantages that they are inexpensive, don't require a permit, can be openly carried and used...

Bernard Nietschmann, "Defending the Reefs" (1995)

...The Indians are very expert in delineating countries upon bark, with wood coal mixed with bear's grease, and which even the women do with great precision.

John Long, *Voyages and Travels* (1791)

A little instruction in the elements of chartography – a little practice in the use of the compass and the spirit level, a topographical map of the town common, an excursion with a road map – would have given me a fat round earth in place of my paper ghost.

Mary Antin, *The Promised Land* (1912)

A traveller would bring his judgment in Question who should despise the Directions of his Map for want of real Roads in it, because here stands a Dott instead of a Town, or a cypher instead of a City, and it must be a long Day's journey to travel thro' two or three inches.

From a letter to *The Spectator,* no. 593, Monday 13 September 1714.

Kim: You suck at drawing, don't you?
Scott: Maps are hard! I could draw it really good if it was a sheep.

Bryan Lee O'Malley, *Scott Pilgrim vs. the World* (2005)

More...

Older map making texts are all about making maps with feet, eyeballs, pens, and paper. Erwin Raisz published the first in English in 1938, *General Cartography.* In 1962 he published *Principles of Cartography.* Both amply exhibit his ability at landform mapping, many of which are still in print. Another classic, very much still being used, is David Greenhood's *Down to Earth: Mapping for Everybody* (1944), republished as *Mapping* (1964). Few of the maps in Wood's *Everything Sings* (2013) were made with a computer (the atlas provides production notes for all its maps). Few parish maps are made with computers.

While not necessary, most of you will be making your maps on computers. Billions and billions of manuals and guides and websites explain how to use GIS software, and web mapping sites are usually easy to use without much guidance. Thoughtful GIS books include Nicholas Chrisman, *Exploring Geographical Information Systems* (2001); Francis Harvey, *A Primer of GIS* (2016); and Paul Longley, Michael Goodchild, David Maguire, and David Rhind, *Geographic Information Science and Systems* (2015).

Sophisticated map making and analysis tools can be free. Check out QGIS (www.qgis.org), a full-function GIS and mapping software package, free and open-source, available for multiple operating systems. For learning programming for web maps, get out of this paper book and onto the web. Of particular coolness is the University of Kentucky's *New Maps Plus* (newmapsplus.uky.edu).

For an engaging overview of maps and graphs in action, helping to figure stuff out, see Howard Wainer, *Graphic Discovery* (2005).

Mark Monmonier's thought-provoking books are full of stories about how maps do their work in the world: *Cartographies of Danger: Mapping Hazards in America* (1997), *Air Apparent: How Meteorologists Learned to Map, Predict, and Dramatize Weather* (1999), *Bushmanders and Bullwinkles: How Politicians Manipulate Electronic Maps and Census Data to Win Elections* (2001), *Spying with Maps: Surveillance Technologies and the Future of Privacy* (2002), *From Squaw Tit to Whorehouse Meadow: How Maps Name, Claim, and Inflame* (2007), *Coast Lines: How Mapmakers Frame the World and Chart Environmental Change* (2008), *No Dig, No Fly, No Go: How Maps Restrict and Control* (2010), and *Lake Effect: Tales of Large Lakes, Arctic Winds, and Recurrent Snows* (2012).

Sources: Jack Kerouac map reproduced by permission of SLL/Sterling Lord Literistic, Inc. Copyright by John Sampas, Literary Representative. Sketch mapping photo courtesy of the *Participatory Avenues* website (iapad.org). Dr. Verghese's AIDS data from Abraham Verghese, *My Own Country: A Doctor's Story* (Vintage, 1994) and "Urbs in Rure: Human Immunodeficiency Virus Infection in Rural Tennessee" (*Journal of Infectious Diseases*, 160:6). Workflow content developed with Nat Case (www.incasellc.com). Gantt Chart recreated from www.freepik.com. Credit score map and data from Andrew van Dam. "Why the South has such low credit scores." *Washington Post,* Feb. 17, 2023.

How do you flatten,
shrink, and locate data?

CHAPTER 5 Geographic Framework

We have to mess with our spherical earth to get it flat. Through a process called "map projection" the curved surface of the earth is flattened. We are used to seeing the flat earth on maps, but what if we flattened a human body? Artists Lilla Locurto and Bill Outcault scanned the entire surface of their bodies with a 3D object scanner. They brought the data into GeoCart map projection software and projected it. We have clearly become accustomed to the radical transformation involved in projecting the earth.

Map Projections

Our earth's surface is curved. Most maps are flat. Transforming the curved surface to a flat surface is called map projection. Projected maps are flat, compact, portable, useful, and always distorted. Any curved surface gets distorted when you flatten it.

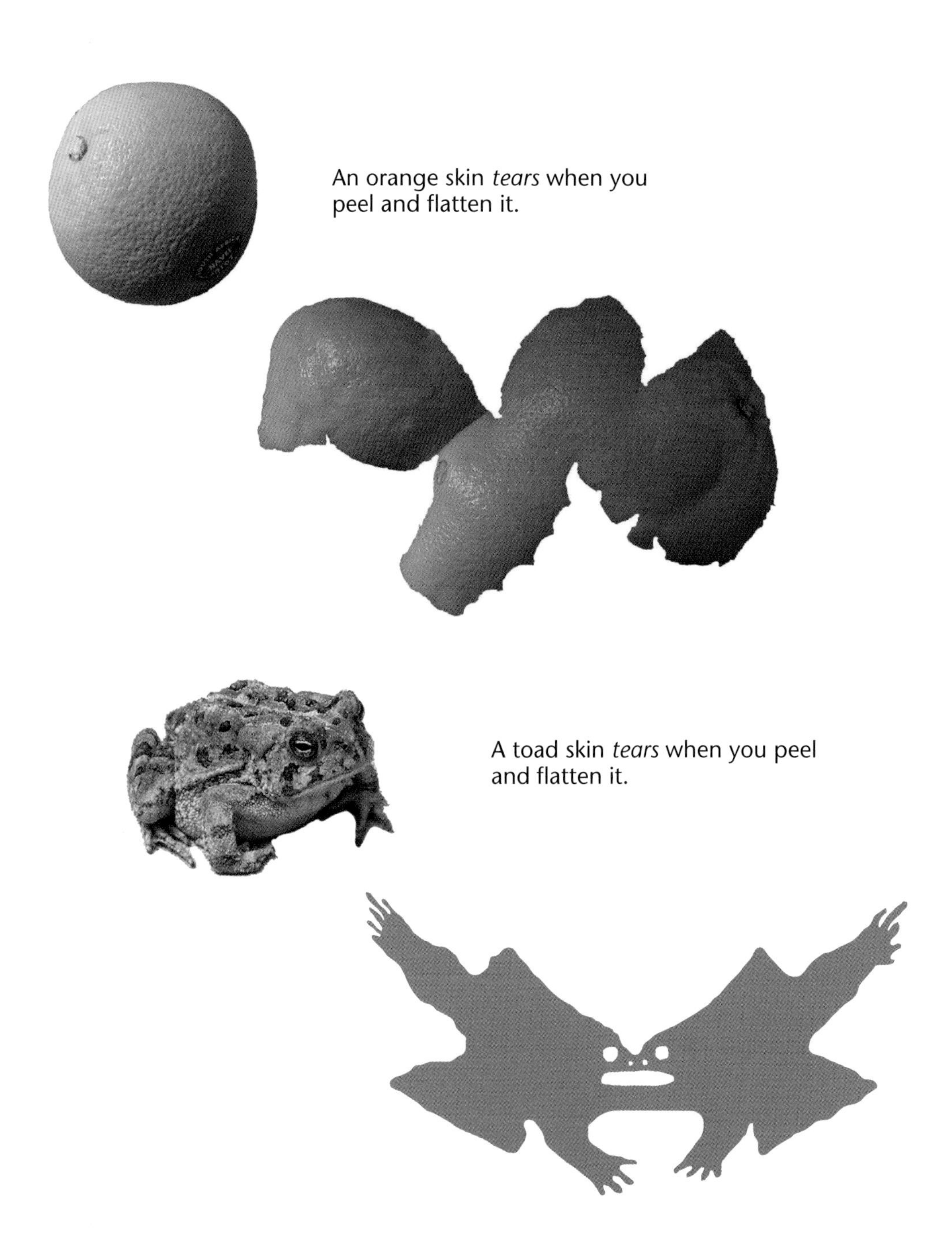

An orange skin *tears* when you peel and flatten it.

A toad skin *tears* when you peel and flatten it.

Strike flat the thick rotundity o' th' world!
William Shakespeare, *King Lear*

The surface of the earth *tears* when you peel and flatten it. Peel a globe and you'll get globe gores (below).

Most map projections stretch and distort the earth to "fill in" the tears. The Mercator projection (bottom) preserves angles, and so shapes in limited areas, but it greatly distorts sizes. Note the size of Greenland on the globe as compared to the Mercator.

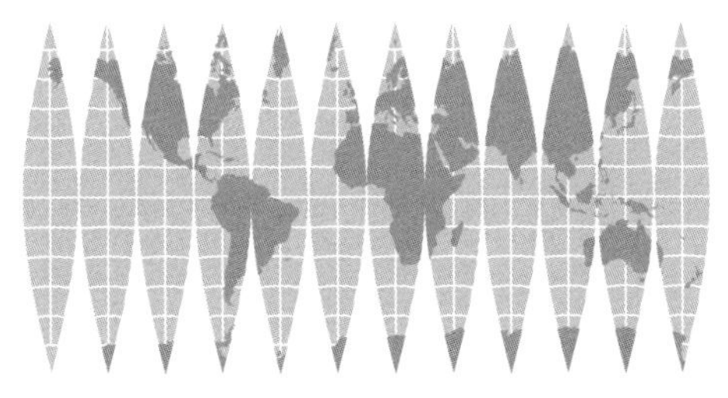

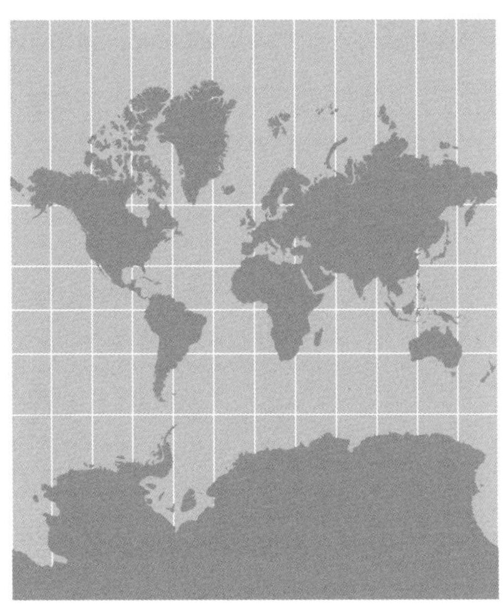

Thinking about Map Projections

There are many ways to think about map projections beyond flattened bodies, oranges, and toads. How is the earth's surface distorted? How do map projections distort your data? What do particular map projections preserve from the spherical earth?

Visualizing Distortions

In the 19th century, Nicolas Auguste Tissot developed an indicatrix to evaluate map projection distortion. Imagine infinitely small ellipses placed at regular intervals on the curved surface of the earth. Imagine these ellipses being projected along with the earth's surface. When scaled to be visible, changes in the ellipses show the location and quality of distortions on the projected map.

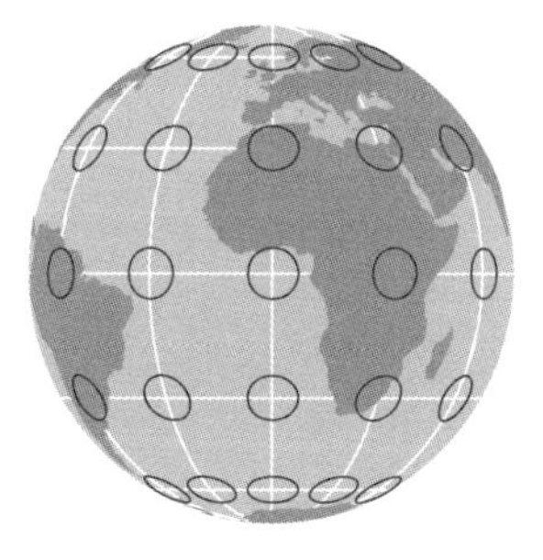

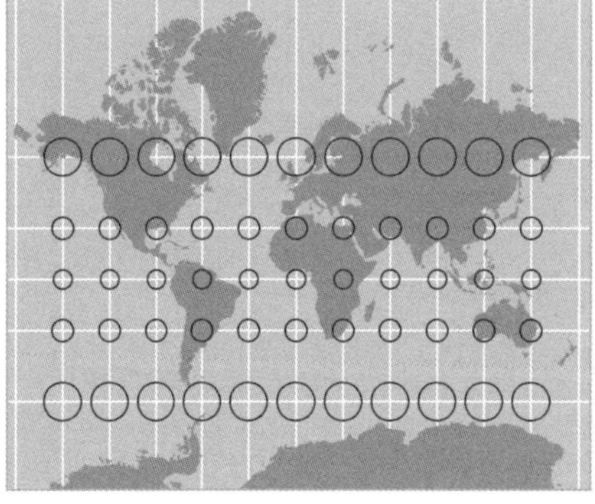

Mercator map projection:
Preserves shapes, changes areas.

Tissot's ellipses change area as you move north and south of the equator on the Mercator map projection (left). The more enlarged the *ellipse,* the more exaggerated the *areas* of the land masses. Ellipse *shapes* are not distorted.

Tissot's ellipses change shape over the surface of this area-preserving map (below). The more distorted the *ellipses,* the more distorted the *shapes* of the land masses. Ellipse *areas* are not changed.

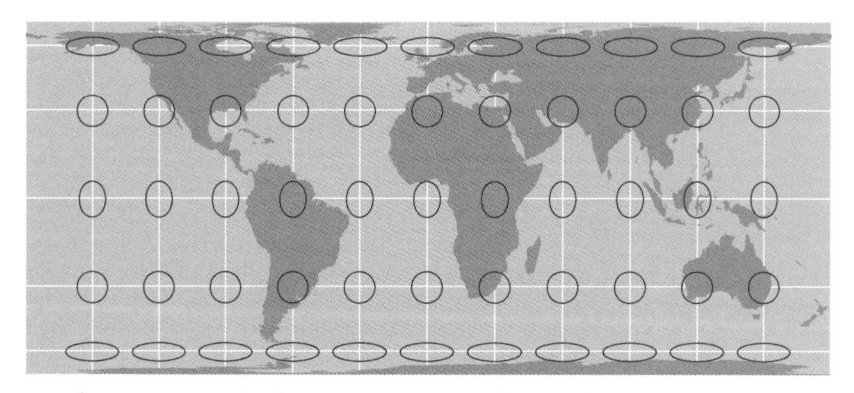

Equal-area map projection: Preserves areas, distorts shapes.

Distorting Data

Mappable data are always associated with a location on the earth's surface. That is, mappable data are always tied to the grid. Because this grid gets distorted when it's projected from the curved surface of the earth to the flat surface of the map, the data tied to the grid are distorted too. Projections matter because of what they do to our data! It's important that what map projections do to our data clarifies, not muddies, it.

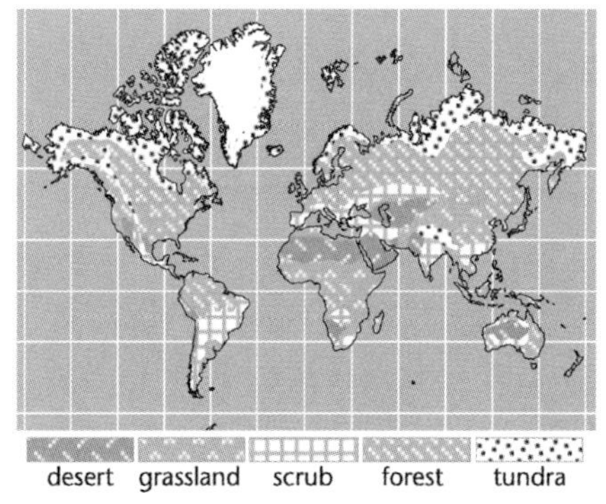

Mercator map projection:
Preserves shapes, distorts areas.

A map of biomes on a Mercator (left) distorts the data. Northern biomes are greatly expanded in area compared to those near the equator. This suggests the global predominance of northern biomes, which is incorrect.

A map projection that preserves areas (below) suggests the limited global extent of northern biomes. But now shapes of continents are distorted! Understand the trade-offs of different projections: with areal data, it's better to distort shapes, not areas.

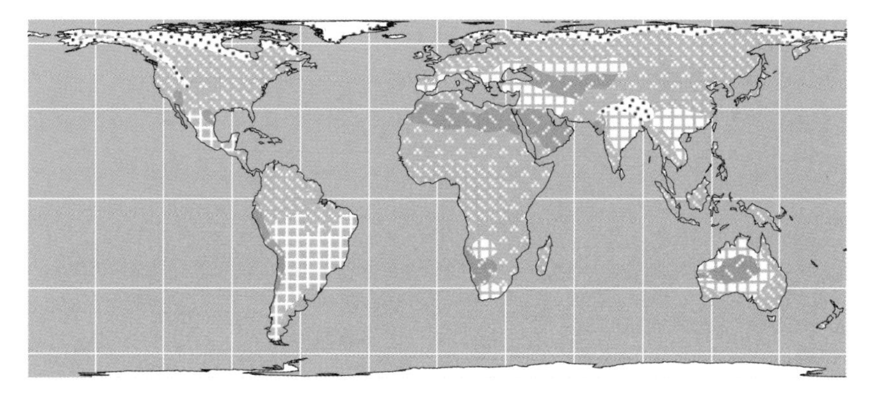

Equal-area map projection: Preserves areas, distorts shapes.

What Map Projections Preserve

No map projection preserves the attributes of a globe, which does preserve the earth's relative sizes, shapes, distances, and directions. Map projections can preserve one or two of these attributes of the globe, but not all four together. Select a map projection that makes the best sense for your data.

Preserving Sizes or Area

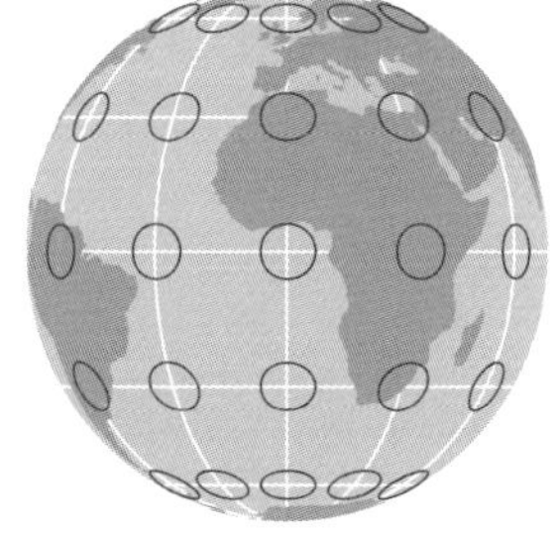

Some projections preserve area. This means that areas of the same size on the globe have the same relative size on the map. Such size, or area-preserving (equal-area) map projections are a good default for maps showing area data.

Mollweide projection: Oval shape, preserves area. Rounded map shape suggests the round earth. The Mollweide can be recentered to minimize shape distortions of regions of greatest interest.

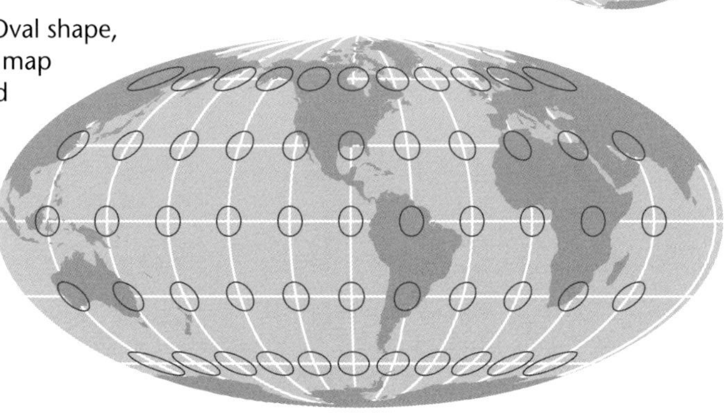

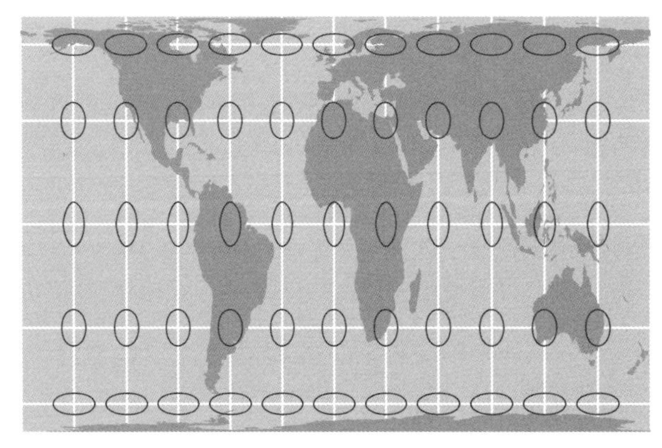

Peters (Gall-Peters) projection: To some map experts, what garlic is to vampires. This equal-area map projection's straight grid makes north-south relationships straightforward. As with any rectangular map projection, it fits into page layouts.

This is a good projection for illustrating the shape distortions inherent in equal-area map projections and the area distortions inherent in shape-preserving projections. Also good as a symbol of affinity with the Global South.

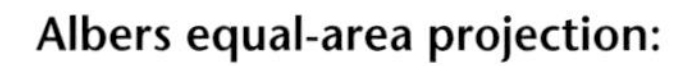

Albers equal-area projection: A common equal-area map projection. Poor for world scale maps because of shape distortion and peculiar form. However, recentering on an area of interest (the U.S., below) and selecting part of the earth (continent, country) results in an equal-area map with minimal shape distortion.

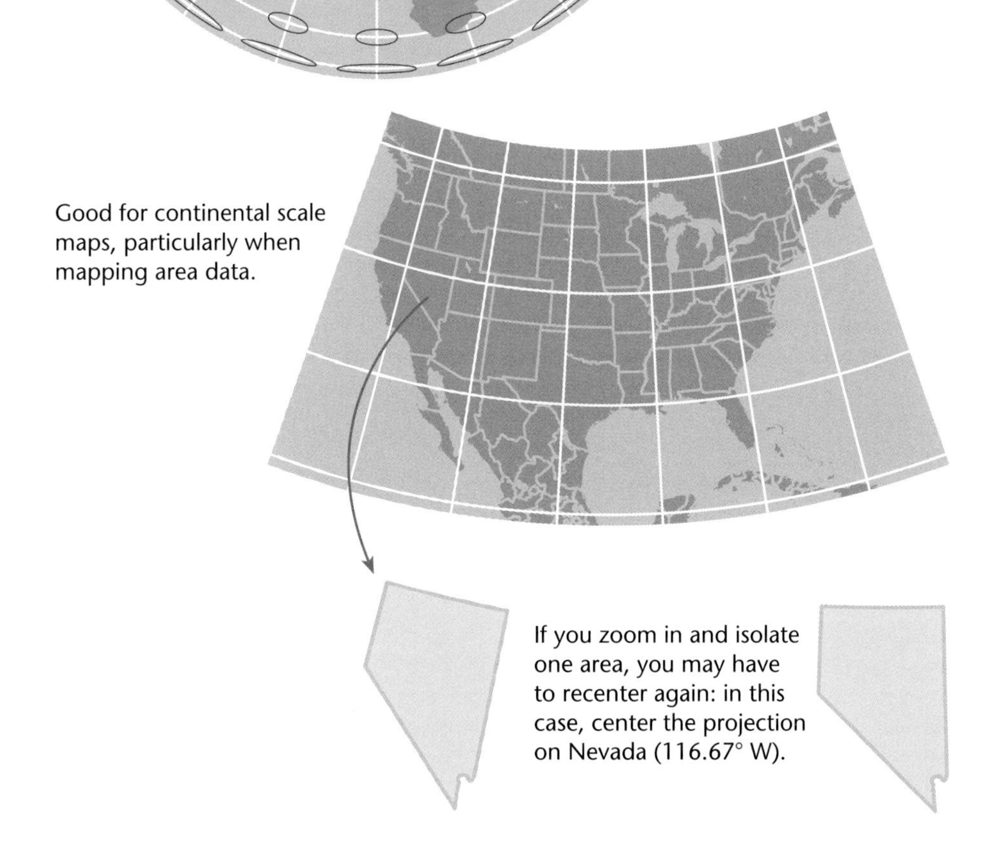

Good for continental scale maps, particularly when mapping area data.

If you zoom in and isolate one area, you may have to recenter again: in this case, center the projection on Nevada (116.67° W).

Preserving Shape (Angles)

Preserving shape (or angles) on a map projection means the projection is *conformal.* Conformal map projections preserve angles (around points) and therefore shape in small areas. As long as you are away from areas of high distortion, the shapes of continents look OK in comparison to shapes on the globe. Areas are, however, distorted. Conformal map projections are good for mapping regions and continents, especially when statistical data are involved. At sub-global scales, distortions of area and shape are not as evident.

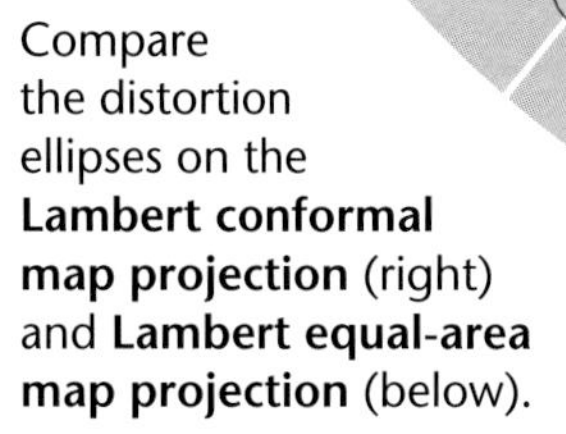

Compare the distortion ellipses on the **Lambert conformal map projection** (right) and **Lambert equal-area map projection** (below).

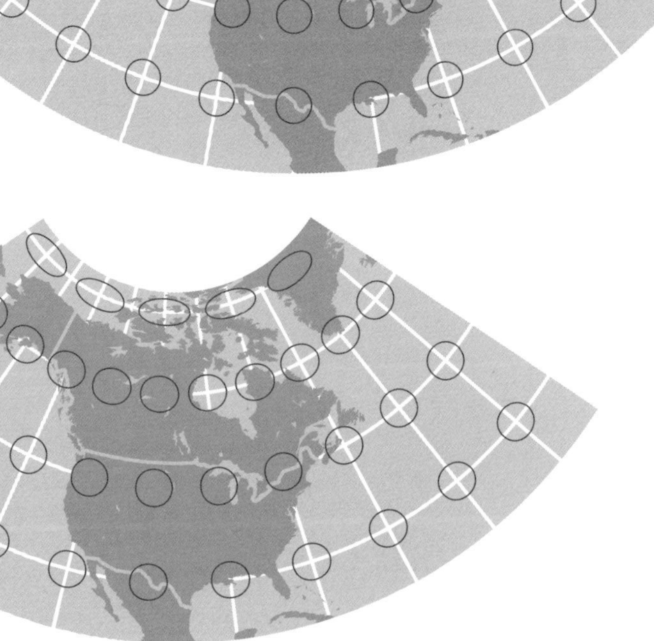

Mercator projection: One of the few conformal world projections. Its distortions of sizes are nasty, and it is a poor choice for a world map. Good for equatorial maps (below) where the area distortion is small. A Mercator with the distorted north and south lopped off was chosen for the Voyager map. Also good for maps of very small areas (below, right). A modified version of the Mercator is used for most web maps, because north is always up, eliminating angular distortion of streets and other features as you move away from the center of the projection.

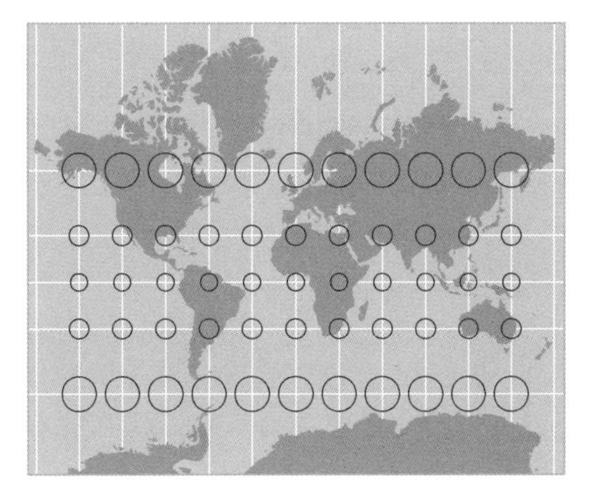

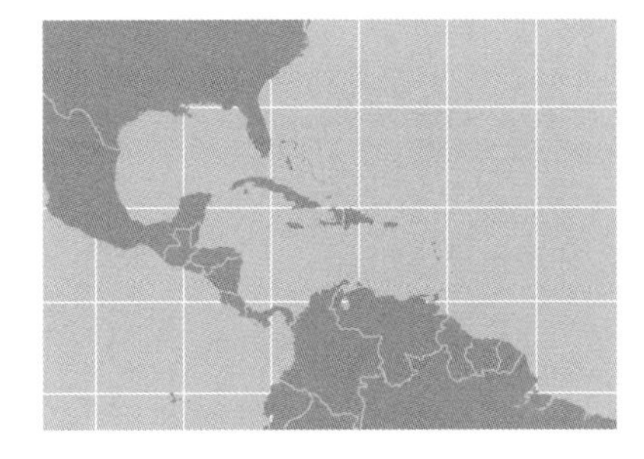

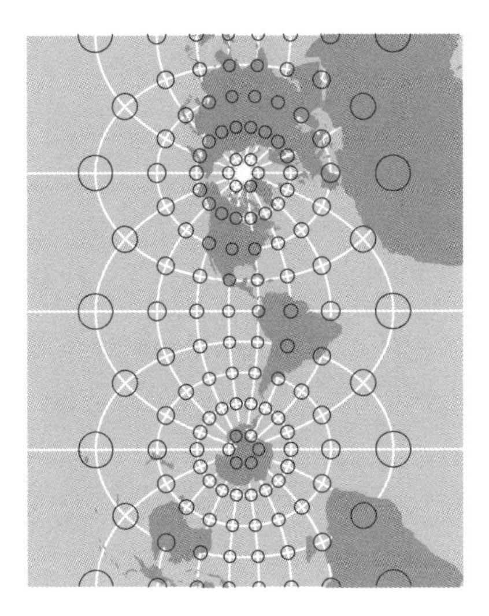

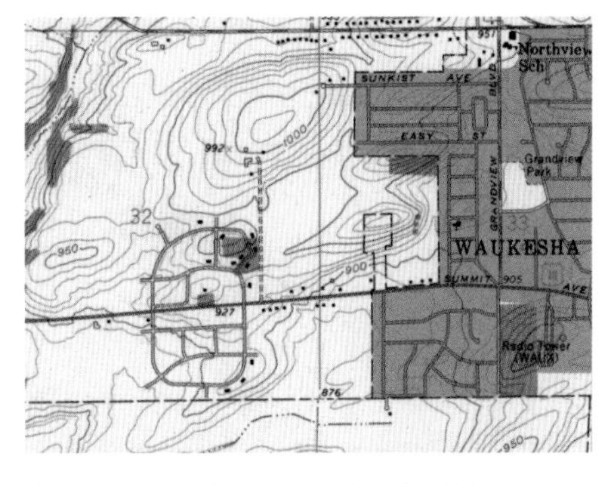

Transverse Mercator: On the Mercator projection, scale is true along the equator. When that projection is recentered sideways along a meridian (or line of longitude), scale is true along that meridian. This recentered projection is known as the Transverse Mercator and is the basis for the Universal Transverse Mercator coordinate system. When areas a few square miles in size are mapped using this projection, they are effectively free of all distortion.

Preserving Distance, Direction

Stretch a piece of string between two points on a globe, and you will get the shortest distance between the points (a great circle). Some projections preserve such distance relations: a straight line between two points on the map is the shortest distance between those two points on the earth. Distance relations cannot be preserved on equal-area maps. Direction can be preserved on area-, shape-, or distance-preserving map projections.

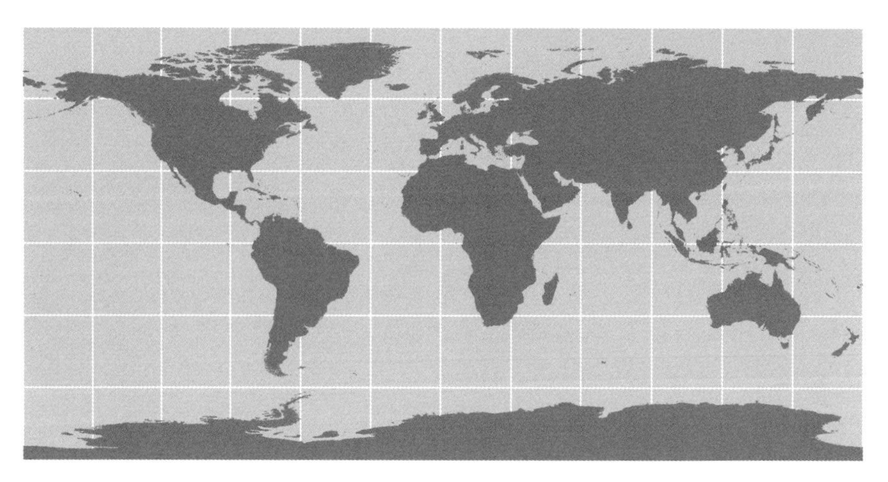

The Geographic Coordinate System, aka the **Geographic** or **Equirectangular projection**, is similar to the **Plate Carrée projection**. Invented by Marinus of Tyre in 100 AD, it "maps meridians to equally spaced vertical straight lines, and circles of latitude to evenly spread horizontal straight lines."

The Geographic Coordinate System is a common default for GIS software and some internet mapping sites, where it distorts the data mapped on it. It preserves nothing but distance, principally for north-south measurements.

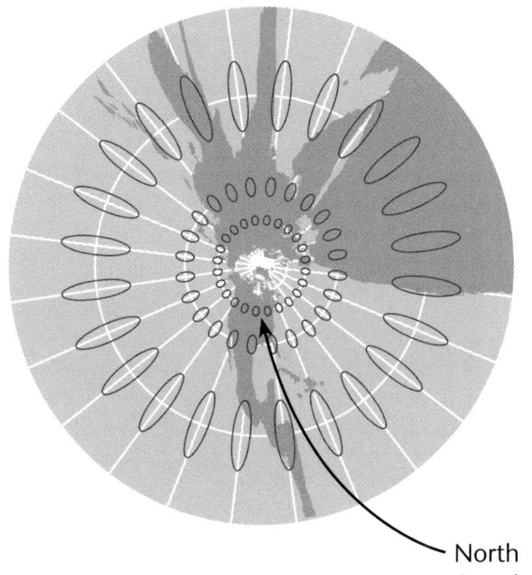

Gnomonic projection: A straight line anywhere on a Gnomonic projection is a great circle route, the shortest distance between two points. Terrifying distortions of area and shape and the inability to show more than half the earth at a time limit other uses of this projection.

Azimuthal equidistant projection: Planar (azimuthal) map projections preserve directions (azimuths) from their center to all other points. The azimuthal equidistant projection also preserves distance. A straight line from the projection center to any other point represents accurate distance in addition to correct direction and the shortest route. Great for travel agencies when centered on their city.

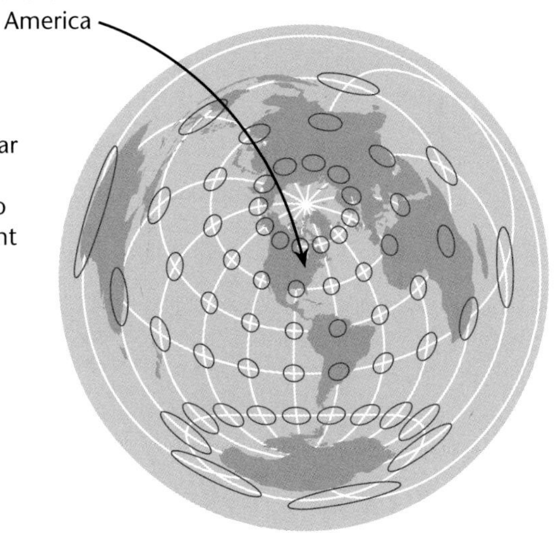

Poor for showing anything but distances from a particular point. Areas are wildly exaggerated and shapes distorted as you move from the center.

Preserving Interruptions

Globe gores, peeled from a globe and flattened, are akin to interrupted map projections. Interrupted map projections minimize distortions on the uninterrupted part of the map, and are typically used on maps of the entire earth.

Interrupted map projections are commonly used for maps of global statistical data. They are also used as icons (Berghaus "star") and as the basis of "cut and assemble" do-it-yourself globes (Fuller).

A Berghaus centered on the north pole is equidistant north of the equator. The Fuller has constant scale along the edges of all 20 of the triangular pieces. Within each of these triangles, area and shape are well preserved.

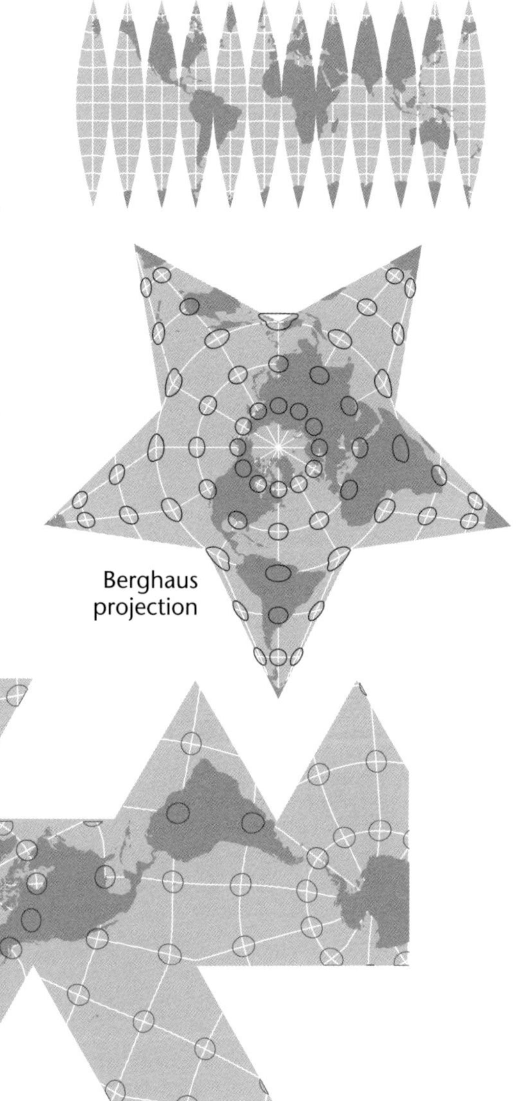

Berghaus projection

Fuller projection

Goode's homolosine projection: Goode's is a common interrupted map projection used for world maps of statistical data. The projection does not distort areas, and shape distortions in the uninterrupted areas of the map are minimized.

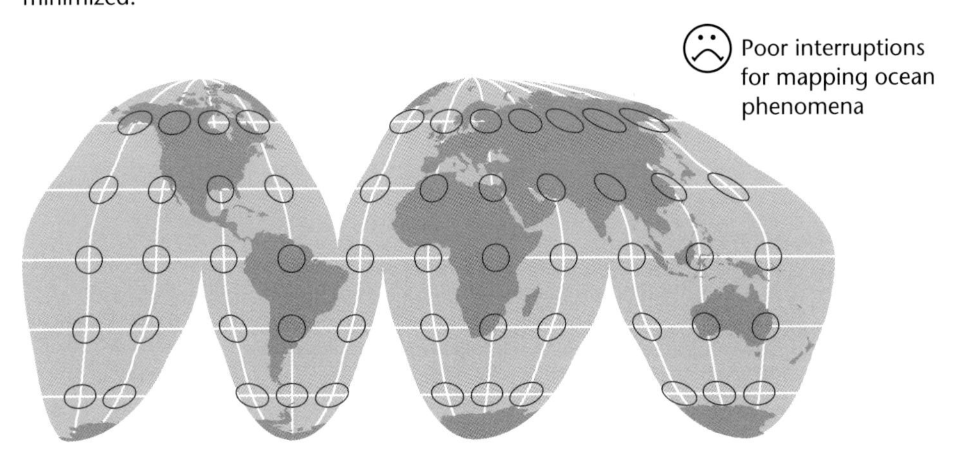

Interruptions can be moved. A map for ocean phenomena can interrupt land areas, for example.

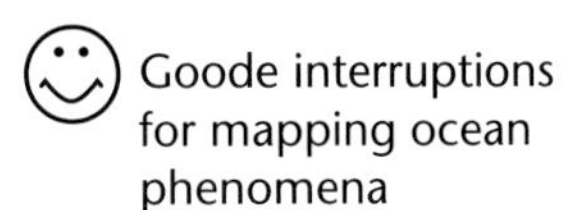

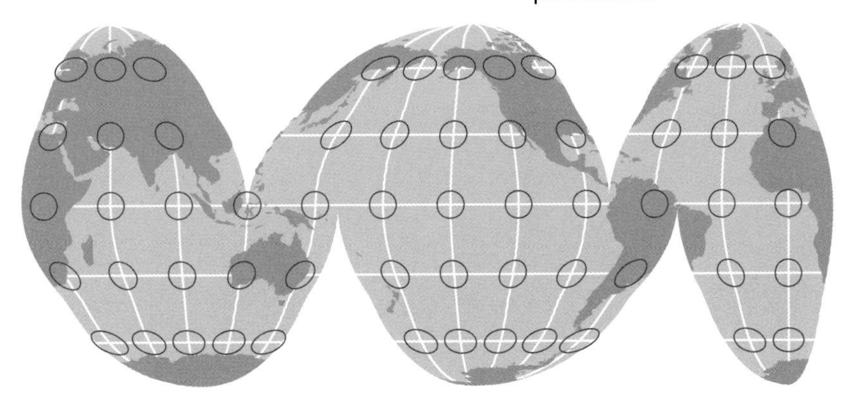

Preserving Everything, Almost

Map projection is most visible at a global scale, where distortions of areas and shape are most evident. Area-preserving projections often badly distort shapes; and shape-preserving projections, area. But there is an alternative – a map projection that does not distort anything!

But that doesn't exist. Crap.

However, instead consider some compromise map projections that distort both area and shape a bit, but neither too badly. They preserve everything, almost.

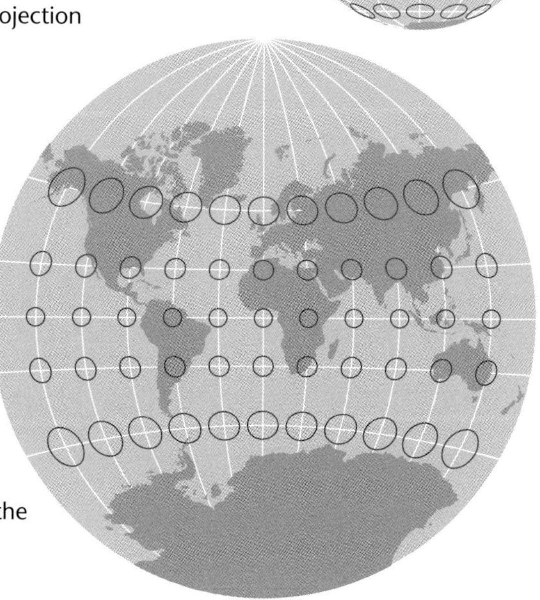

The Van der Grinten projection does not preserve shape or area, but minimizes their distortions in all but polar regions. Usually the polar regions are lopped off and the map presented as a rectangle.

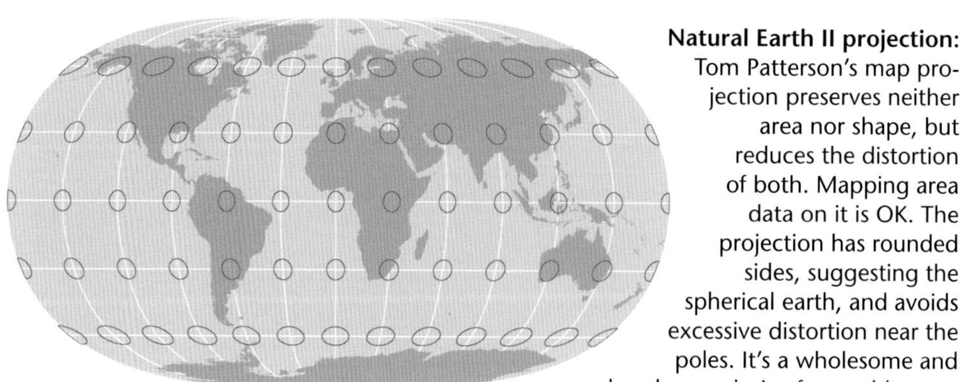

Natural Earth II projection: Tom Patterson's map projection preserves neither area nor shape, but reduces the distortion of both. Mapping area data on it is OK. The projection has rounded sides, suggesting the spherical earth, and avoids excessive distortion near the poles. It's a wholesome and handsome choice for world maps.

Winkel Tripel projection: A compromise map projection used by the National Geographic Society. It was chosen in part due to its modest distortion, but also because it fills out a 24″x36″ poster map better than similar compromise projections.

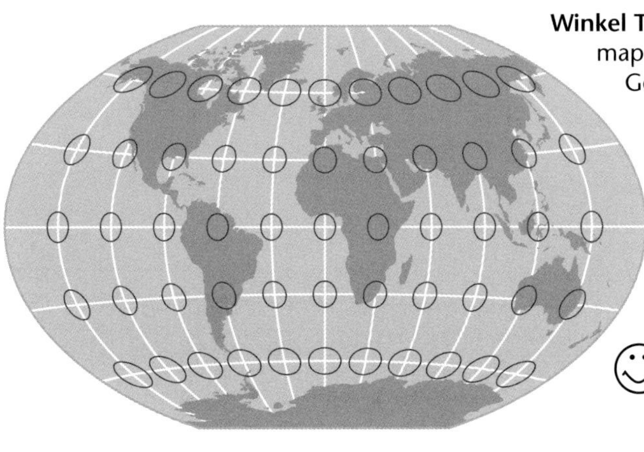

Good for a general world map and for mapping global phenomena

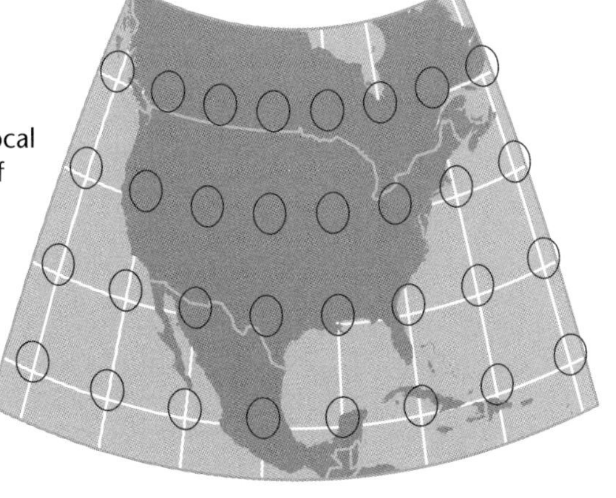

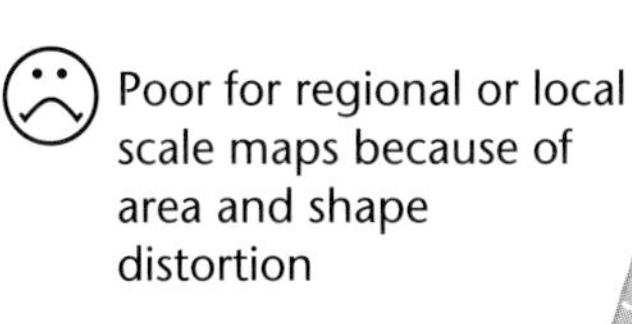

Poor for regional or local scale maps because of area and shape distortion

Map Scale

The earth is big. Maps are small. Map scale describes the difference, verbally, visually, or with numbers. Map scale will be determined by your goals for your map. Map scale affects how much of the earth, and how much detail, can be shown on a map. Map scale terminology is counterintuitive, but think of it this way: how big is a house on the map? If it's small, it's a small scale map. Large, it's a large scale map.

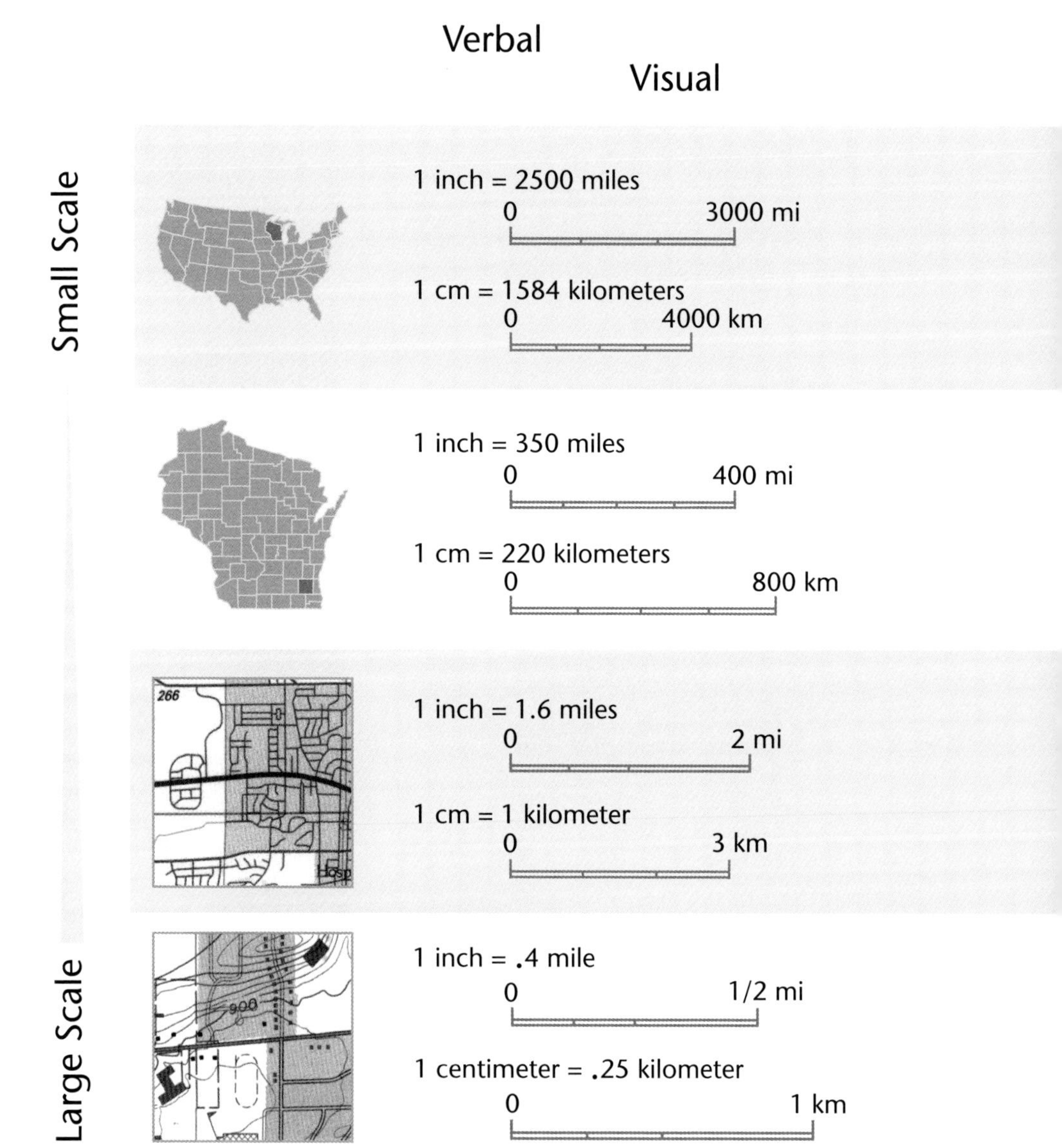

Numerical

Small Scale

1 : 155,000,000

A representative fraction (RF) shows the proportion between map distance and earth distance for any unit of measure. 1 inch on the map is 155 million inches on the earth. 1 cm on the map is 155 million cm on the earth.

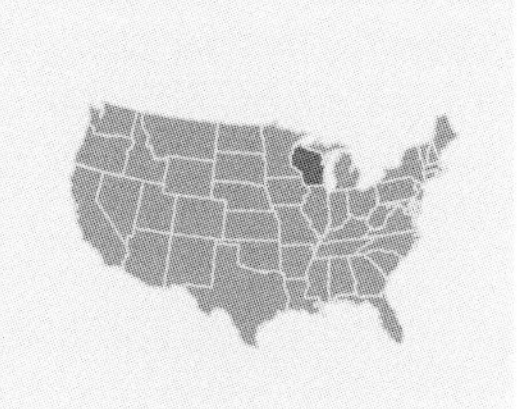

1 : 22,000,000

Distortions from map projections become less visually noticeable at regional and local scales. These distortions may become evident when combining map layers with different projections in GIS.

1 : 100,000

Divide a representative fraction:

1 / 100,000 = .00001
1 / 24,000 = .00004

The former is smaller than the latter: thus 1:100,000 is smaller scale than 1:24,000 (larger scale).

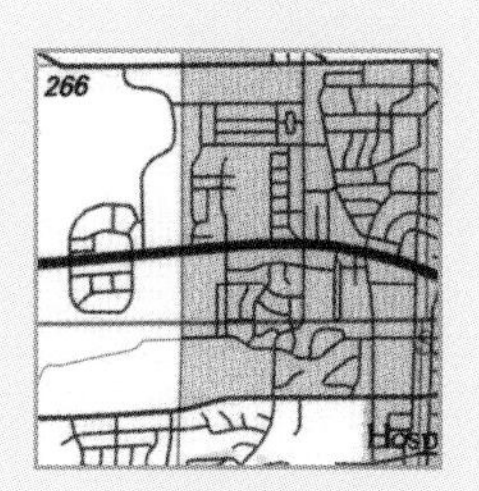

1 : 24,000

Larger-scale maps show more detail, but of a limited area. Map projection distortions are less evident and distance is relatively accurate over the entire map.

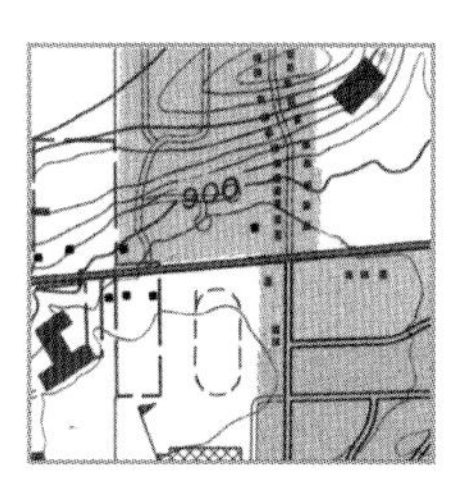

Large Scale

Earth's Shape and Georeferencing

The earth is a geoid, an imperfect 3D object. An ellipsoid is a slightly squished-down sphere that approximates our geoid-shaped earth. Mapping things requires a system for locating those things (georeferencing), typically based on a pair of coordinates. Latitude and longitude are based on our 3D earth. Other systems operate in a projected 2D world.

Ellipsoids and Datums

There are a diversity of ellipsoids that can serve as an approximation of our earth. Many were created to best fit the earth in specific countries. Maps and data sets based on different ellipsoids will not work together.

A datum is based on a single common point shared by the geoid (our imperfect earth) and a particular ellipsoid. All spatial relationships (locations, directions, scales) derive from this single point. Maps and data sets based on different datums will not work together.

Ellipsoids, datums, projections, and georeferencing systems combine in strange ways with great diversity. Expect to convert your geographic data sets in order to get them to work together.

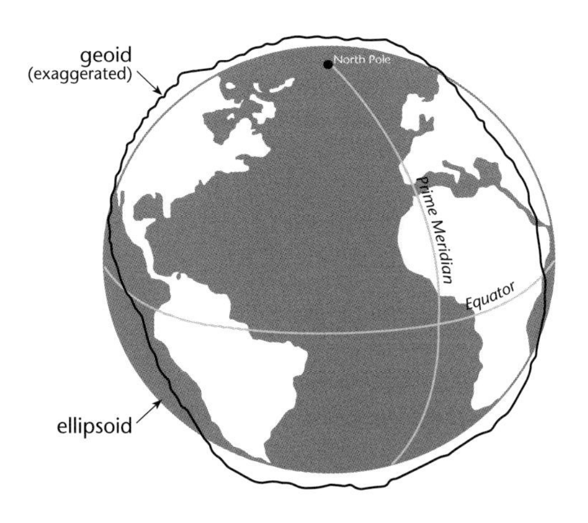

The World Geodetic System of 1984 (WGS84) is the modern standard ellipsoid. The North American Datum of 1983 (NAD83), closely approximating WGS84, has been widely adopted in North America.

Map Coordinates

Map coordinates – also known as georeferences – typically consist of a pair of numbers or letters that locate data, tying them to the grid. Geographic data are distinguished from other data by the fact that they can be located. There are many different map coordinate systems and means of georeferencing. You need to pay attention to which system is associated with your data. Coordinate systems can be converted from one to another, and often have to be when making maps with GIS.

Where is (0, 0)? Where, on earth, should the origin (0, 0) be? If Washington, DC, is the origin, then all other locations are in relation to Washington. Different coordinate systems have different origins.

Area covered? How much of the earth is covered by the coordinate system? Coordinate systems may cover all or only part of the earth.

Flat or spherical? Coordinate systems covering part of the earth assume a flat earth, to take advantage of easier planar geometry. Coordinate systems covering the entire earth assume spherical geometry.

Units? Coordinate systems can be in English units (feet), metric units (meters), or degrees. Different coordinate systems have different units of measurement.

Latitude and Longitude

Latitude and longitude cover the entire earth with one system and a single origin. It's used when you need a single coordinate system for our 3D earth. In this system, locations are specified in degrees, which can be subdivided. There are 60 minutes in 1 degree and 60 seconds in 1 minute. Decimal degrees are increasingly common.

The equator is the origin for **latitude**. Lines of latitude are called parallels. Parallels run east-west, measuring 90° north and 90° south of the equator. Parallels never converge: one degree of latitude is always 69 miles or 111 km.

Greenwich, England, is the origin (prime meridian) for **longitude**. Lines of longitude are called meridians. Meridians run north-south, measuring 180° east and 180° west of the prime meridian. Meridians converge at the poles. One degree of longitude at the equator is 69 miles or 111 kilometers. One degree of longitude at the poles is 0 miles or km (a point).

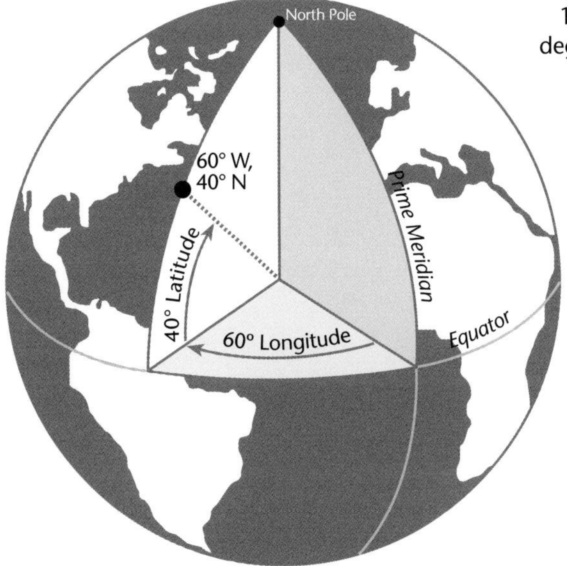

The single origin (0, 0) is off the coast of Africa. Coordinates fall into one of four quadrants to the N/S (latitude) and E/W (longitude) of this origin.

Latitude and longitude can operate in a 3D or 2D world. Map projections flatten and distort the grid of latitude and longitude.

Universal Transverse Mercator (UTM)

The Universal Transverse Mercator (UTM) is a projected coordinate system. UTM, based on the "transverse" (sideways) Mercator projection, covers most of the earth, which is divided into 60 zones, each 6° wide, running from 84° north to 80° south. Planar geometry (a flat earth) makes computations easy. UTM is measured in meters. A point is located in terms of how many meters east and north it is from the origin. UTM is used by environmental scientists, the military, and any other professionals who work at a regional or local scale but need their maps to coordinate with maps of other areas on the earth.

Zone 17

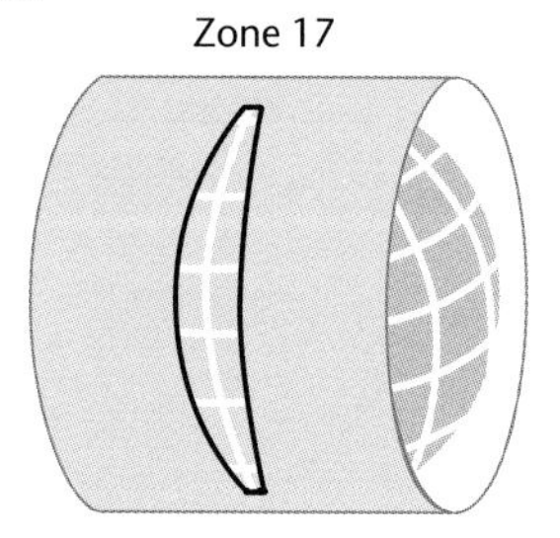

UTM zones are widest at the equator and narrow toward the poles. The poles use the universal stereographic coordinate system. Ten UTM zones (10 to 19) cover the continental U.S.

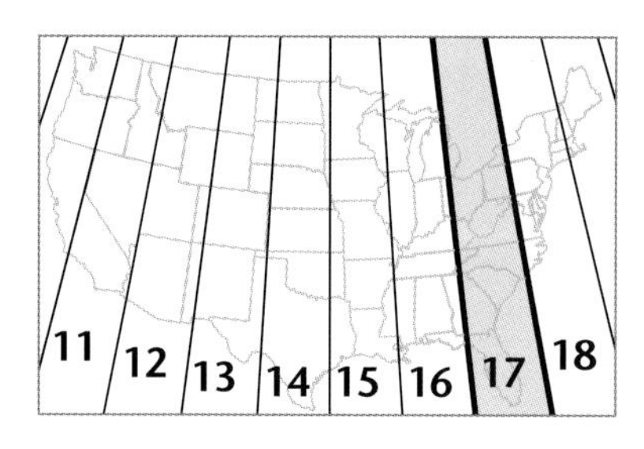

Each 6°-wide zone has a north and south zone. The central meridian of each zone is assigned a value of 500,000 meters (not tied to any earth feature; thus east-west measurements are called *false eastings*). North-south measurements are in terms of the equator. In the south zone, the equator is assigned a value of 10,000,000 meters to avoid negative coordinates.

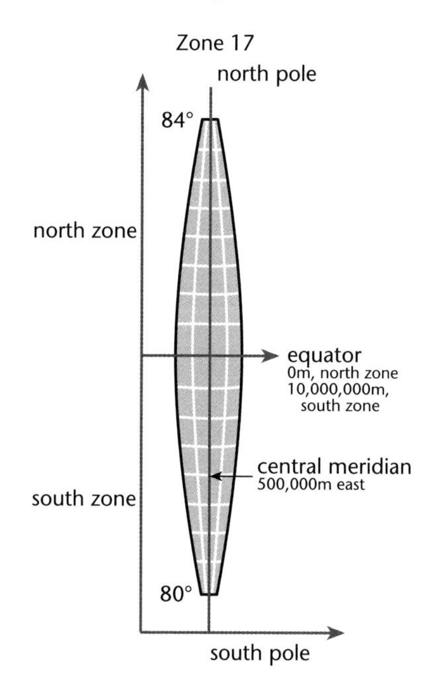

State Plane Coordinate System (SPCS)

The **State Plane Coordinate System** (SPCS) is also based on a flat 2D earth. SPCS is used only in the United States, which is divided into over a hundred areas, each with its own coordinate system. Since each area is relatively small, distortions from projection are minimal. The most recent version of SPCS is based on the NAD83 datum.

SPCS is measured in feet, meters, or both. A point is located in terms of how many units east and north it is from the origin. Where the false origin is set varies. Some states use a central meridian as a false *easting*, as with UTM. Others establish the false origin outside of the bounds of the state (but only coordinates within state boundaries are used).

SPCS is used by planners, urban utilities, and environmental engineers. Similar coordinate systems are used in other parts of the world.

Small U.S. states have a single SPCS zone; larger states (excepting Montana) are divided into several zones. No SPCS zone crosses a state boundary.

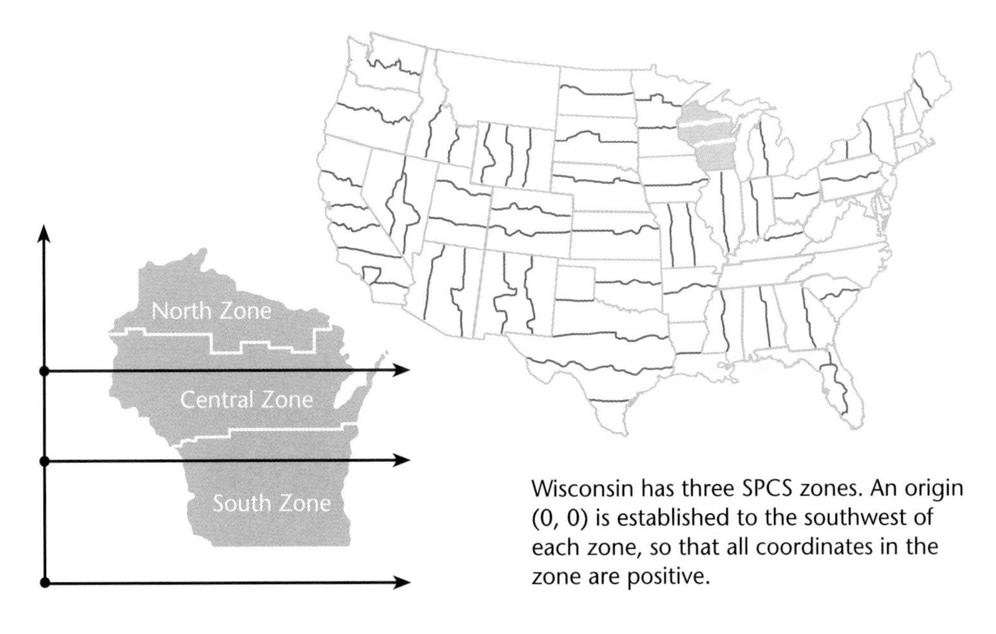

Wisconsin has three SPCS zones. An origin (0, 0) is established to the southwest of each zone, so that all coordinates in the zone are positive.

But Do You Really Need a Map Projection?

You may have noticed as you worked your way through this chapter that almost every example involved a map of the world or a big part of it. And if you're doing mapping like that, then a critical understanding of map projections and their distortions is important.

But many maps don't require you to think about map projections. A sketch map providing directions to a friend's house has no projection. The fictional maps of Tolkien's Middle Earth or for *Game of Thrones* have no projection. Maps in tourist guides? That Google map on your phone you use for navigation every day? They may have a projection, but given the detailed scale and way you're using the map, you need not fret about it. Map projections are worth knowing about – they're important! – but it's also important to know that they don't always matter.

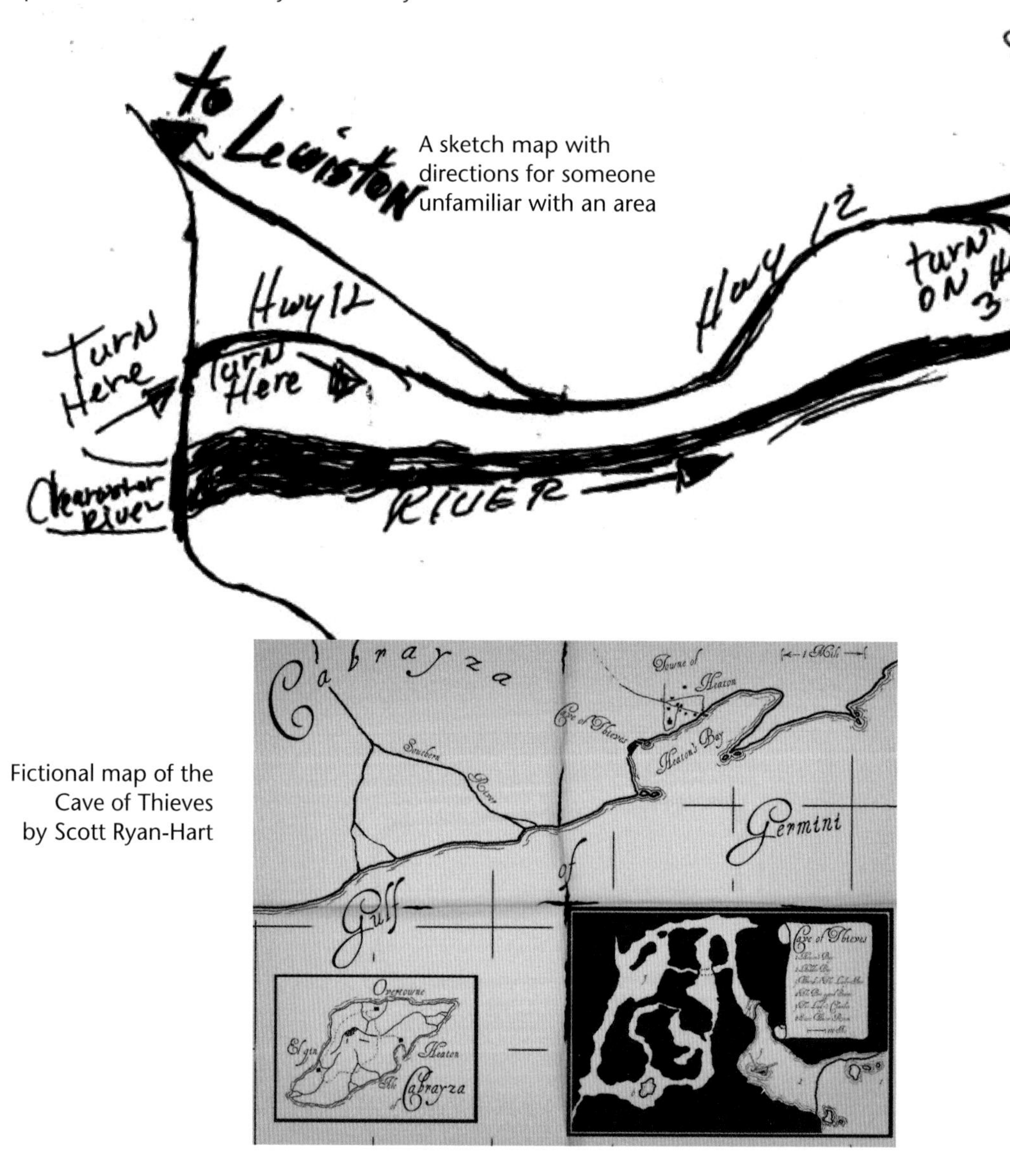

A sketch map with directions for someone unfamiliar with an area

Fictional map of the Cave of Thieves by Scott Ryan-Hart

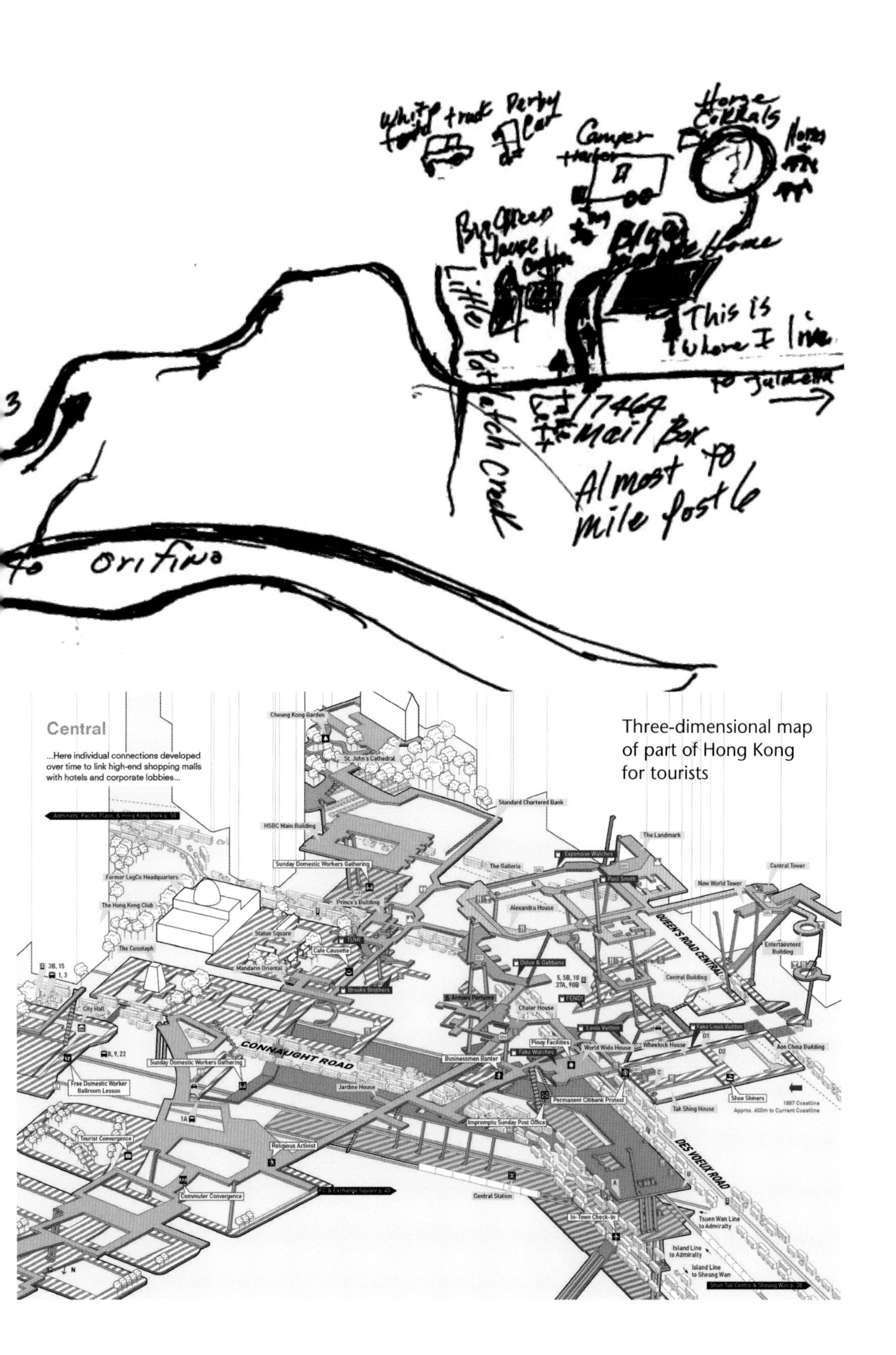

Three-dimensional map of part of Hong Kong for tourists

And the map stutters inarticulate lines —

R.H.S. Crossman, *Rediscovery* (1929)

In the shadowed art of cartography each map is a cursed whisper, charting realms where sanity fears to tread. Gazing upon these cryptic lines invites madness, for within them writhe unspeakable horrors and the yawning void of ancient, unseen worlds.

Arvids Alberts Veidemans-Ozolins

According to the map we've only gone 4 inches.

Harry Dunne, *Dumb and Dumber* (1994)

Men do some things better, like reading maps, because only the male mind can conceive of one inch equaling one hundred miles

Roseanne Barr

More...

A highly accessible introduction to map projections is Denis Wood, Ward Kaiser, and Bob Abramms, *Seeing through Maps: The Power of Images to Shape Our World View* (2006).

The best book on map projections is John Snyder's *Flattening the Earth* (1997). Snyder's approach is historical and exhaustive. The treatment is technical but not intimidating. Want to know about the diverse ecosystem of map coordinate systems? See Clifford Mugnier's *Coordinate Systems of the World: Datums and Grids* (2023).

A great handbook describing and illustrating dozens of map projections is *An Album of Map Projections,* published by John Snyder and Philip Voxland (U.S. Geological Survey Professional Paper #1453, 1989). See also Tau Rho Alpha, Janis Detterman, and James Morley, *Atlas of Oblique Maps* (U.S. Geological Survey Miscellaneous Investigations Series #I-1799, 1988), for a collection of very cool oblique maps. Also see *Working with Map Projections: A Guide to Their Selection* by Fritz Kessler and Sarah Battersby (2019). For neonate mappers, check out *Map Projections for Babies* by Dan Ford (2021). By "babies" Ford means infants, not whiney map makers.

Every cartography textbook has a chapter or two on map projections and coordinate systems. Check out any of the previously cited cartography texts for more information than you will ever need.

Sources: Body projection image courtesy of Lilla Locurto and Bill Outcault. The majority of map projections in this chapter were generated in GeoCart software, and a few in ESRI's ArcGIS. The flat toad is redrawn from Edward Tufte, *The Visual Display of Quantitative Information* (Graphics Press, 1983). The map of vegetation on the Mercator map projection is redrawn from a map in Anne Spirn's *The Granite Garden: Urban Nature and Human Design* (Basic Books, 1984). The latitude and longitude earth was redrawn from David Greenhood's *Down to Earth: Mapping for Everybody* (Holiday House, 1951). The state plane coordinate and Universal Transverse Mercator diagrams were redrawn from Philip Muehrcke and Juliana Muehrcke, *Map Use* (JP Publications, 1998), and Kraak and Ormeling's *Cartography* (Pearson, 1996). The sketch map is from Denis Wood's collection. Map of the Cave of Thieves reproduced courtesy of Scott Ryan-Hart. The tourist map of Hong Kong is from *Cities Without Ground: A Hong Kong Guidebook* (Oro Editions, 2012).

Uh ... why is the Voyager story *ending* at the *beginning* of the map?

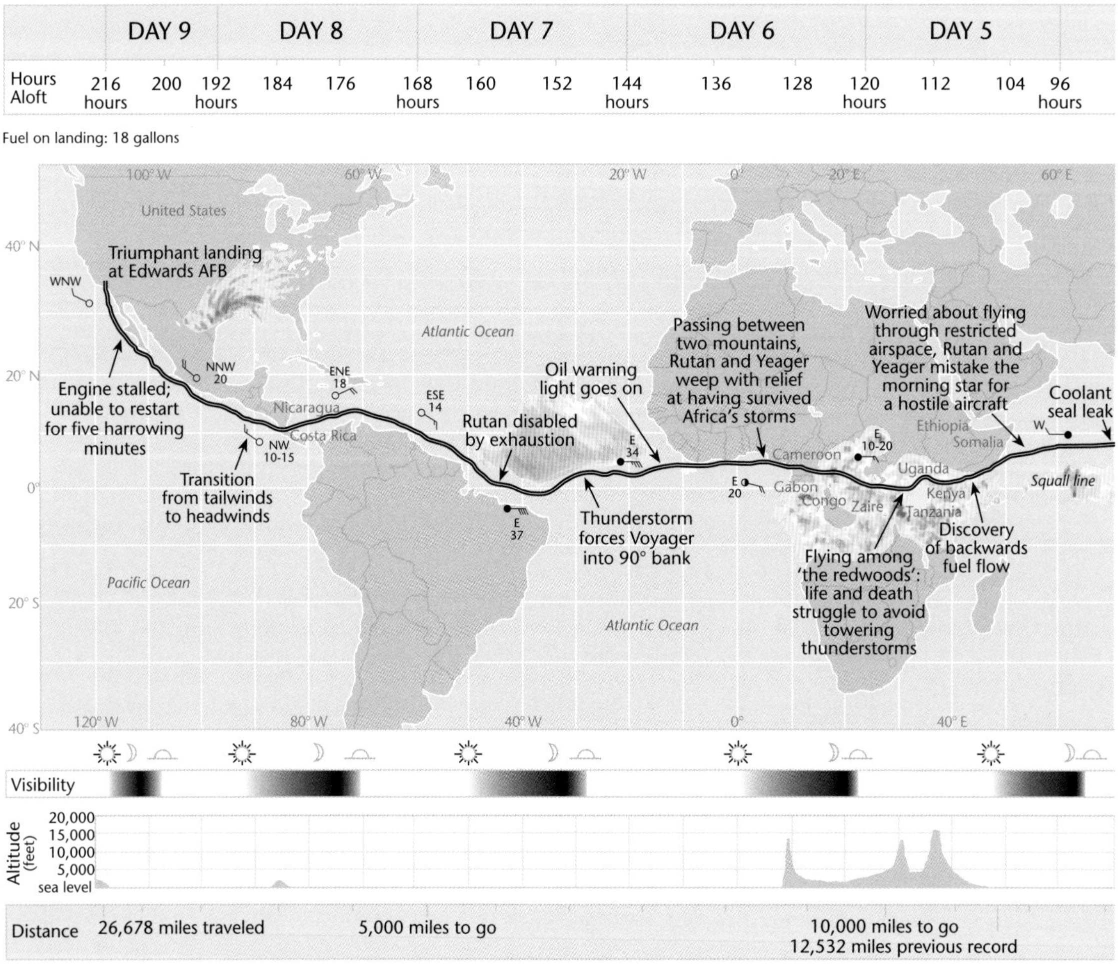

	DAY 9		DAY 8		DAY 7			DAY 6			DAY 5		
Hours Aloft	216 hours	200	192 hours	184	176	168 hours	160	152	144 hours	136	128	120 hours	112	104	96 hours

Fuel on landing: 18 gallons

Triumphant landing at Edwards AFB

Engine stalled; unable to restart for five harrowing minutes

Transition from tailwinds to headwinds

Rutan disabled by exhaustion

Oil warning light goes on

Thunderstorm forces Voyager into 90° bank

Passing between two mountains, Rutan and Yeager weep with relief at having survived Africa's storms

Flying among 'the redwoods': life and death struggle to avoid towering thunderstorms

Worried about flying through restricted airspace, Rutan and Yeager mistake the morning star for a hostile aircraft

Discovery of backwards fuel flow

Coolant seal leak

Squall line

United States, Nicaragua, Costa Rica, Atlantic Ocean, Pacific Ocean, Cameroon, Gabon, Congo, Zaire, Uganda, Kenya, Tanzania, Ethiopia, Somalia

WNW; NNW 20; NW 10-15; ENE 18; ESE 14; E 37; E 34; E 20; E 10-20; W

40° N, 20° N, 0°, 20° S, 40° S; 100° W, 60° W, 20° W, 0°, 20° E, 60° E; 120° W, 80° W, 40° W, 0°, 40° E

Visibility

Altitude (feet): 20,000; 15,000; 10,000; 5,000; sea level

Distance: 26,678 miles traveled; 5,000 miles to go; 10,000 miles to go; 12,532 miles previous record

Flight data courtesy of Len Snellman and Larry Burch, Voyager meteorologists
Mapped by David DiBiase and John Krygier, Department of Geography, University of Wisconsin-Madison, 1987

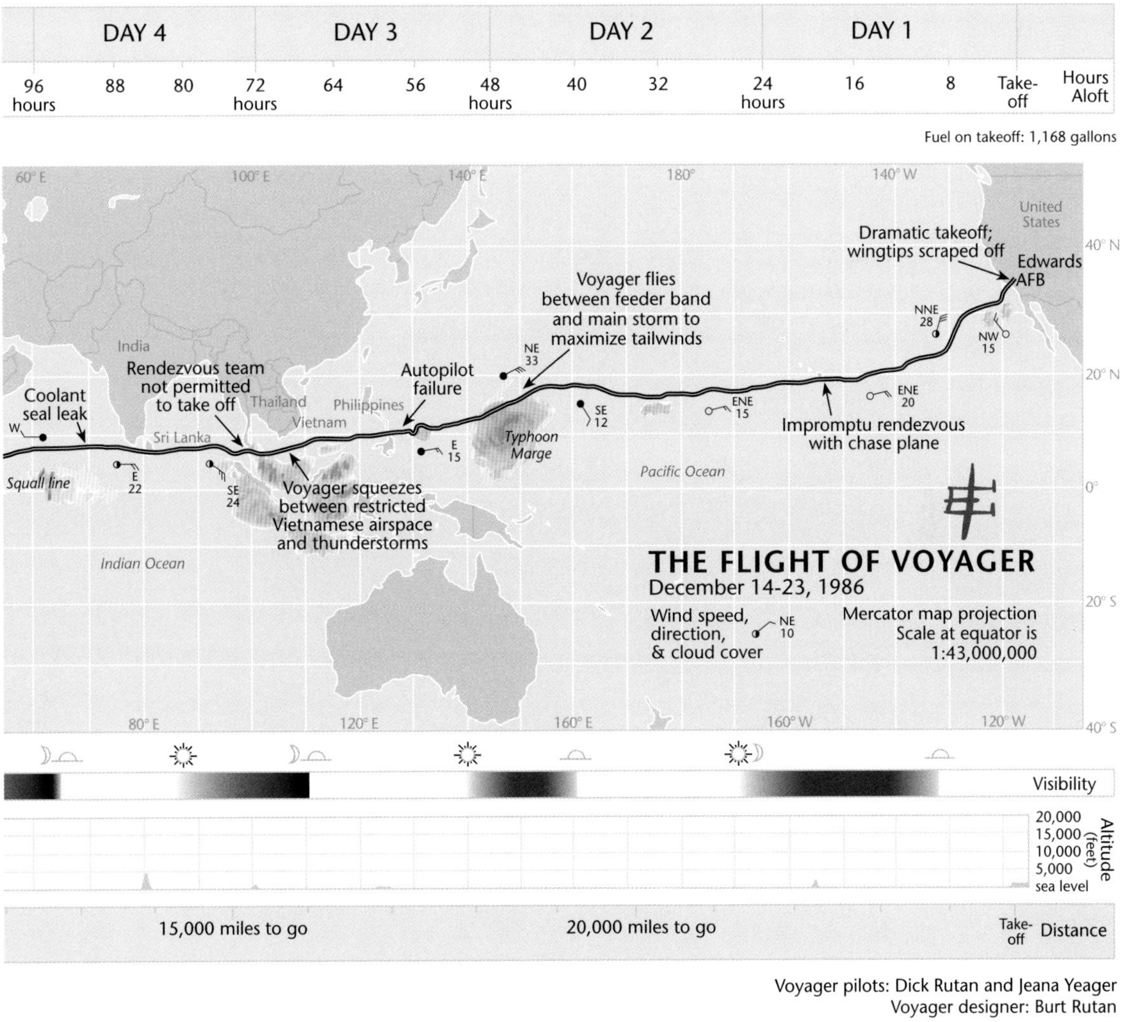

DAY 4 DAY 3 DAY 2 DAY 1

96 hours 88 80 72 hours 64 56 48 hours 40 32 24 hours 16 8 Take-off Hours Aloft

Fuel on takeoff: 1,168 gallons

Dramatic takeoff; wingtips scraped off

Edwards AFB

Voyager flies between feeder band and main storm to maximize tailwinds

Autopilot failure

Rendezvous team not permitted to take off

Coolant seal leak

Impromptu rendezvous with chase plane

Voyager squeezes between restricted Vietnamese airspace and thunderstorms

Typhoon Marge

Squall line

India

Thailand

Philippines

Vietnam

Sri Lanka

Pacific Ocean

Indian Ocean

United States

THE FLIGHT OF VOYAGER

December 14-23, 1986

Wind speed, direction, & cloud cover

Mercator map projection
Scale at equator is 1:43,000,000

Visibility

Altitude (feet): 20,000 15,000 10,000 5,000 sea level

15,000 miles to go 20,000 miles to go Take-off Distance

Voyager pilots: Dick Rutan and Jeana Yeager
Voyager designer: Burt Rutan

What's up with that?

There was a problem. Voyager's route began in California and headed west, sending the reader backwards across the endpapers – violating reading habits.

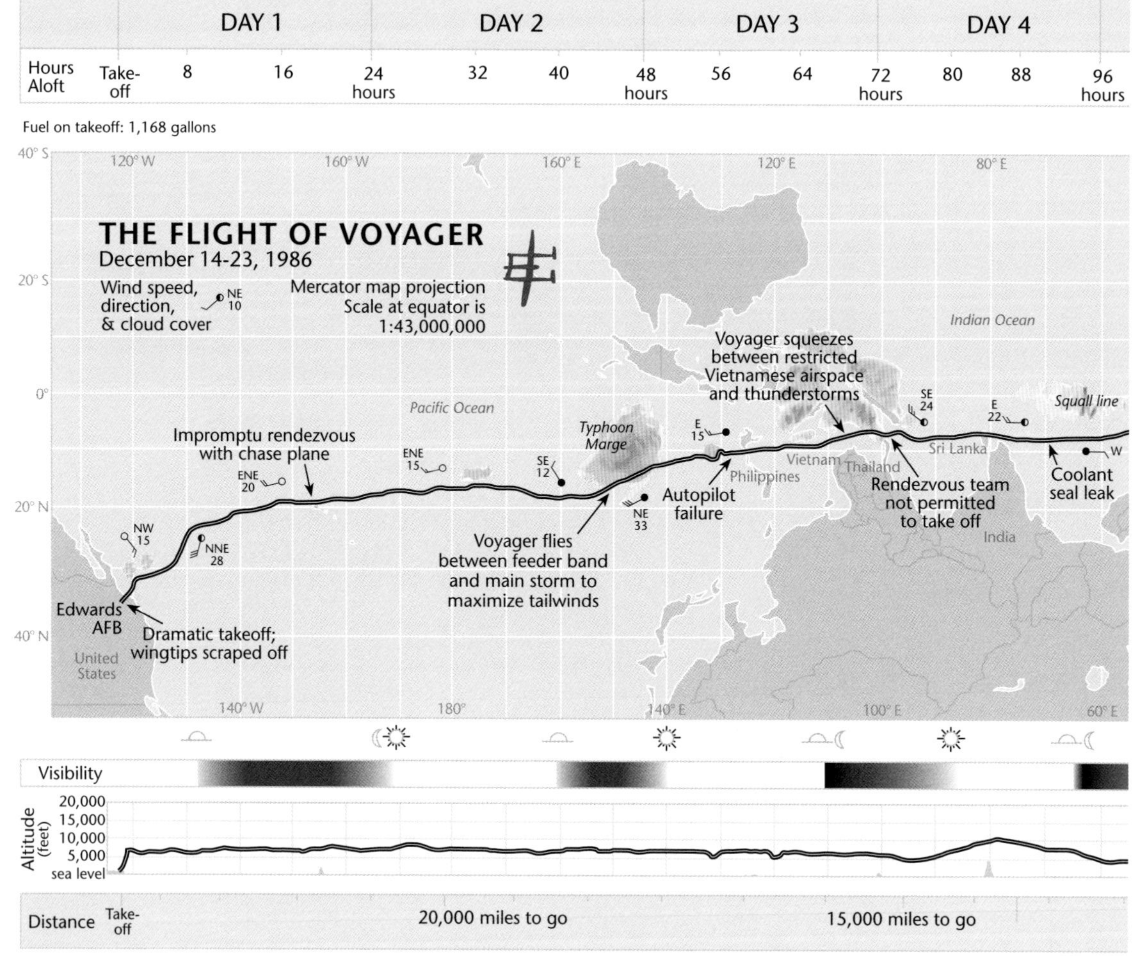

Flight data courtesy of Len Snellman and Larry Burch, Voyager meteorologists
Mapped by David DiBiase and John Krygier, Department of Geography, University of Wisconsin-Madison 1987

Thankfully, north isn't really up. The data suggest south up, and we suggested south up for the map. "No!" shrieked the publisher. "You can't have south up!" And we didn't – in the book. But now, here it is.

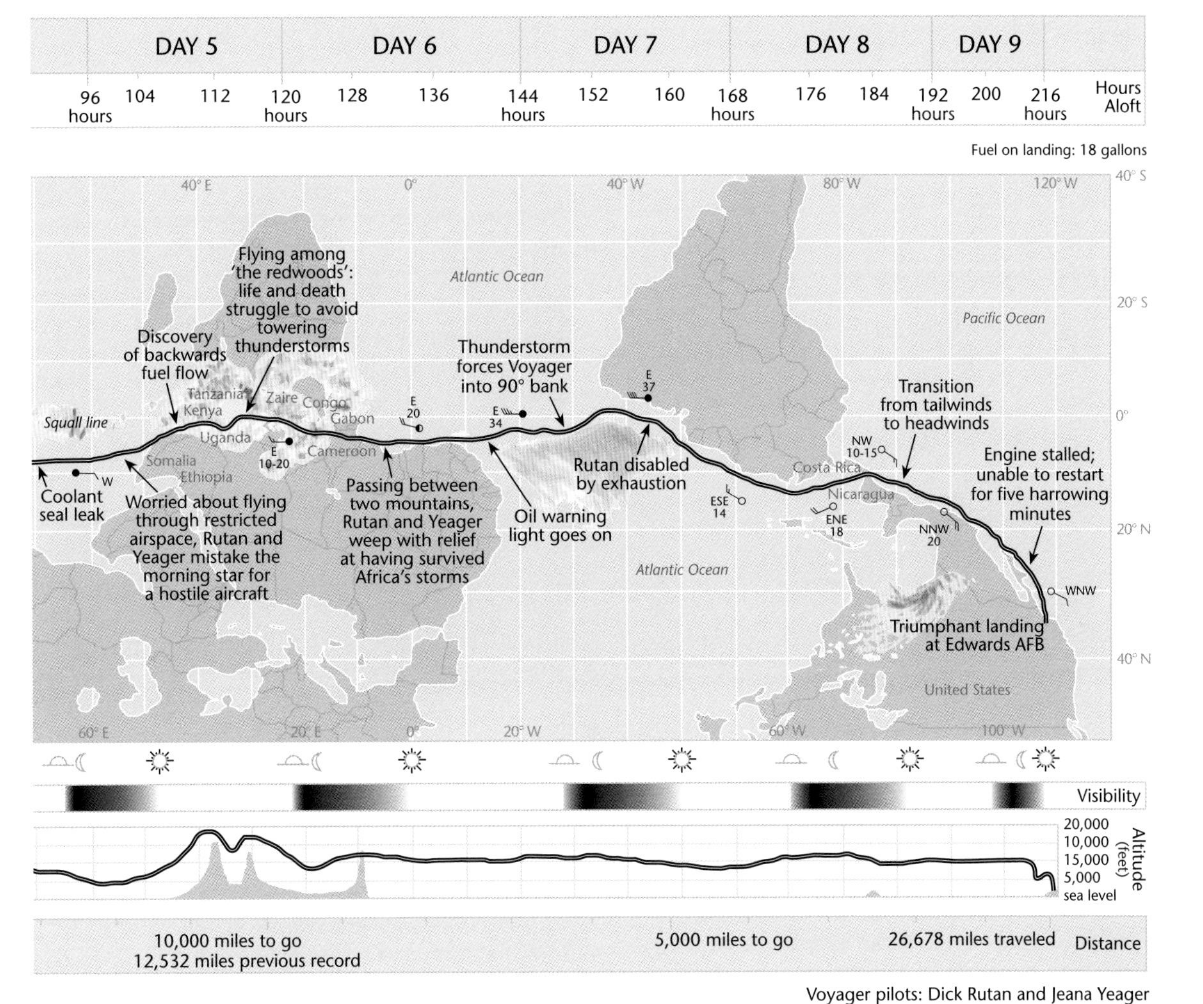

CHAPTER 6 The Big Picture of Map Design

Map design is tough. What's the point of your map? What kind of data do you have? What tools are you using? What's your geographic framework? With answers to these questions in mind, intelligently design the diverse pieces of your map into a coherent whole. One "big picture" approach to map design is borrowed from advertising: layout and visual arrangement. *The medium is the message*. Another approach is Edward Tufte's "graphical excellence." *The data are the message*. Both approaches played a role in designing the Voyager map, as they should with any mapping project.

Map Pieces

The "big picture" of map design in part requires arranging and rearranging your map with its associated map pieces: title, scale, explanatory text, legend, directional indicator, border, sources and credits, and insets and locator maps. None are absolutely required! Your intent for the map will determine which are used and how they are displayed on your map.

Title

Titles should, if possible, include

What: the topic of the map
Where: the geographic area
When: temporal information

Title type size, in general, should be two to three times the size of the type on the map itself and bolder. A subtitle, in smaller type, is good for complex map subjects.

 title content:

Population Change

 title size:

Population Change

 title content and size:

Population Change in Ohio
By county, 1900-2020

Scale

Maps from local to continental scale should include a scale, especially if your map's readers need to make measurements on the map. Verbal and visual scales are more intuitive, numerical scales more flexible (units can be metric or English). If your map's users might reduce or increase the size of the map, a visual scale is best (it will remain accurate even if scaled).

mi

1 inch = 1000 mi

1 : 1,200,000

Small-scale maps (of the entire earth or a substantial portion of it) should not include a simple visual scale, because such maps always contain substantial scale variations.

Explanatory Text

Explanatory text can be vital to the success of your map. You can't express everything you need your map's readers to understand with the map itself. Use text blocks on the map to communicate information about the map content, its broader context, and your goals.

On a historical map, include a paragraph setting the historical context, and also include blocks of text on the map that tell what happened at important locations.

Explain your interpretation of your map's patterns with text: tell your map's readers (in addition to showing them with the map) what you think about the mapped data.

On a choropleth map showing changes in average income over the past decade (by county in a state), explain your interpretation that suburban counties are becoming richer, and urban and rural counties poorer, due to recent tax cuts.

The readers of your map may agree or disagree with your interpretation, but your interpretation and intent will be clearly communicated.

Legend

Map legends vary greatly but should include any map symbol you think may not be familiar to your audience. The legend is the key to interpreting the map.

However, don't insult your map's readers by including obvious symbols in the legend. You need not preface the legend with "Legend" or "Key," as most map readers know that without being told.

Directional Indicator

Use a directional indicator if

The map is not oriented north
The map is of an area unfamiliar to your intended audience

Directional indicators can often be left off the map if the orientation is obvious (unless your audience is stupid). They should be left off map projections where north varies. If included, avoid appallingly large and ugly directional indicators.

Border

A border, or neatline, drawn around your map and its pieces may draw everything together. Try your map without borders. It gives a more open feel to the design. If a border is used, make it narrow, possibly in gray: noticeable but not distracting.

Sources, Credits, Etc.

Maps may include all relevant metadata, including

Data sources and citations
Map maker and date
Organization and logos
Disclaimers and legal information
Map series information
Copyright and use issues
Map projection and coordinate system

Including projection and coordinate system information is important if you think someone may need that information to combine your map with other data in GIS.

Insets and Locator Maps

Choosing a single map scale is difficult. At a smaller scale, data in one area on the map are too dense. At a larger scale, the dense data can be distinguished, but the geographic extent is limited.

An inset that jumps to a larger scale (less area, more detail) helps the map viewer to understand areas on the map where data are more dense and difficult to distinguish at the scale of the main map. An inset that jumps to a smaller scale (more area, less detail) helps the map viewer contextualize the area of the main map. An orthographic map projection is commonly used as a smaller-scale global locator map.

Larger-scale maps benefit greatly by including a smaller-scale locator map.

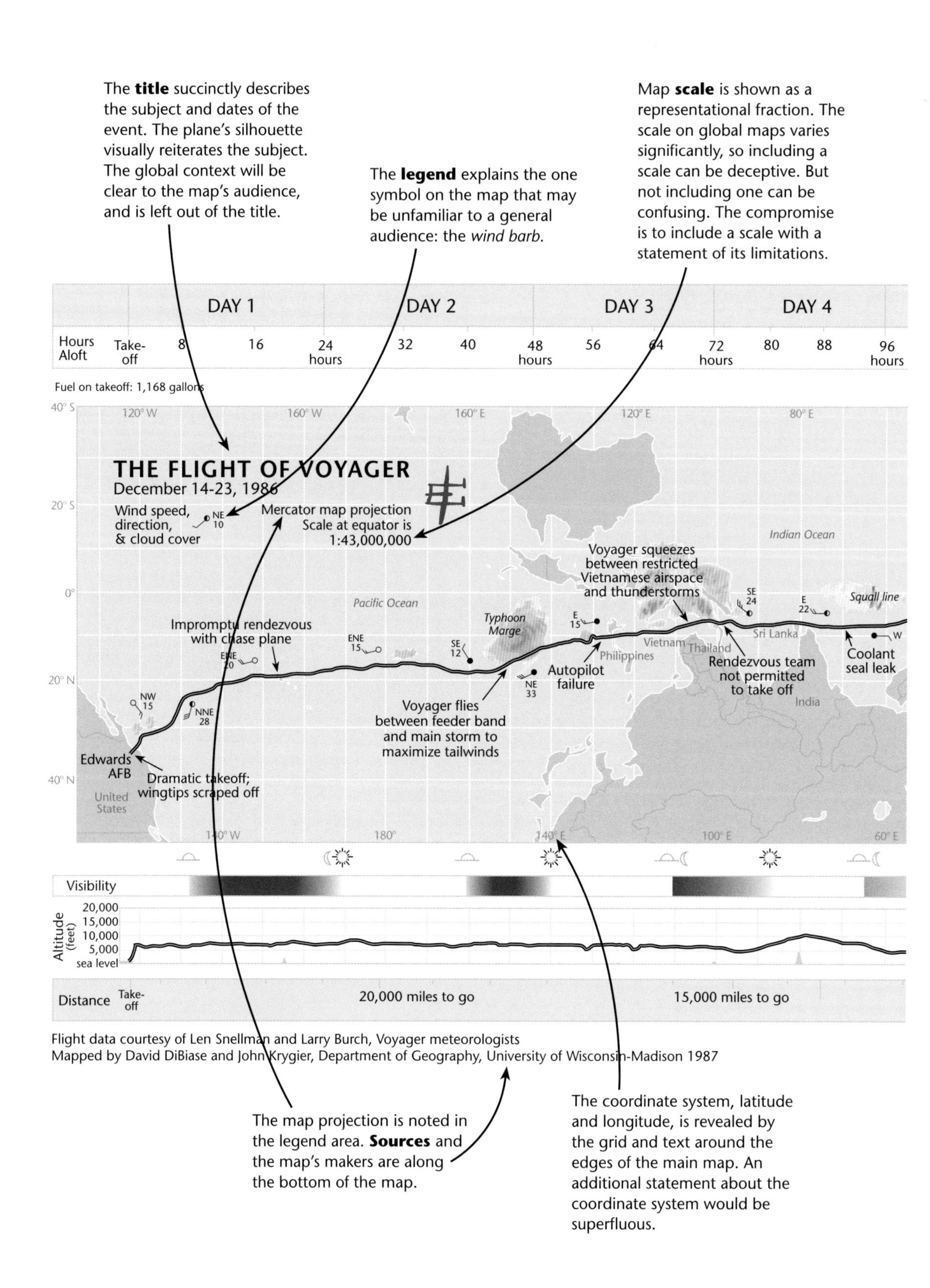
The **title** succinctly describes the subject and dates of the event. The plane's silhouette visually reiterates the subject. The global context will be clear to the map's audience, and is left out of the title.
The **legend** explains the one symbol on the map that may be unfamiliar to a general audience: the *wind barb*.
Map **scale** is shown as a representational fraction. The scale on global maps varies significantly, so including a scale can be deceptive. But not including one can be confusing. The compromise is to include a scale with a statement of its limitations.
DAY 1
DAY 2
DAY 3
DAY 4
Hours Aloft
Take-off
8
16
24 hours
32
40
48 hours
56
64
72 hours
80
88
96 hours
Fuel on takeoff: 1,168 gallons
40° S
20° S
0°
20° N
40° N
120° W
160° W
160° E
120° E
80° E
140° W
180°
140° E
100° E
60° E
THE FLIGHT OF VOYAGER
December 14-23, 1986
Wind speed, direction, & cloud cover
NE 10
Mercator map projection
Scale at equator is 1:43,000,000
Indian Ocean
Pacific Ocean
Voyager squeezes between restricted Vietnamese airspace and thunderstorms
Squall line
Typhoon Marge
Impromptu rendezvous with chase plane
Vietnam
Thailand
Sri Lanka
Philippines
India
Autopilot failure
Rendezvous team not permitted to take off
Coolant seal leak
Voyager flies between feeder band and main storm to maximize tailwinds
Edwards AFB
Dramatic takeoff; wingtips scraped off
United States
Visibility
Altitude (feet)
20,000
15,000
10,000
5,000
sea level
Distance
Take-off
20,000 miles to go
15,000 miles to go
Flight data courtesy of Len Snellman and Larry Burch, Voyager meteorologists
Mapped by David DiBiase and John Krygier, Department of Geography, University of Wisconsin-Madison 1987
The map projection is noted in the legend area. **Sources** and the map's makers are along the bottom of the map.
The coordinate system, latitude and longitude, is revealed by the grid and text around the edges of the main map. An additional statement about the coordinate system would be superfluous.

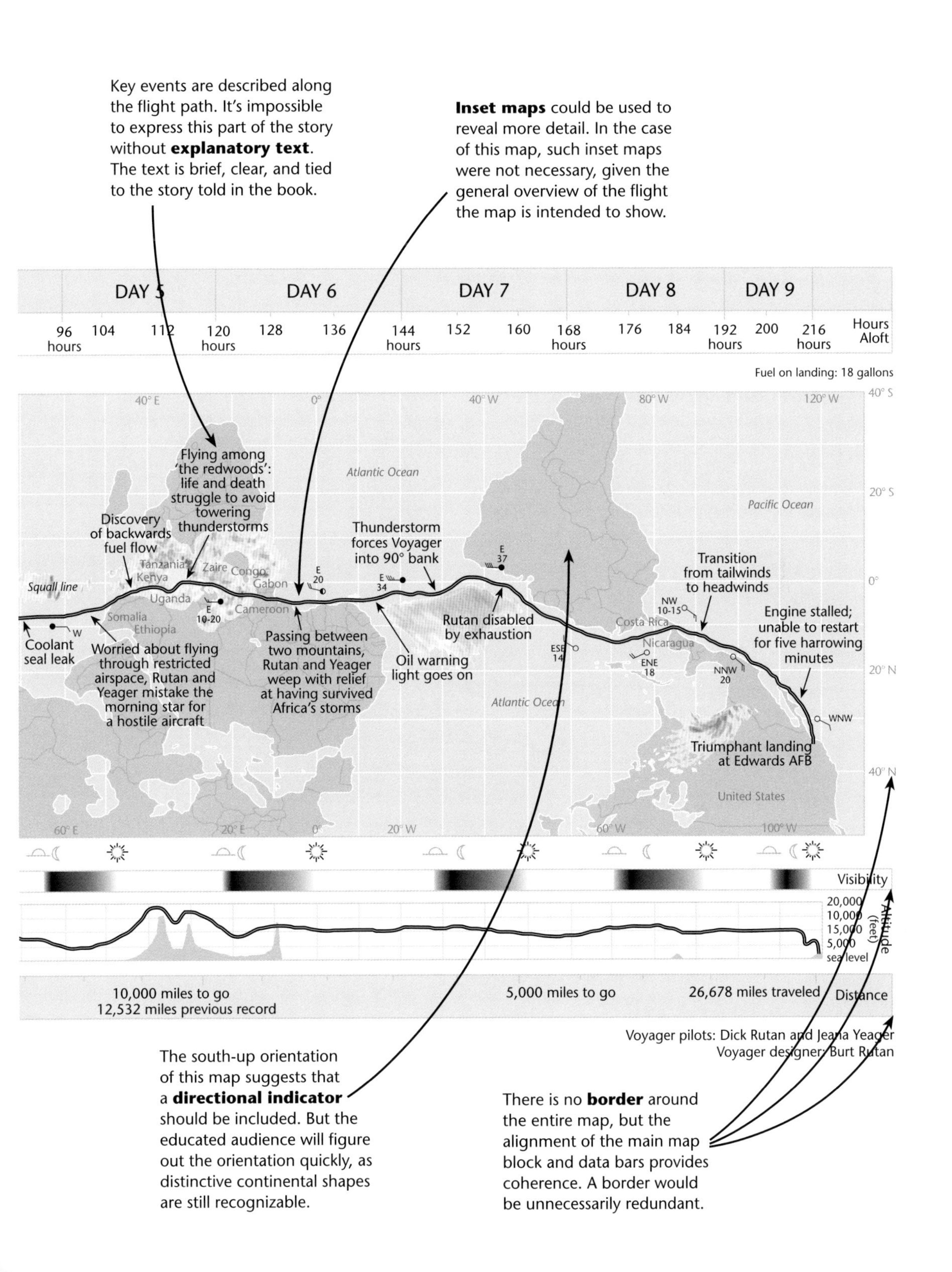

Key events are described along the flight path. It's impossible to express this part of the story without **explanatory text**. The text is brief, clear, and tied to the story told in the book.
Inset maps could be used to reveal more detail. In the case of this map, such inset maps were not necessary, given the general overview of the flight the map is intended to show.
DAY 5
DAY 6
DAY 7
DAY 8
DAY 9
96 hours
104
112
120 hours
128
136
144 hours
152
160
168 hours
176
184
192 hours
200
216 hours
Hours Aloft
Fuel on landing: 18 gallons
40° E
0°
40° W
80° W
120° W
40° S
20° S
0°
20° N
40° N
Atlantic Ocean
Pacific Ocean
Flying among 'the redwoods': life and death struggle to avoid towering thunderstorms
Discovery of backwards fuel flow
Thunderstorm forces Voyager into 90° bank
Transition from tailwinds to headwinds
Engine stalled; unable to restart for five harrowing minutes
Squall line
Tanzania
Kenya
Zaire
Congo
Gabon
Uganda
Somalia
Ethiopia
Cameroon
Costa Rica
Nicaragua
E 20
E 34
E 37
E 10-20
NW 10-15
ESE 14
ENE 18
NNW 20
WNW
W
Rutan disabled by exhaustion
Coolant seal leak
Worried about flying through restricted airspace, Rutan and Yeager mistake the morning star for a hostile aircraft
Passing between two mountains, Rutan and Yeager weep with relief at having survived Africa's storms
Oil warning light goes on
Atlantic Ocean
Triumphant landing at Edwards AFB
United States
60° E
20° E
0°
20° W
60° W
100° W
Visibility
20,000
10,000
15,000
5,000
sea level
Altitude (feet)
10,000 miles to go
12,532 miles previous record
5,000 miles to go
26,678 miles traveled
Distance
Voyager pilots: Dick Rutan and Jeana Yeager
Voyager designer: Burt Rutan
The south-up orientation of this map suggests that a **directional indicator** should be included. But the educated audience will figure out the orientation quickly, as distinctive continental shapes are still recognizable.
There is no **border** around the entire map, but the alignment of the main map block and data bars provides coherence. A border would be unnecessarily redundant.

Thinking about the Big Picture

Abstract map diagrams like the small ones on the following page assume that map content doesn't matter. Map content matters more than anything. Interesting content trumps bad design every time. But why not strive for both content and design? Diagrams like these can draw your attention to the many ways in which map pieces can be arranged and rearranged to enhance your goals for your map. Tufte's "graphical excellence" serves as a complementary approach, driven by your data, to the big picture of map design.

Roving Eyes

Eyes move over maps, so can we guide the way they move by design? George Jenks pioneered eye movement studies of maps. In 1973 he published the eye scans below. Arrows indicate the start of the path of the subject's eyes over the map. Jenks's study and subsequent research show that people's eye movements over maps are erratic, individualistic, and nearly impossible to predict.

But where there's nothing else to hang on to, assuming a predictable eye movement path is one strategy that can help the map maker focus his or her attention on the way the map and its pieces are coming together. The big picture of map design is experimentation guided by intuition and evaluation. Design strategies include playing with paths, the visual center, balance, symmetry, sight-lines, and grids.

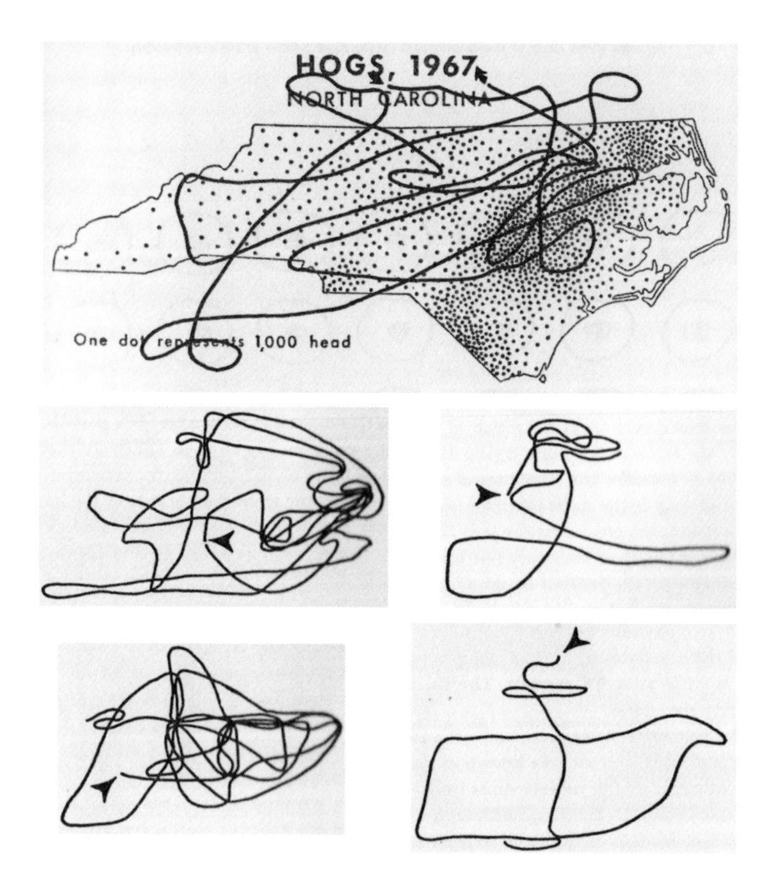

Visual Arrangement

Assume that reading a map follows a **path**. Arrange map pieces so that those to be seen first – such as the title – intuitively lead to the map pieces you wish your viewer to see next. Where the viewer starts varies culturally, and can be shaped by numbering and other visual cues.

The **visual center** of a map is slightly above the actual center. Centering implies importance. Try positioning map pieces so that the most important are near the visual center of the map. The map reader may focus near this center and assume the elements there are most important.

Balance can be assessed when all your map pieces are in place. Map pieces vary in weight: some seem heavier, others lighter. Does your arrangement seem off-balance? Unless you want to suggest a lack of balance to your readers, rearrange your map pieces for better balance.

Symmetry can be thought of as balance around a central vertical axis. Symmetrical balance (left) feels traditional, conservative, and cautious. Asymmetry (right) depends on off-center weights and balances. It typically feels modern, progressive, complex, and more creative.

Sight-lines are invisible horizontal or vertical lines that touch the top, bottom, or sides of map elements. Minimizing the number of sight-lines reduces disjointedness and stabilizes and enhances map layout. This allows your map's readers to focus on the map's subject.

Symmetrical grids (left) are based on two central axes and top, bottom, and side margins. **Asymmetrical grids** (right) are more complex, but still depend on the visual center while maintaining top, bottom, and side margins.

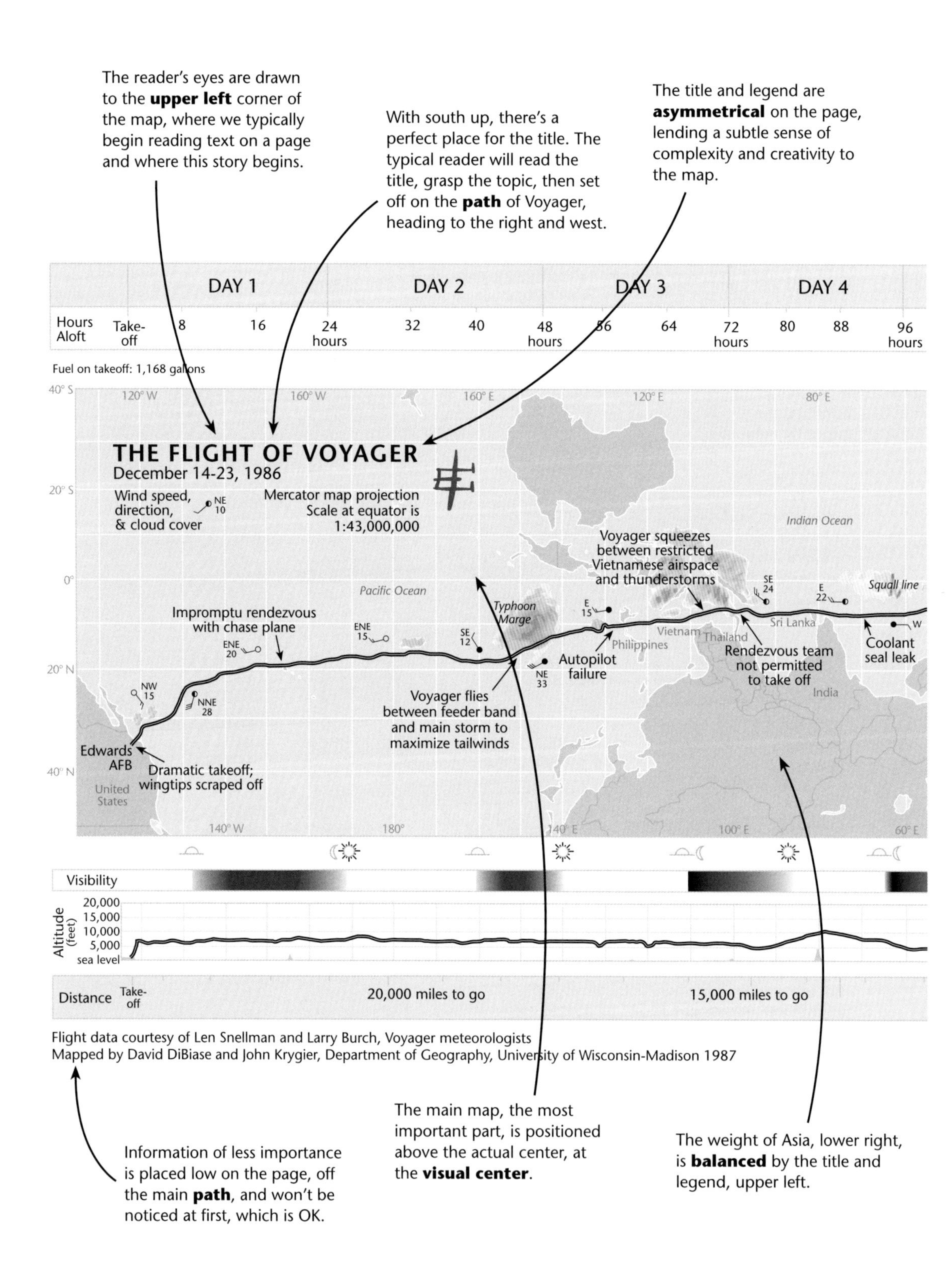

The reader's eyes are drawn to the **upper left** corner of the map, where we typically begin reading text on a page and where this story begins.
With south up, there's a perfect place for the title. The typical reader will read the title, grasp the topic, then set off on the **path** of Voyager, heading to the right and west.
The title and legend are **asymmetrical** on the page, lending a subtle sense of complexity and creativity to the map.
DAY 1
DAY 2
DAY 3
DAY 4
Hours Aloft
Take-off
8
16
24 hours
32
40
48 hours
56
64
72 hours
80
88
96 hours
Fuel on takeoff: 1,168 gallons
40° S
20° S
0°
20° N
40° N
120° W
160° W
160° E
120° E
80° E
THE FLIGHT OF VOYAGER
December 14-23, 1986
Wind speed, direction, & cloud cover
NE 10
Mercator map projection
Scale at equator is 1:43,000,000
Indian Ocean
Voyager squeezes between restricted Vietnamese airspace and thunderstorms
Pacific Ocean
Squall line
Typhoon Marge
Impromptu rendezvous with chase plane
ENE 15
ENE 20
SE 12
E 15
SE 24
E 22
W
Sri Lanka
Vietnam
Thailand
Philippines
Autopilot failure
Rendezvous team not permitted to take off
Coolant seal leak
NE 33
India
NW 15
NNE 28
Voyager flies between feeder band and main storm to maximize tailwinds
Edwards AFB
Dramatic takeoff; wingtips scraped off
United States
140° W
180°
140° E
100° E
60° E
Visibility
Altitude (feet)
20,000
15,000
10,000
5,000
sea level
Distance
Take-off
20,000 miles to go
15,000 miles to go
Flight data courtesy of Len Snellman and Larry Burch, Voyager meteorologists
Mapped by David DiBiase and John Krygier, Department of Geography, University of Wisconsin-Madison 1987
Information of less importance is placed low on the page, off the main **path**, and won't be noticed at first, which is OK.
The main map, the most important part, is positioned above the actual center, at the **visual center**.
The weight of Asia, lower right, is **balanced** by the title and legend, upper left.

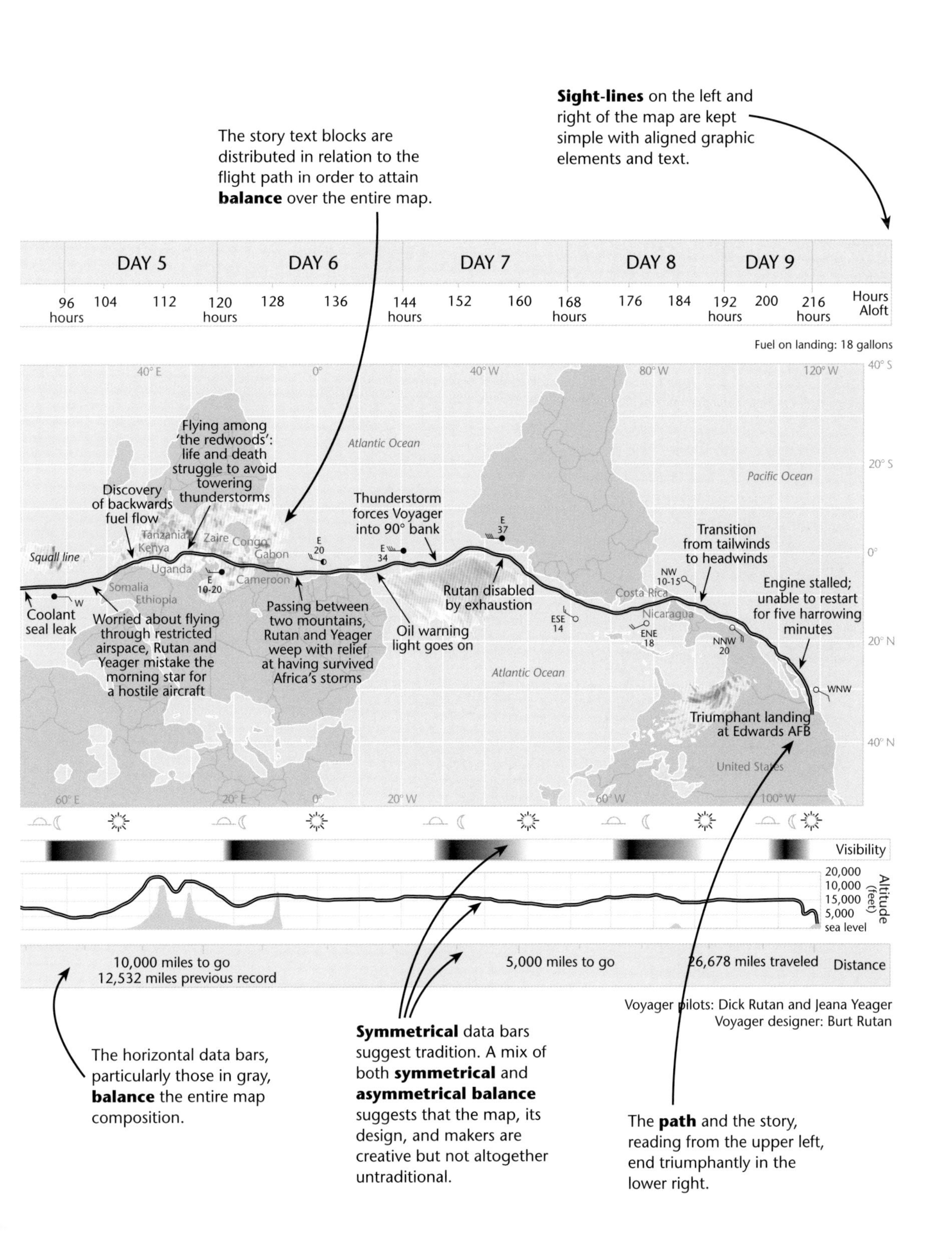

Sight-lines on the left and right of the map are kept simple with aligned graphic elements and text.
The story text blocks are distributed in relation to the flight path in order to attain balance over the entire map.
DAY 5
DAY 6
DAY 7
DAY 8
DAY 9
96 hours
104
112
120 hours
128
136
144 hours
152
160
168 hours
176
184
192 hours
200
216 hours
Hours Aloft
Fuel on landing: 18 gallons
40° E
0°
40° W
80° W
120° W
40° S
20° S
0°
20° N
40° N
Atlantic Ocean
Pacific Ocean
Flying among 'the redwoods': life and death struggle to avoid towering thunderstorms
Discovery of backwards fuel flow
Thunderstorm forces Voyager into 90° bank
Transition from tailwinds to headwinds
Squall line
Tanzania
Kenya
Zaire
Congo
Gabon
Uganda
Cameroon
Somalia
Ethiopia
Costa Rica
Nicaragua
Engine stalled; unable to restart for five harrowing minutes
Coolant seal leak
Worried about flying through restricted airspace, Rutan and Yeager mistake the morning star for a hostile aircraft
Passing between two mountains, Rutan and Yeager weep with relief at having survived Africa's storms
Oil warning light goes on
Rutan disabled by exhaustion
Atlantic Ocean
Triumphant landing at Edwards AFB
United States
60° E
20° E
0°
20° W
60° W
100° W
Visibility
20,000
10,000
15,000
5,000
sea level
Altitude (feet)
10,000 miles to go
12,532 miles previous record
5,000 miles to go
26,678 miles traveled
Distance
Voyager pilots: Dick Rutan and Jeana Yeager
Voyager designer: Burt Rutan
The horizontal data bars, particularly those in gray, balance the entire map composition.
Symmetrical data bars suggest tradition. A mix of both symmetrical and asymmetrical balance suggests that the map, its design, and makers are creative but not altogether untraditional.
The path and the story, reading from the upper left, end triumphantly in the lower right.

Graphical Excellence

A different way to think about the big picture
of map design are the ideas of Edward Tufte.
Unlike the emphasis on the purely visual encouraged by path, center, balance, symmetry,
sight-lines, and grids, Tufte is concerned
with interesting data and complex ideas
presented with clarity and intelligence.
Below are 23 "Tufteisms" from his
books. They are intended to help
you think about your map
design choices.
They are not rules.
There are no rules.

Revise and edit.
Forgo chartjunk.
Erase non-data-ink.
To clarify, add detail.
Erase redundant data-ink.
Maximize the data-ink ratio.
Above all else, show the data.
The revelation of the complex.
Showing complexity is hard work.
Show data variation, not design variation.
Graphics must not quote data out of context.
Graphical excellence is nearly always multivariate.
Graphical excellence requires telling the truth about the data.
If the numbers are boring, then you've got the wrong numbers.
Write out explanations of the data on the graphic itself. Label important events in the data.
Graphical excellence consists of complex ideas communicated with clarity precision, and efficiency.
Clear, detailed, and thorough labeling should be used to defeat graphical distortion and ambiguity.
The number of information-carrying (variable) dimensions depicted should not exceed the number
Graphical excellence is the well-designed presentation of interesting data – a matter of substance,
In time-series displays of money, deflated and standardized units of monetary measurement are nearly
Graphical excellence is that which gives to the viewer the greatest number of ideas in the shortest
If the nature of the data suggests the shape of the graphic, follow that suggestion. Otherwise, move
The representation of numbers, as physically measured on the surface of the graphic itself, should

of dimensions in the data.
of statistics, and of design.
always better than nominal units.
time with the least ink in the smallest space.
toward horizontal graphics about 50 percent wider than tall.
be directly proportional to the numerical quantities represented.

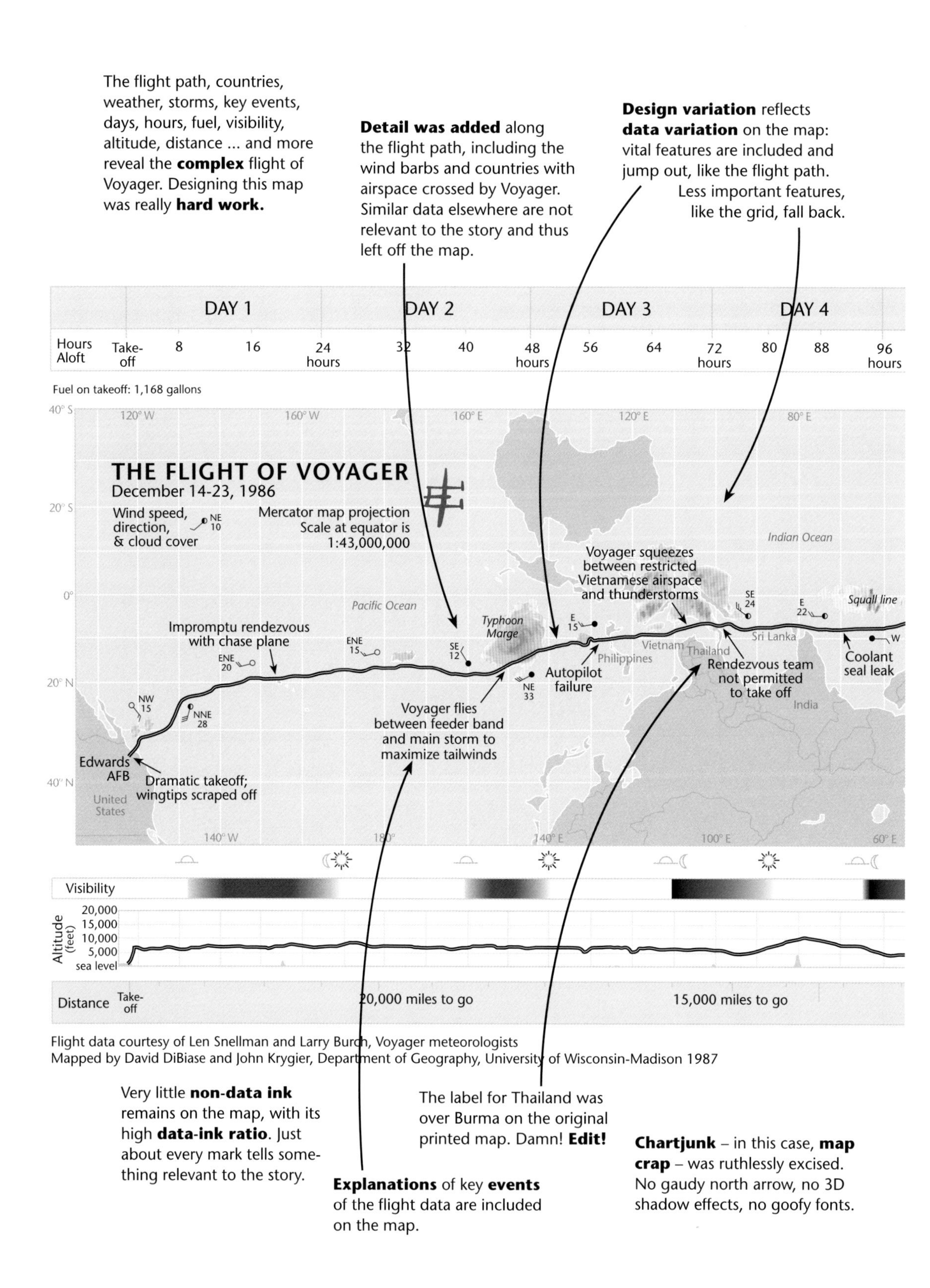

The flight path, countries, weather, storms, key events, days, hours, fuel, visibility, altitude, distance ... and more reveal the **complex** flight of Voyager. Designing this map was really **hard work.**
Detail was added along the flight path, including the wind barbs and countries with airspace crossed by Voyager. Similar data elsewhere are not relevant to the story and thus left off the map.
Design variation reflects **data variation** on the map: vital features are included and jump out, like the flight path. Less important features, like the grid, fall back.
DAY 1
DAY 2
DAY 3
DAY 4
Hours Aloft
Take-off
8
16
24 hours
32
40
48 hours
56
64
72 hours
80
88
96 hours
Fuel on takeoff: 1,168 gallons
40° S
20° S
0°
20° N
40° N
120° W
160° W
160° E
120° E
80° E
140° W
180°
140° E
100° E
60° E
THE FLIGHT OF VOYAGER
December 14-23, 1986
Wind speed, direction, & cloud cover
NE 10
Mercator map projection
Scale at equator is 1:43,000,000
Indian Ocean
Pacific Ocean
Voyager squeezes between restricted Vietnamese airspace and thunderstorms
Squall line
Typhoon Marge
Impromptu rendezvous with chase plane
ENE 20
ENE 15
SE 12
E 15
SE 24
E 22
W
Vietnam
Thailand
Philippines
Sri Lanka
India
Coolant seal leak
Rendezvous team not permitted to take off
Autopilot failure
NE 33
NW 15
NNE 28
Voyager flies between feeder band and main storm to maximize tailwinds
Edwards AFB
Dramatic takeoff; wingtips scraped off
United States
Visibility
Altitude (feet)
20,000
15,000
10,000
5,000
sea level
Distance
Take-off
20,000 miles to go
15,000 miles to go
Flight data courtesy of Len Snellman and Larry Burch, Voyager meteorologists
Mapped by David DiBiase and John Krygier, Department of Geography, University of Wisconsin-Madison 1987
Very little **non-data ink** remains on the map, with its high **data-ink ratio**. Just about every mark tells something relevant to the story.
The label for Thailand was over Burma on the original printed map. Damn! **Edit!**
Chartjunk – in this case, **map crap** – was ruthlessly excised. No gaudy north arrow, no 3D shadow effects, no goofy fonts.
Explanations of key **events** of the flight data are included on the map.

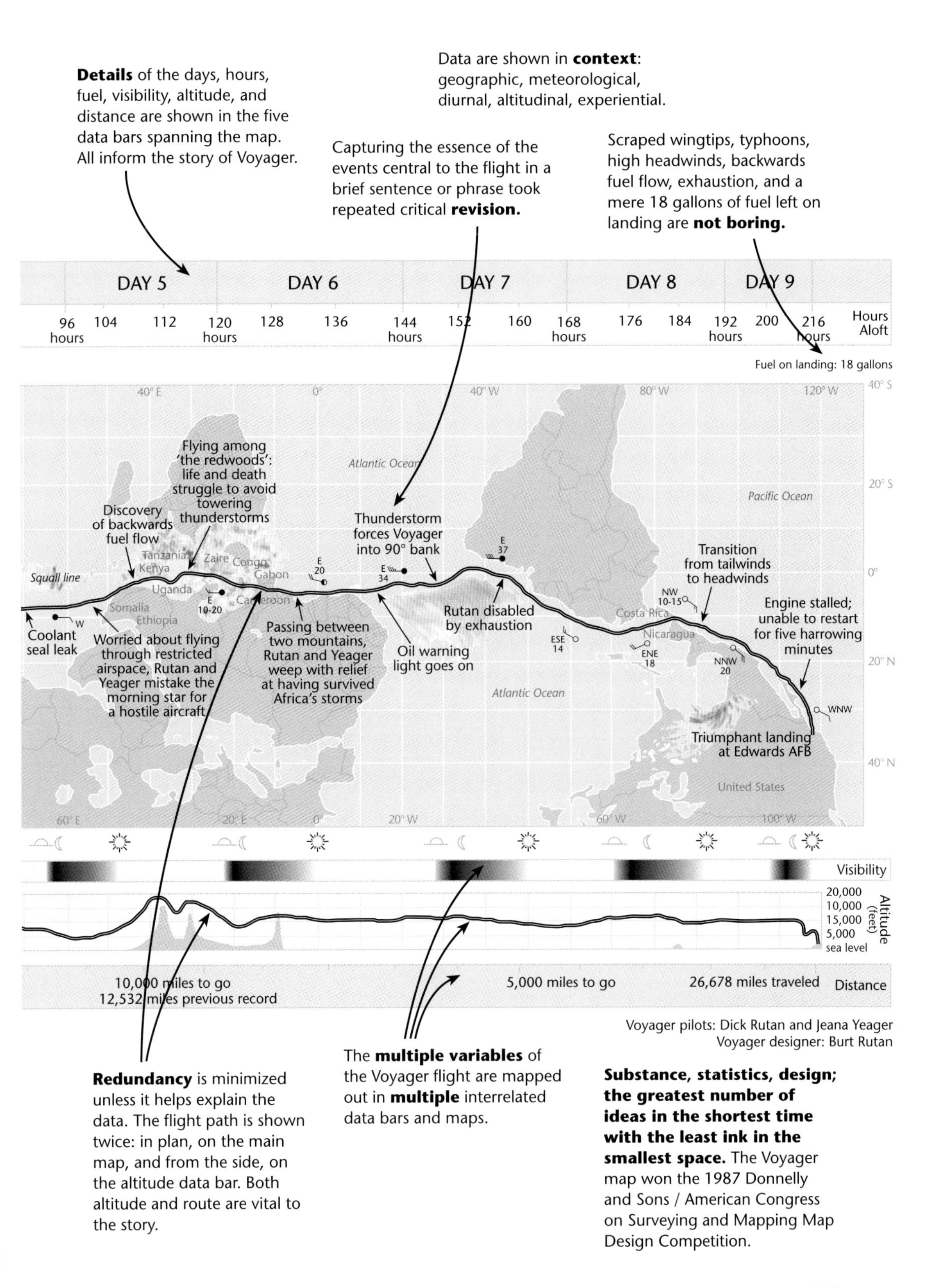

Details of the days, hours, fuel, visibility, altitude, and distance are shown in the five data bars spanning the map. All inform the story of Voyager.
Data are shown in **context**: geographic, meteorological, diurnal, altitudinal, experiential.
Capturing the essence of the events central to the flight in a brief sentence or phrase took repeated critical **revision.**
Scraped wingtips, typhoons, high headwinds, backwards fuel flow, exhaustion, and a mere 18 gallons of fuel left on landing are **not boring.**
DAY 5
DAY 6
DAY 7
DAY 8
DAY 9
96 hours
104
112
120 hours
128
136
144 hours
152
160
168 hours
176
184
192 hours
200
216 hours
Hours Aloft
Fuel on landing: 18 gallons
40° E
0°
40° W
80° W
120° W
40° S
20° S
0°
20° N
40° N
Atlantic Ocean
Pacific Ocean
Flying among 'the redwoods': life and death struggle to avoid towering thunderstorms
Discovery of backwards fuel flow
Thunderstorm forces Voyager into 90° bank
Transition from tailwinds to headwinds
Squall line
Tanzania
Kenya
Zaire
Congo
Gabon
Uganda
Somalia
Ethiopia
Cameroon
Costa Rica
Nicaragua
Engine stalled; unable to restart for five harrowing minutes
Coolant seal leak
Worried about flying through restricted airspace, Rutan and Yeager mistake the morning star for a hostile aircraft
Passing between two mountains, Rutan and Yeager weep with relief at having survived Africa's storms
Oil warning light goes on
Rutan disabled by exhaustion
Atlantic Ocean
Triumphant landing at Edwards AFB
United States
60° E
20° E
0°
20° W
60° W
100° W
Visibility
20,000
10,000
15,000
5,000
sea level
Altitude (feet)
10,000 miles to go
12,532 miles previous record
5,000 miles to go
26,678 miles traveled
Distance
Voyager pilots: Dick Rutan and Jeana Yeager
Voyager designer: Burt Rutan
Redundancy is minimized unless it helps explain the data. The flight path is shown twice: in plan, on the main map, and from the side, on the altitude data bar. Both altitude and route are vital to the story.
The **multiple variables** of the Voyager flight are mapped out in **multiple** interrelated data bars and maps.
Substance, statistics, design; the greatest number of ideas in the shortest time with the least ink in the smallest space. The Voyager map won the 1987 Donnelly and Sons / American Congress on Surveying and Mapping Map Design Competition.

I knew every page in that atlas by heart. How many days and nights I had lingered over its old faded maps, following the blue rivers from the mountains to the sea, wondering what the little towns really looked like, and how wide were the sprawling lakes! I had a lot of fun with that atlas, traveling, in my mind, all over the world.

Hugh Lofting, *The Voyages of Dr. Doolittle* (1922)

When our maps do not fit the territory, when we act as if our inferences are factual knowledge, we prepare ourselves for a world that isn't there. If this happens often enough, the inevitable result is frustration and an ever-increasing tendency to warp the territory to fit our maps. We see what we want to see, and the more we see it, the more likely we are to reinforce this distorted perception, in the familiar circular and spiral feedback pattern.

Harry L. Weinberg, *Levels of Knowing and Existence* (1959)

Open the daunting map beneath —

Ralph Waldo Emerson, *Monadnoc* (1847)

You know how Venice looks upon the map –
Isles that the mainland hardly can let go?

Robert Browning, *Stafford* (1837)

More...

Philosophers and others are beginning to take a serious interest in information graphics. John Bender and Michael Marrinan's *The Culture of Diagram* (2010) comes at it from a very academic historical perspective that's nonetheless valuable. James Elkins's *How to Use Your Eyes* (2000) is an anthology of graphics, each of which he discusses. Almost anything by Elkins is interesting. Carolyn Handa's *Visual Rhetoric in a Digital World* (2004) is a critical source book of readings. Denis Wood and John Fels apply this kind of thinking to maps in *The Natures of Maps* (2008). Sandra Rendgen's *Information Graphics* (2012) is a saucy picture book with a pretty cool introduction. Also see *Data Visualization Made Simple* by Kristen Sosulski (2018), *Data Visualization: A Practical Introduction* by Kieran Healy (2018), and *Better Data Visualizations: A Guide for Scholars, Researchers, and Wonks* by Jonathan Schwabish (2021).

Any of Edward Tufte's books on information graphics is a must-see for the big picture of map design: *The Visual Display of Quantitative Information* (2001), *Envisioning Information* (1990), *Visual Explanations: Images and Quantities, Evidence and Narrative* (1997), *Beautiful Evidence* (2006), and *Seeing with Fresh Eyes: Meaning, Space, Data, Truth* (2020).

Henry Beck's map of the London Underground is a design classic. The tale of its design is the subject of Ken Garland's *Mr. Beck's Underground Map* (1994). Design is not just a matter of tasteful layouts: it has profound public and political dimensions. Nate Burgos's discussion of iconic modernist designer Herbert Bayer's *World Geo-Graphic Atlas* is worth a look. Janet Abrams and Peter Hall's anthology *Else/Where: Mapping* (2006) collects maps designed by utilizing a bewildering range of cutting-edge design perspectives.

Check out the Visionary Press's *Information Graphic Visionaries* book series at visionarypress.com for a growing series of historical tomes on information design and graphics.

The annual North American Cartographic Information Society (nacis.org) meeting has many events devoted to map design. The affiliated *Atlas of Design* (atlasofdesign.org) is the best source for the best map designs, issued in six volumes so far, every volume worth owning.

Sources: George Jenks's eye movement illustrations were published in his "Visual Integration in Thematic Mapping: Fact or Fiction?" *International Yearbook of Cartography* 13, 1973.

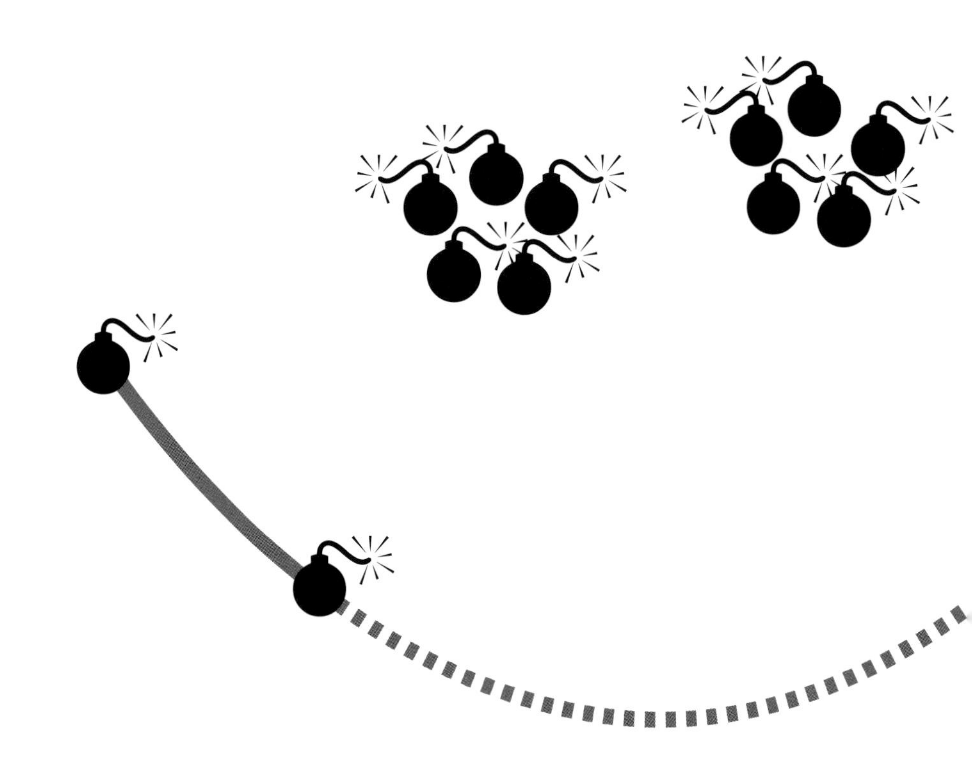

What jumps out at you?

Geo-Smiley Terror Spree

Luke Helder, a university student from Minnesota, went on a 2-week spree of bombings throughout the Midwestern U.S. in an attempt to create a giant smiley face across the nation.

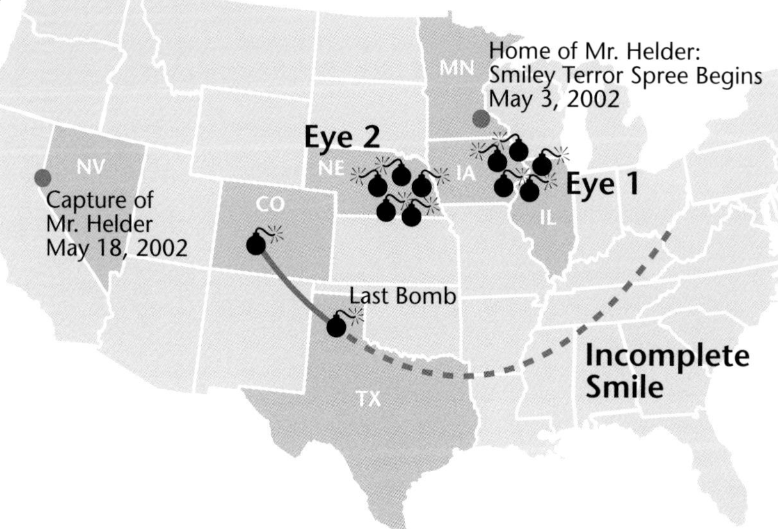

This is a crime that could only make sense in a map-immersed society like ours, since it is a crime that makes sense only if mapped, and then only if mapped at the the scale of the entire U.S. Zoom in, say, on the bombings that make up the eyes,

... and the pattern disappears.

The pattern jumps out only when the relevant elements are ***emphasized.***

CHAPTER 7 The Inner Workings of Map Design

Each map piece has its own internal structure, a structure that in some ways mimics that of the map as a whole. This is especially true of the map itself, that part of the whole display that justifies calling the whole thing a map. This piece differs from the others in that the space of the map proper is geographic whereas that of the big picture is merely graphic. Here then other sorts of rhetorical strategies come into play, ways of focusing the map reader's attention that also play a role in the "big picture" but are of the essence here: figure-ground and visual difference.

Thinking about Visual Differences

Our visual lives are full of depth and contrast. Maps too can be designed with depth and contrast, enhancing their ability to effectively communicate. The perceptual effect called figure-ground is behind our ability to see visual depth. Purposeful use of figure-ground helps create effective maps. Keep in mind that figure-ground plays a meaningful role in what gets relegated to the background or left off the map entirely.

Figure-Ground

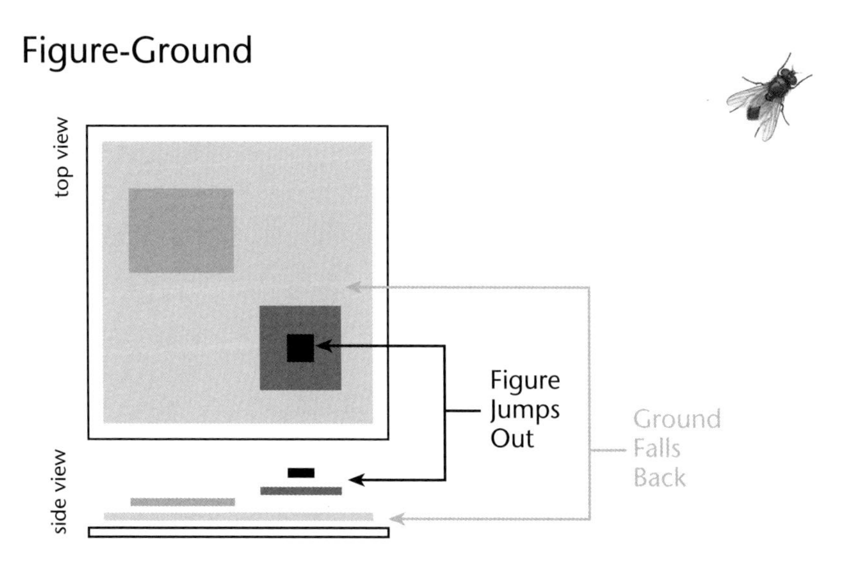

On a map, some parts stand out – figure – some fall back – ground – and others fall off the map altogether.

A successful figure-ground strategy on a map reveals what's most important first; these elements jump out. Less important elements are less visually noticeable and fall toward the back. Figure-ground relationships clearly communicate what you want your map to emphasize.

Figure on maps

- Most important map elements
- More important meaning
- Distinct form and shape
- Jumps out

Ground on maps

- Least important map elements
- Less important meaning
- Indistinct form and shape
- Falls back

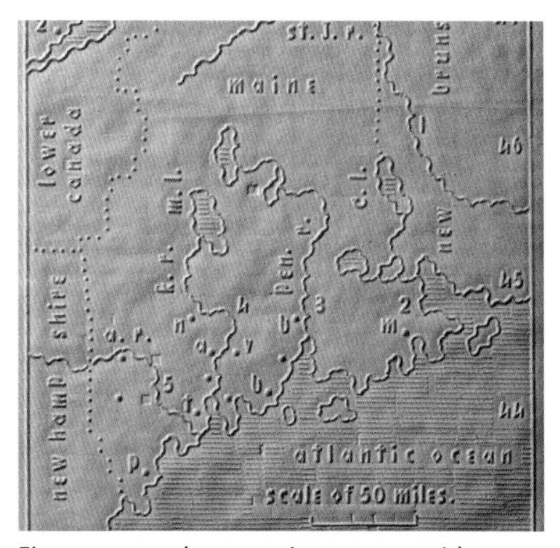

Figure-ground perception occurs with hearing, taste, smell, touch, and vision. Raised symbol maps for the blind are effective for communicating tactile figure-ground distinctions. The raised symbols literally stand out as figure over the smooth map background.

Her Figure on Our Ground

The famous Ditchley portrait of Queen Elizabeth I standing on a map of England reminds us that more is at stake in figure-ground relationships than elevating an object above the background.

On maps, the less powerful are relegated to the background or left off the map altogether, erased from our consciousness. Early maps of the Americas, Australia, and Africa communicated that there was much land available for colonization. The map silenced the voices of those already living there. It did this by burying the inhabitants below the ground.

Neither Figure Nor Ground

A typical web map reveals a system of tubes for driving around and buying stuff. Streams, wetlands, vegetation, bike trails, and walking trails take time to find. You can't show everything, but what you do show, what stands out as figure, reveals your intent for your map and your goals.

While creating your map, consider what should jump out, what should fall back, and, importantly, the implications of what falls off the map altogether.

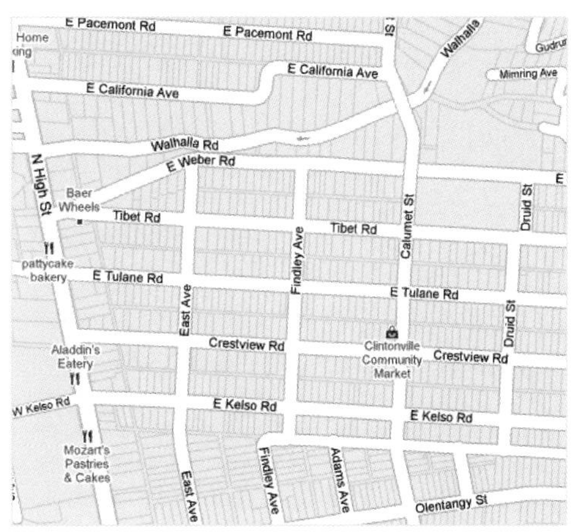

Enhancing Visual Differences

There are diverse ways to add depth to flat maps to help the map reader see the point of your map. The abstract examples below provide ideas for establishing figure-ground relationships on your map and can be combined on your map. Not all will work on every map.

Weaker Figure-Ground | Stronger Figure-Ground

Visual difference: Noticeable visual differences separate figure from ground. To focus attention on the important areas on your map, make them visually distinct from less important areas.

Detail: Figure has more detail than ground. To focus attention on the most important area on your map, generalize and reduce detail in less important areas.

Edges: Sharp, defined edges separate figure from ground. Conversely, weak edges move less important elements on the map from figure to ground. Gray or white (reversed-out) lines and type weaken edges and move less important information to a lower visual level.

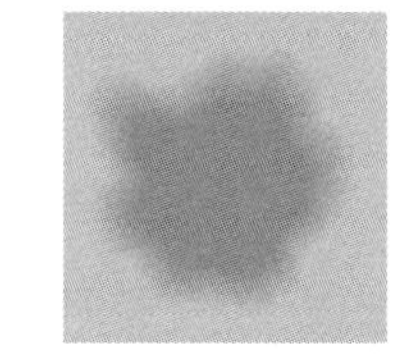

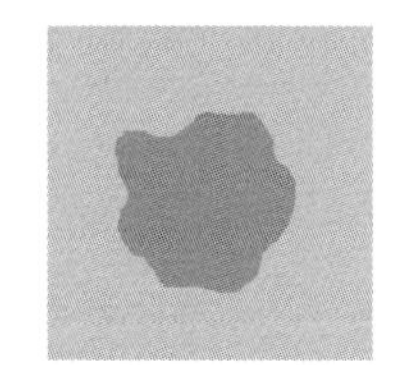

Texture: Isolated coarse textures tend to stand out as figure and move to higher visual levels.

Layering: Visual depth is enhanced when the ground appears to continue behind the figure. Grids of latitude and longitude can be visually manipulated to focus attention on the most important area of your map by appearing to run behind the most important parts.

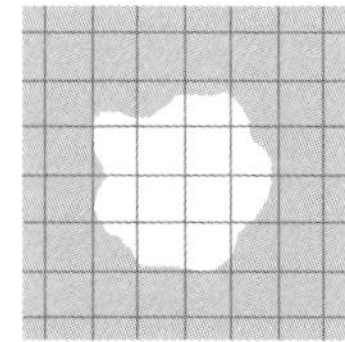

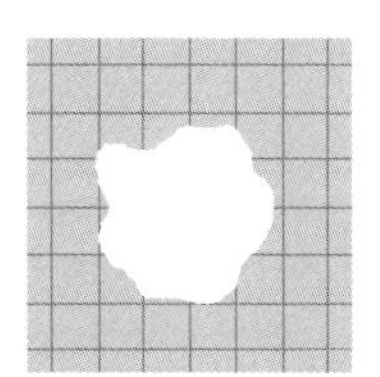

Shape and size: Figure has shape and size. Map elements with simple closed shapes tend to be seen as figure. However, complex shapes also draw attention and tend toward figure. Larger symbols tend toward stronger figure.

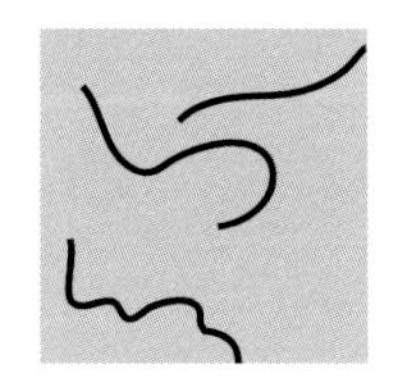

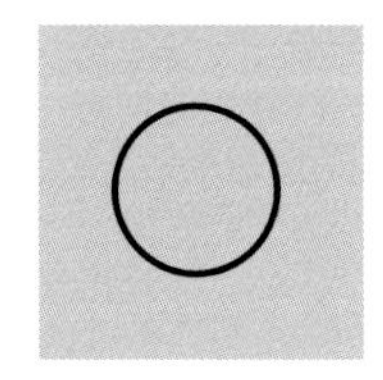

Closure: Closed objects tend to jump out from the ground.

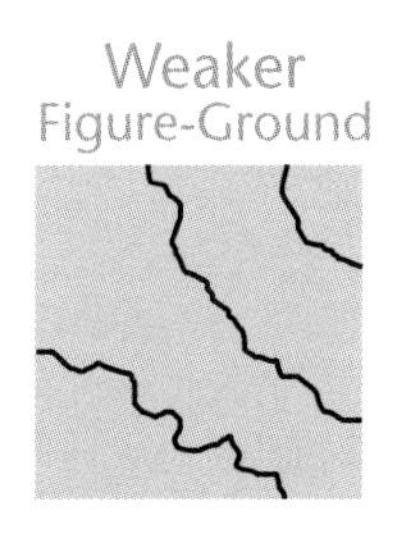

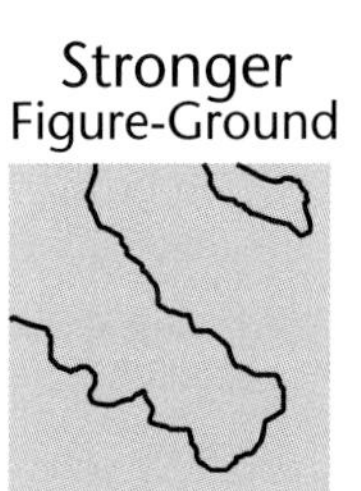

Proximity: Objects close together tend to stand out as figure.

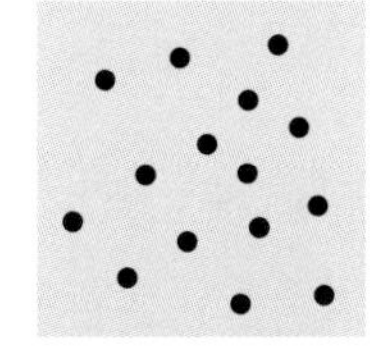

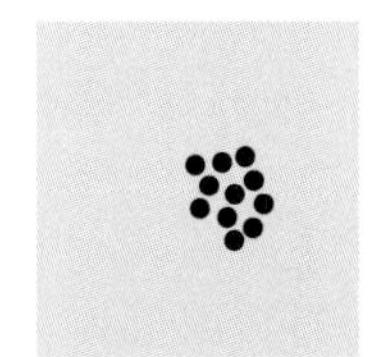

Simplicity: Simple objects tend to form stronger figure.

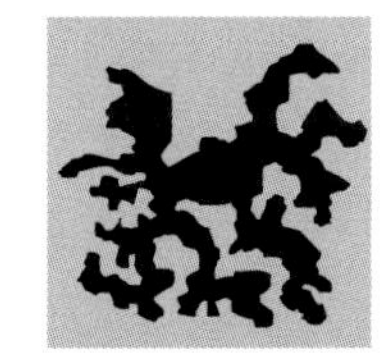

Direction: Objects with the same orientation, heading in the same direction, tend to form figure.

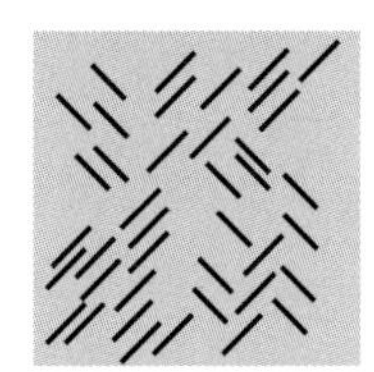

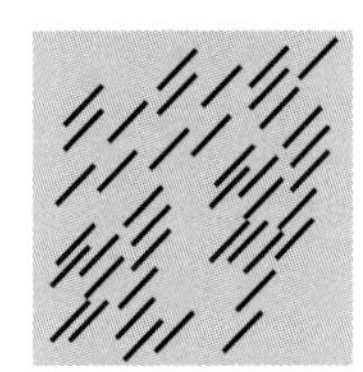

Familiarity: Objects with familiar, recognizable shapes jump out as figure.

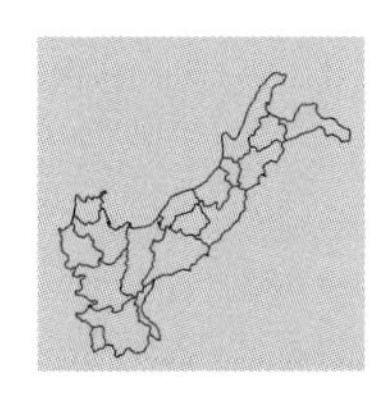

Color: Strong figure is created by intense colors, reds, and highly contrasting hues (yellow-black, white-blue). Complementary hues (red-green, blue-orange) create ambiguous figure-ground.

Visualizing Visual Differences

A design guide shows ways you can visually manipulate point, line, and area symbols on a map to achieve visual depth. Create a design guide, like the one below, based on common elements in your work. Print or view the guide in the final medium, and use it to help design your map. Subtle differences are noticeable, but if you want the map reader to see differences on your map, make sure there are substantial differences in the symbols you choose.

A Visual Design Guide

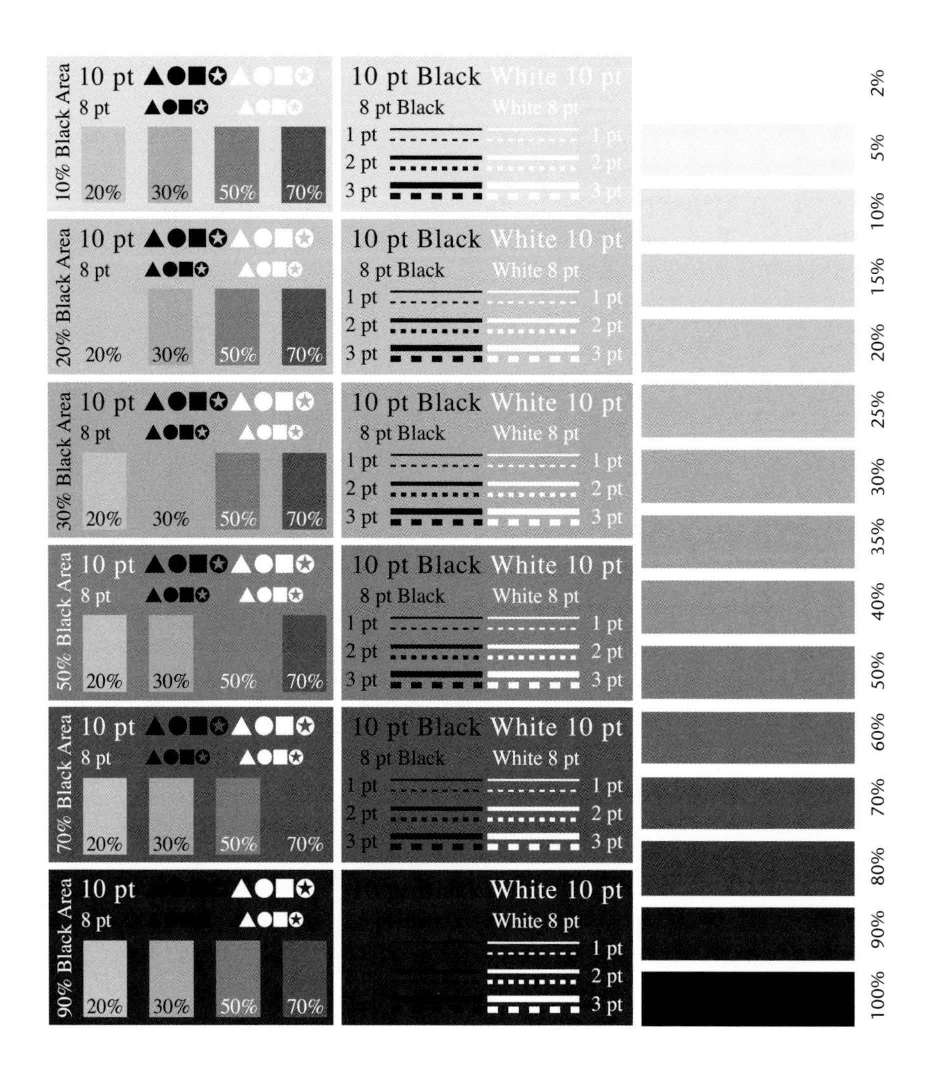

Symbol Differentiation

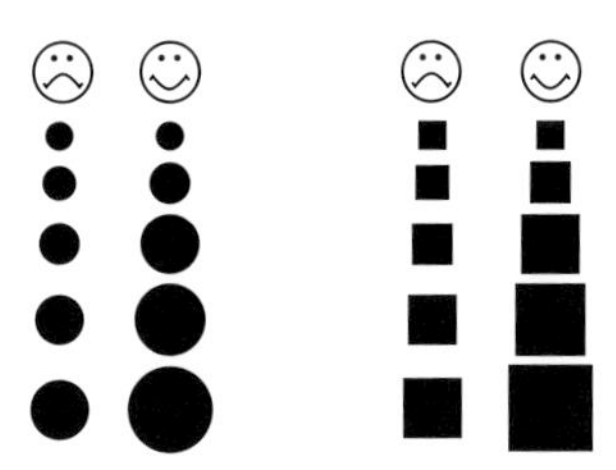

To differentiate symbols, ensure that they are different enough to notice.

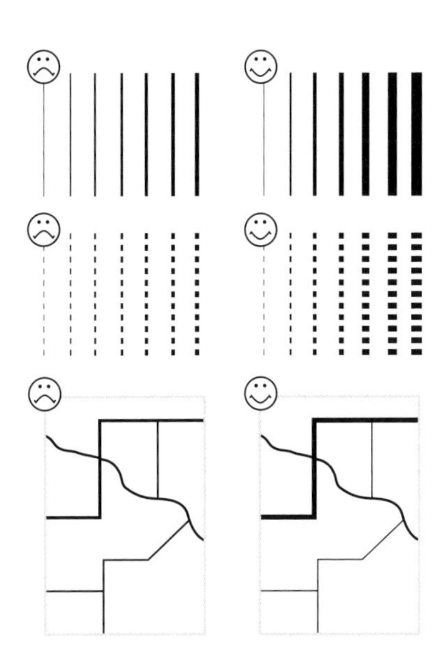

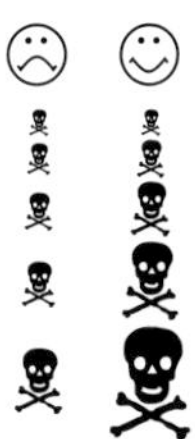

Trust your eye and err in the direction of too much difference.

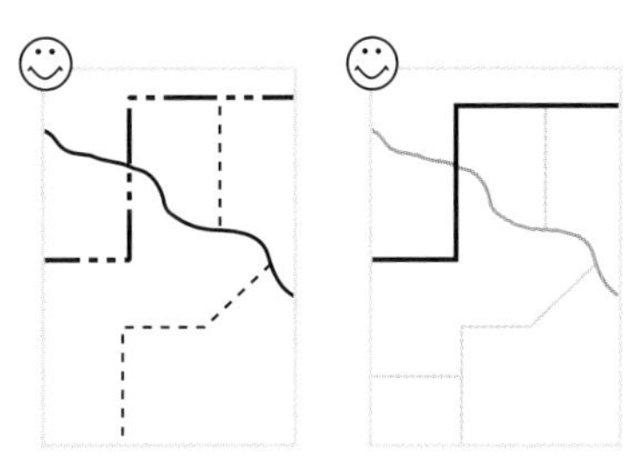

Without **visual differences** among the symbols on the map you have a nasty, unintelligible mess.

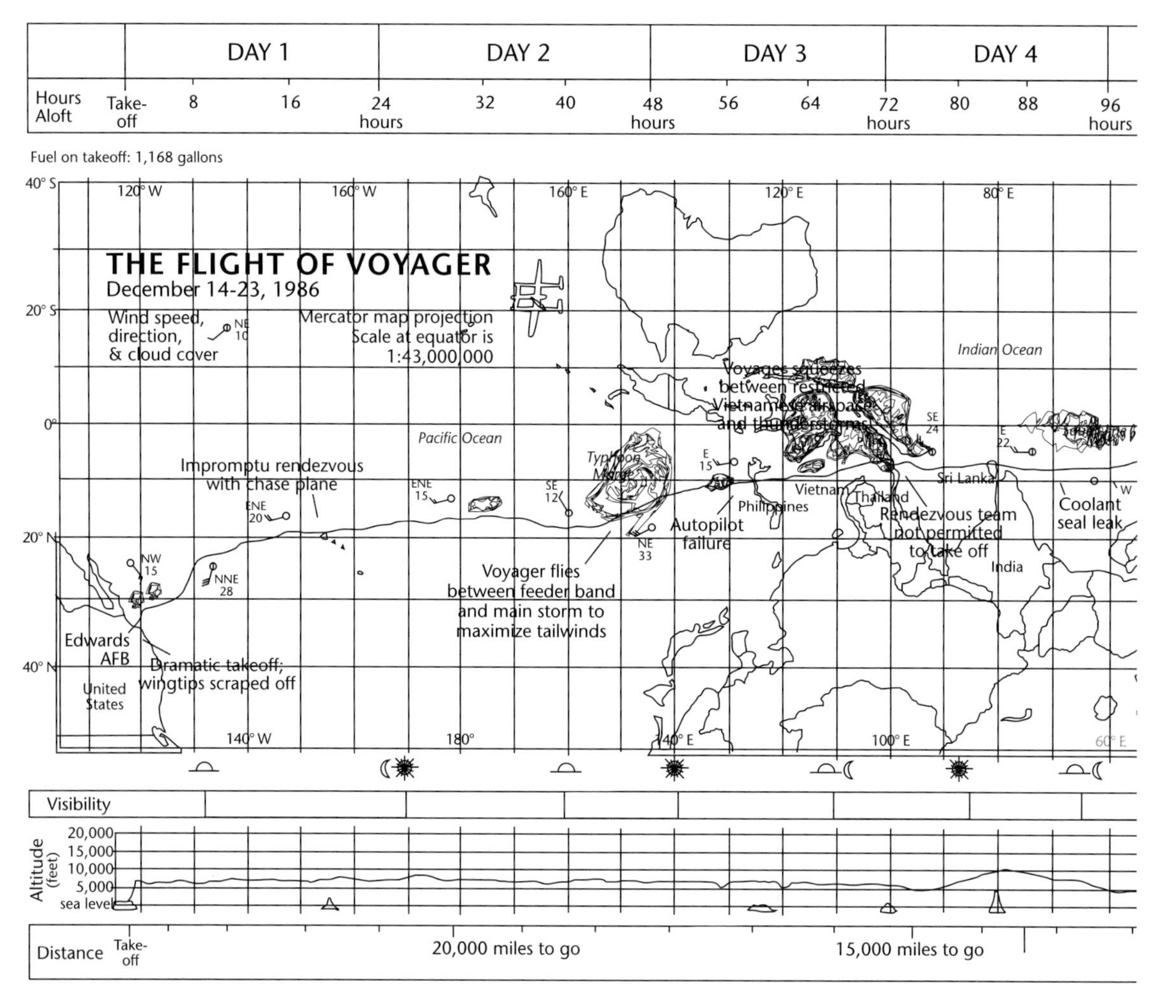

Flight data courtesy of Len Snellman and Larry Burch, Voyager meteorologists
Mapped by David DiBiase and John Krygier, Department of Geography, University of Wisconsin-Madison, 1987

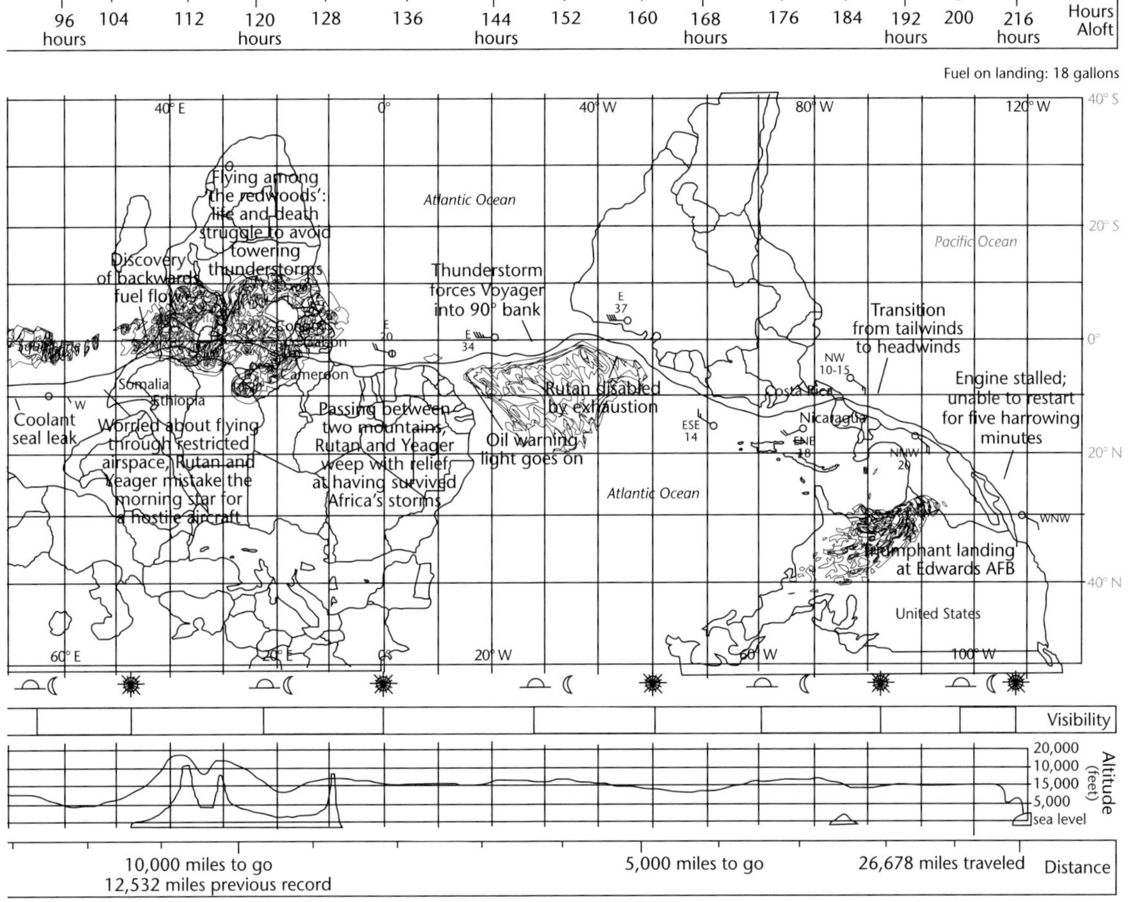

Voyager pilots: Dick Rutan and Jeana Yeager
Voyager designer: Burt Rutan

So, add **visual differences** – driven by your data and the goals for your map...

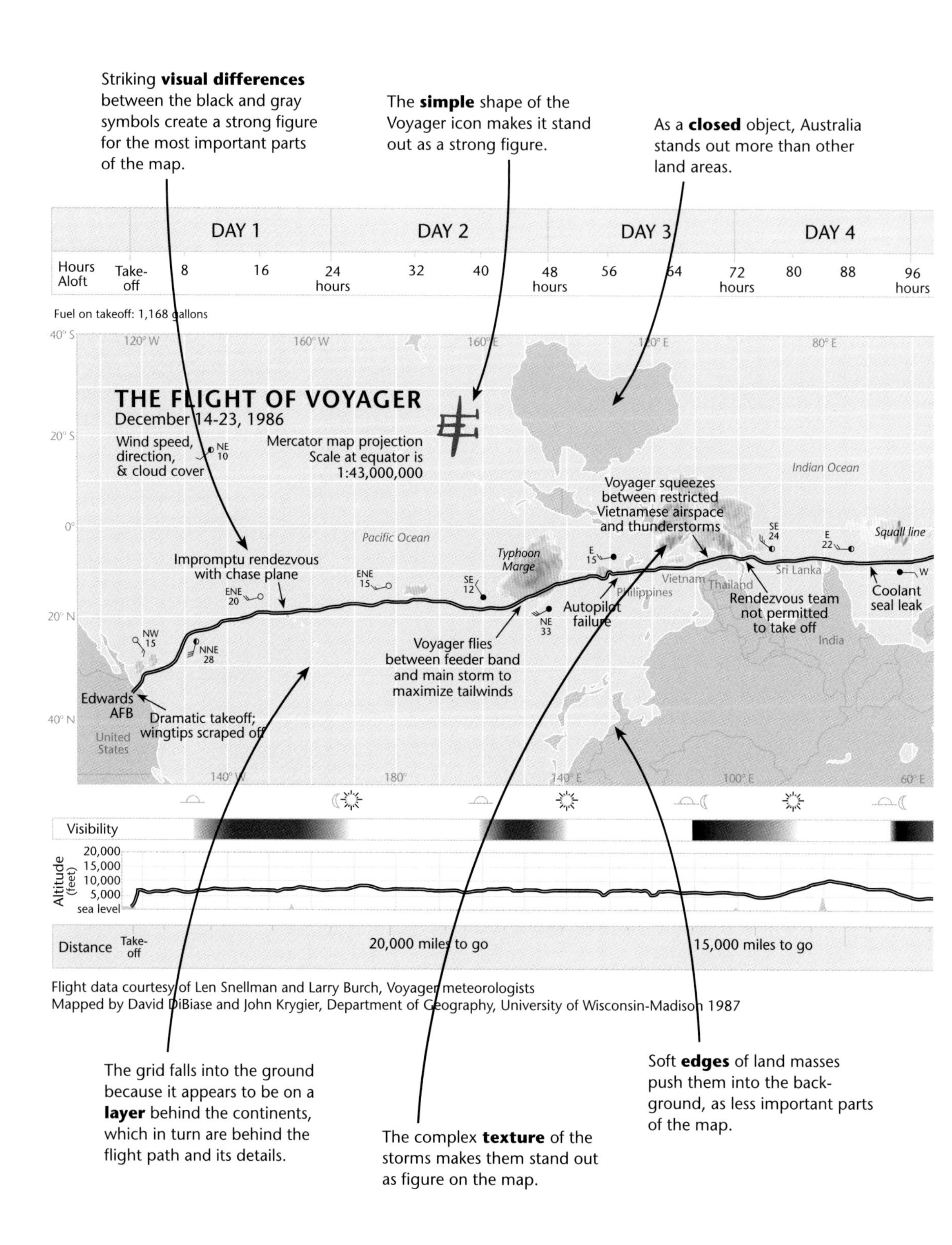

Striking **visual differences** between the black and gray symbols create a strong figure for the most important parts of the map.
The **simple** shape of the Voyager icon makes it stand out as a strong figure.
As a **closed** object, Australia stands out more than other land areas.
DAY 1
DAY 2
DAY 3
DAY 4
Hours Aloft
Take-off
8
16
24 hours
32
40
48 hours
56
64
72 hours
80
88
96 hours
Fuel on takeoff: 1,168 gallons
40° S
20° S
0°
20° N
40° N
120° W
160° W
160° E
120° E
80° E
140° W
180°
140° E
100° E
60° E
THE FLIGHT OF VOYAGER
December 14-23, 1986
Wind speed, direction, & cloud cover
NE 10
Mercator map projection
Scale at equator is 1:43,000,000
Indian Ocean
Pacific Ocean
Voyager squeezes between restricted Vietnamese airspace and thunderstorms
Squall line
Typhoon Marge
Impromptu rendezvous with chase plane
ENE 15
SE 12
E 15
SE 24
E 22
W
ENE 20
NE 33
Autopilot failure
Vietnam
Thailand
Philippines
Sri Lanka
Rendezvous team not permitted to take off
Coolant seal leak
India
NW 15
NNE 28
Voyager flies between feeder band and main storm to maximize tailwinds
Edwards AFB
Dramatic takeoff; wingtips scraped off
United States
Visibility
Altitude (feet)
20,000
15,000
10,000
5,000
sea level
Distance
Take-off
20,000 miles to go
15,000 miles to go
Flight data courtesy of Len Snellman and Larry Burch, Voyager meteorologists
Mapped by David DiBiase and John Krygier, Department of Geography, University of Wisconsin-Madison 1987
The grid falls into the ground because it appears to be on a **layer** behind the continents, which in turn are behind the flight path and its details.
The complex **texture** of the storms makes them stand out as figure on the map.
Soft **edges** of land masses push them into the background, as less important parts of the map.

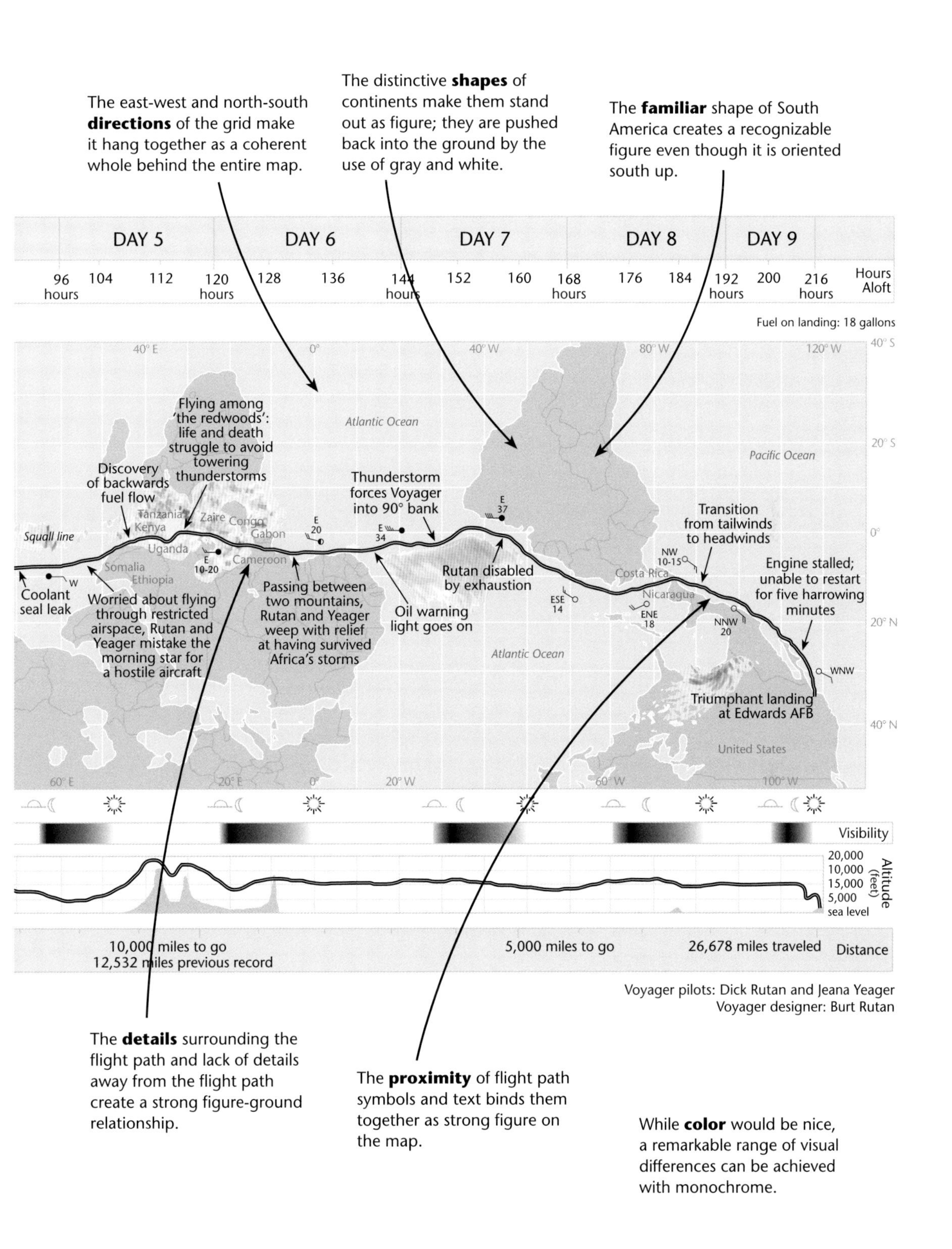

The east-west and north-south **directions** of the grid make it hang together as a coherent whole behind the entire map.
The distinctive **shapes** of continents make them stand out as figure; they are pushed back into the ground by the use of gray and white.
The **familiar** shape of South America creates a recognizable figure even though it is oriented south up.
DAY 5
DAY 6
DAY 7
DAY 8
DAY 9
96 hours
104
112
120 hours
128
136
144 hours
152
160
168 hours
176
184
192 hours
200
216 hours
Hours Aloft
Fuel on landing: 18 gallons
40° E
0°
40° W
80° W
120° W
40° S
20° S
0°
20° N
40° N
Atlantic Ocean
Pacific Ocean
Flying among 'the redwoods': life and death struggle to avoid towering thunderstorms
Discovery of backwards fuel flow
Thunderstorm forces Voyager into 90° bank
Transition from tailwinds to headwinds
Squall line
Tanzania
Kenya
Zaire
Congo
Gabon
Uganda
Somalia
Ethiopia
Cameroon
Costa Rica
Nicaragua
Coolant seal leak
Worried about flying through restricted airspace, Rutan and Yeager mistake the morning star for a hostile aircraft
Passing between two mountains, Rutan and Yeager weep with relief at having survived Africa's storms
Oil warning light goes on
Rutan disabled by exhaustion
Engine stalled; unable to restart for five harrowing minutes
Triumphant landing at Edwards AFB
Atlantic Ocean
United States
60° E
20° E
0°
20° W
60° W
100° W
Visibility
20,000
10,000
15,000
5,000
sea level
Altitude (feet)
10,000 miles to go
12,532 miles previous record
5,000 miles to go
26,678 miles traveled
Distance
Voyager pilots: Dick Rutan and Jeana Yeager
Voyager designer: Burt Rutan
The **details** surrounding the flight path and lack of details away from the flight path create a strong figure-ground relationship.
The **proximity** of flight path symbols and text binds them together as strong figure on the map.
While **color** would be nice, a remarkable range of visual differences can be achieved with monochrome.

[looking at map] 'You are here'... Wow! How do they know? That's so cool!!

Twister, *A Shot In The Park* (2007)

He looked on the map for the Lonely Tree, but did not find it.

"That map is absolutely worthless," said Good Fortune emphatically.

"Perhaps it is only a matter of looking long enough," Christian suggested. "Yesterday I found Durben Mot, and there's a lot of sand here.... We must be here, and the place is called Gatsen Mot, and there's a well near."

"That's wonderful!" cried Good Fortune, "and those who say a man should be sparing of his words are right. Gatsen Mot is Mongolian for Lonely Tree, and I beg your pardon for my hasty speech."

Fritz Muhlenweg, *Big Tiger and Christian* (1954)

Young C——— took the jokes of his companions on his chart and its Indian towns good-naturedly enough, and the map was nailed to a big spruce tree and used for a target for rifle practice...

Frederick Schwatka, *Along Alaska's Great River* (1898)

We have watched mutant creatures crawl from sewers into cold flat starlight and whisper shyly to each other, drawing maps and messages in faecal mud.

China Miéville, *Perdido Street Station* (2000)

More...

One of the best discussions of the internal structure of information graphics is Edward Tufte's in the second half of his book *The Visual Display of Quantitative Information* (2001). Another stimulating treatment is that of Richard Saul Wurman in his *Information Anxiety* (1989), and there's plenty of valuable material in his *Information Architects* (1997), with contributions from 20 important designers. In cartography, check out Borden Dent et al., *Cartography: Thematic Map Design* (2008), with its excellent overview of intellectual and visual hierarchies on maps.

On map design in general as well as the concerns of this chapter, see Cynthia Brewer, *Designing Better Maps: A Guide for GIS Users* (2024) and *Designed Maps: A Sourcebook for GIS Users* (2008).

Focused on graphs and PowerPoint but useful for map makers: Stephen Kosslyn, *Graph Design for the Eye and Mind* (2006) and *Clear and to the Point: Eight Psychological Principles for Compelling PowerPoint Presentations* (2007). But also see Tufte's screed against PowerPoint, *The Cognitive Style of PowerPoint* (2003). Also see *Show Me the Numbers: Designing Tables and Graphs to Enlighten* by Stephen Few (2012)

Sources: The Geo-Smiley Terror Spree map is based on data from a map in *Time* (May 20, 2002). The tactile map is from the David Rumsey map collection, www.davidrumsey.com. The Ditchley portrait of Queen Elizabeth I is from the Wikimedia Commons (commons.wikimedia.org). The section of the chapter on enhancing visual differences is based on Borden Dent's chapter on intellectual and visual hierarchies in his *Cartography: Thematic Map Design* (McGraw-Hill, 2008). The symbol differentiation examples are redrawn from R.W. Anson and F. J. Ormeling, eds., *Basic Cartography* (International Cartographic Association, 1984).

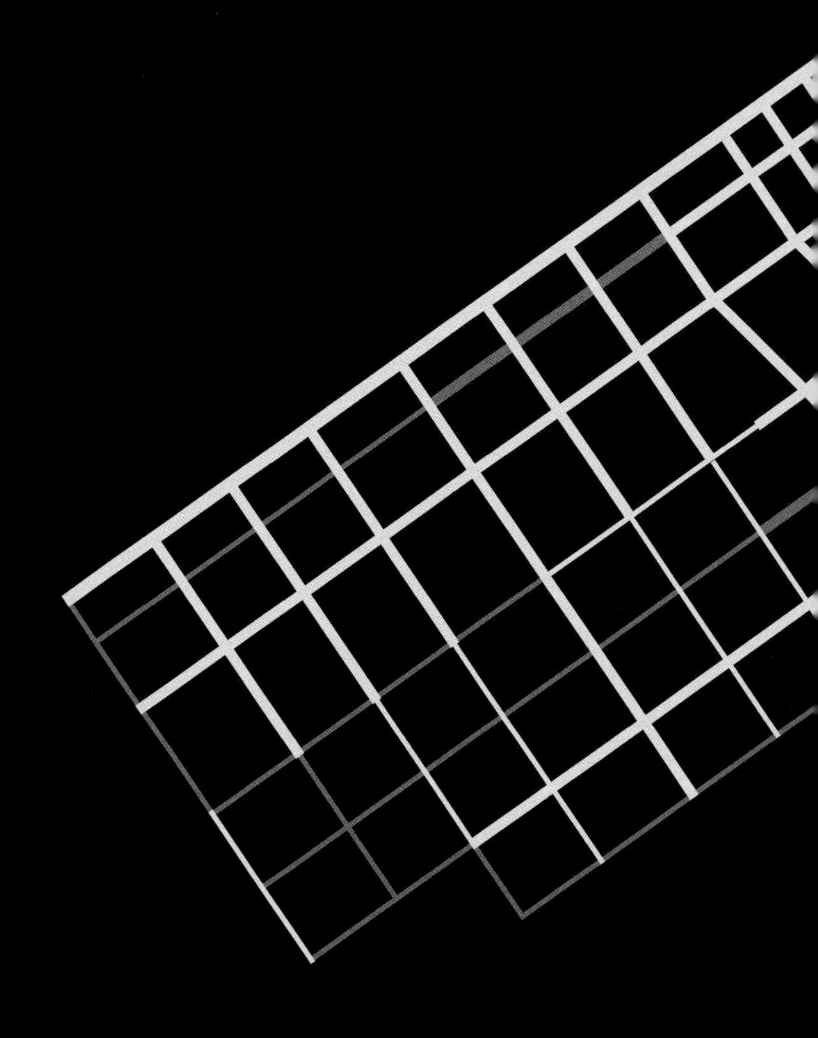

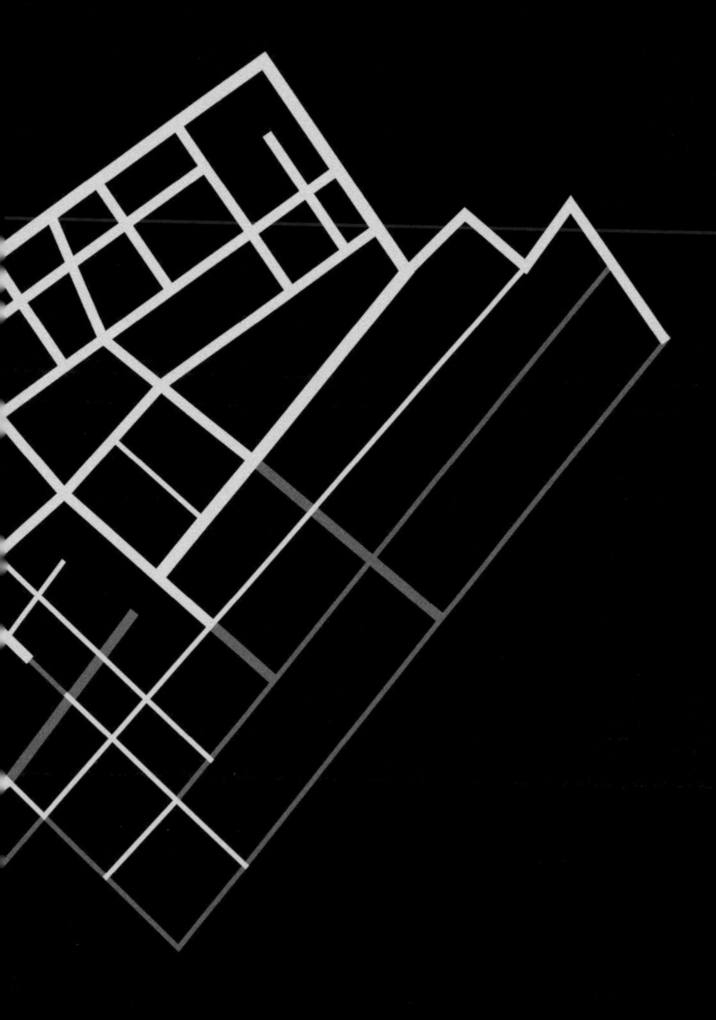

Do you need to move beyond
black and white?

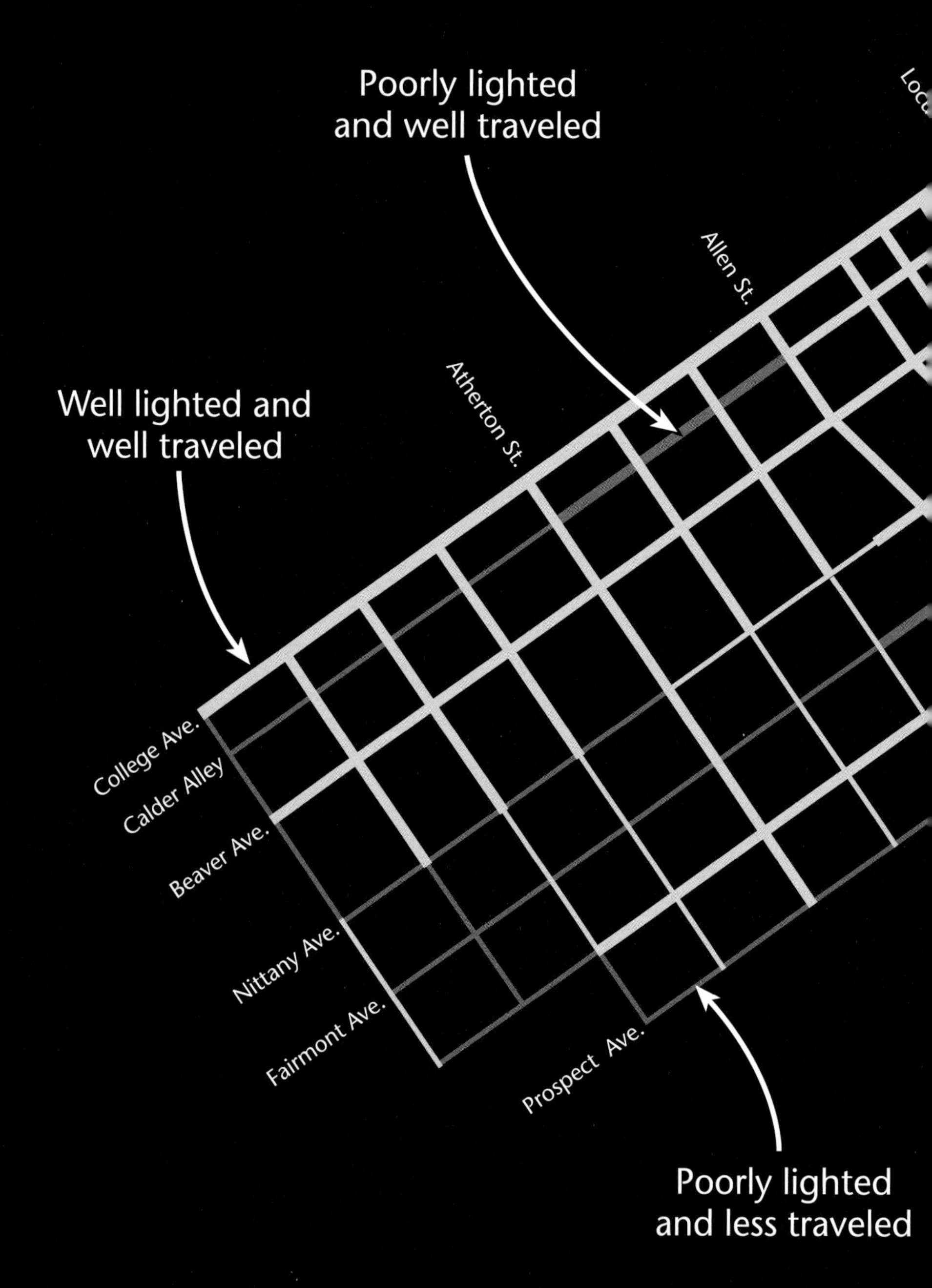
Poorly lighted
and well traveled
Well lighted and
well traveled
Poorly lighted
and less traveled
Allen St.
Atherton St.
College Ave.
Calder Alley
Beaver Ave.
Nittany Ave.
Fairmont Ave.
Prospect Ave.

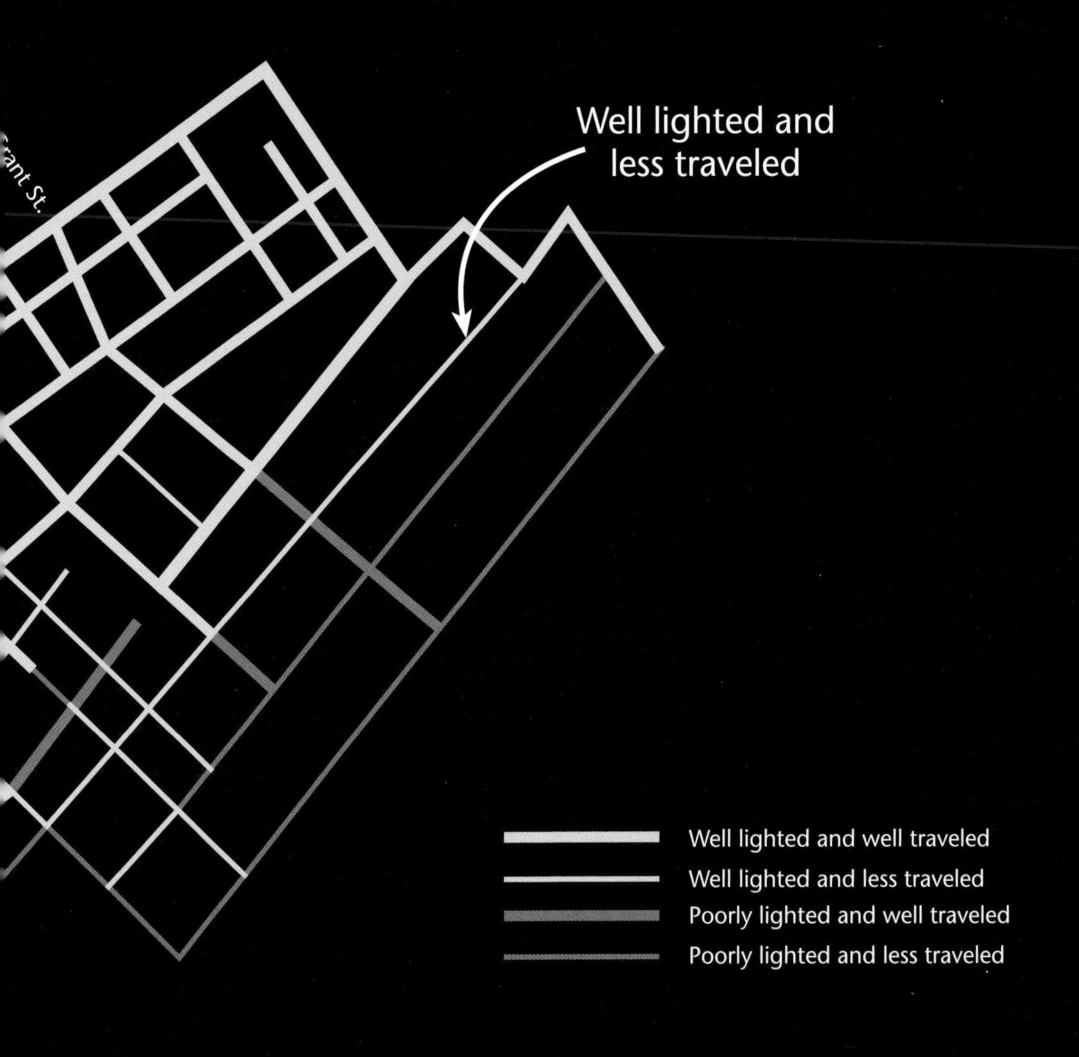

CHAPTER 8 Color on Maps

You don't need color to make excellent maps. This "night" map of State College, Pennsylvania, shows how well streets are lighted at night and how many people are around. The map helps people choose a safe route, and the map wouldn't do that any better if color were used.

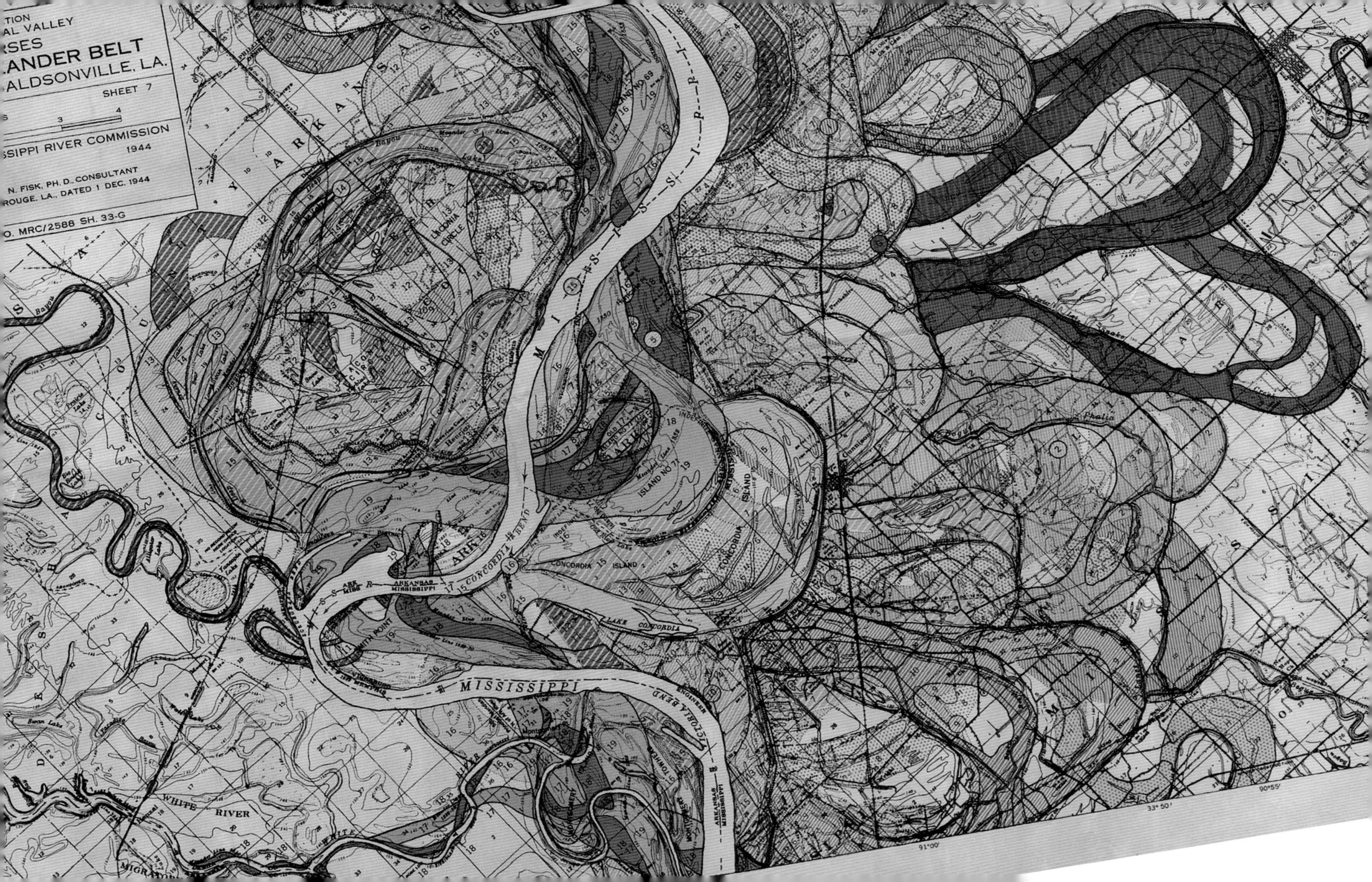

TION
AL VALLEY
RSES
ANDER BELT
ALDSONVILLE, LA.
SHEET 7
SSIPPI RIVER COMMISSION
1944
N. FISK, PH. D., CONSULTANT
ROUGE, LA., DATED 1 DEC. 1944
O. MRC/2588 SH. 33-G
MISSISSIPPI
ARKANSAS
CONCORDIA BEND
CONCORDIA ISLAND
LAKE CONCORDIA
SMITH POINT
LACONIA CIRCLE
ISLAND NO 69
VICTORIA BEND
WHITE RIVER

ARKANSAS
MISSISSIPPI
COAHOMA COUNTY
OLD TOWN BEND
ISLAND NO. 14

Thinking about Color on Maps

Color has a huge impact – positive or negative – on the design of your map. When used well, color vastly extends the effectiveness of your map. When used poorly, it easily draws attention away from your data and your goals for the map. Tufte's idea of graphical excellence, the visual variables, human perception, and symbolic connotations all provide ways to think about appropriate color choices for your map.

Graphical Excellence with Color

"Above all, do no harm" is the adage of Edward Tufte in *Envisioning Information*. Tufte's color Tufteisms, in part drawing from the work of cartographer Eduard Imhof, serve as a guide to excellence of color use on maps.

Graphical excellence is the well-designed presentation of interesting data – a matter of substance, of statistics, and of design.
Use color with an awareness that adjacent colors perceptually modify each other.
Use strong color for important data in small areas against a muted background.
Use color redundancy to reduce perceptual color shifts and ambiguity.
Use color to distinguish and differentiate features on your map.
Use muted color for less important or background data.
Use color to distinguish order in quantitative data.
Use color to mimic the color of phenomena.
Use muted color over large adjacent areas.
Use color to engage your map's viewers.
Use color palettes found in nature.

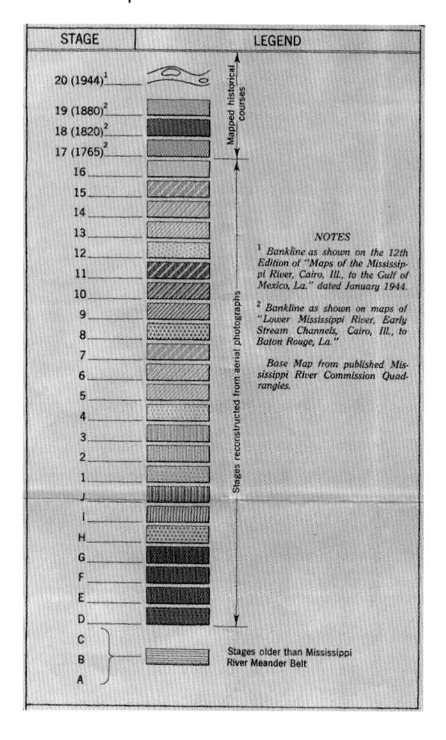

Ancient Courses, Mississippi River Meander Belt, Cape Girardeau, Missouri, to Donaldsonville, Louisiana, Sheet 7

Sheet 7 of the *Ancient Courses of the Mississippi River* map series was published in Harold Fisk's *Geological Investigation of the Alluvial Valley of the Lower Mississippi River* (1944). This spectacular map, expressing engaging data with graphical excellence, reveals changes in the course of the Mississippi River over thousands of years. The map maker differentiates 27 stages of the river. Color (and texture) are used to effectively reveal the tangled knot that is the lower Mississippi. It would be impossible to communicate these complex data without the use of color.

The choice of colors along with the interesting data **engage viewers**, making the subject of fluvial geomorphology seem quite fascinating.

The range of earthy, warm hues used on the map evoke **the phenomena** of ancient river courses.

Color (as well as texture) is chosen to help **distinguish and differentiate** the 27 historical river courses. The challenge is in the sheer number of categories and their complex spatial patterns.

Data of natural phenomena mapped with **natural color palettes** are true to the phenomena, visually engaging, and reveal the complexity of the phenomena.

The **muted** tan background color allows the historical river beds to stand out as the most vital part of this map.

Because the riverbed data are chronological, color value could have been used to **distinguish order**. Instead, the choice was to distinguish qualitative differences, as with the use of color on geologic maps.

Symbolic, Cultural, and Personal Connotations of Color

Understanding the symbolic meanings of color can guide effective color use on maps.

Symbolic Color Connotations: Symbolic connotations subtly shape viewer reactions and should be guided by your goals for your map. Generic Western cultural color connotations include

Blue: water, cool, positive numbers, serenity, purity, depth

Green: vegetation, lowlands, forests, youth, spring, nature, peace

Red: warm, important, negative numbers, action, anger, danger, power, warning

Yellow/tan: dry, lack of vegetation, intermediate elevation, heat

Orange: harvest, fall, abundance, fire, attention, action, warning

Brown: landforms (mountains, hills), contours, earthy, dirty, warm

Purple: dignity, royalty, sorrow, despair, richness, elegant

White: purity, clean, faith, illness, life, clarity, absence, light

Black: mystery, strength, heaviness, death, nighttime, presence

Gray: quiet, reserved, sophisticated, controlled, light, bland, dull

Cultural Color Connotations: The meaning of different colors varies from culture to culture, further complicating the use of color on maps. Check for cultural color connotations if you are mapping for a global audience.

Blue: safe cross-cultural color, because it is the color of the sky, which stands over all peoples

Green: fertility and paganism in Europe, sacred for Muslims, mourning and unhappiness in Asia

Red: Bolsheviks, communists, and other politically left organizations, purity in India

Yellow/tan: peaceful resistance movement associated with Corazón Aquino in Philippines

Orange: pro-Western activists in Ukraine, Protestants in Ireland, sacred Hindu color

Brown: mourning in India, Nazis in West, ceremonial for Australian Aboriginals

Purple: death and crucifixion in Europe, mysticism, prostitution in the Middle East

White: unhappiness in India, mourning in China, royalists and traditionalists in Western world

Black: fascists, anarchists, and other extremists in Western world, death, mourning in West

Gray: corporate culture in the West (also blue), dead and dull in Feng Shui

☹ color conventions

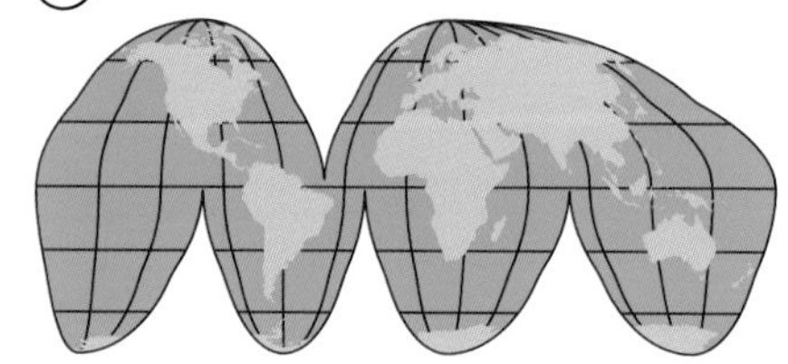

☺ color conventions

Margaret Pearce and Michael Hermann designed a narrative map of the travels and experiences of Samuel de Champlain in Canada. Their goal, to design a map expressing the emotions, voices, and multiple experiences of Champlain, his men, and indigenous peoples: interesting data and complex ideas presented with clarity and intelligence. Graphical excellence with color.

Color and Experience: Different voices and experiences lead to different maps – or embed them in one map. Pearce and Hermann use color hue (right) to map the multiple voices and experiences in the Champlain narrative. Champlain in blue, indigenous people in green, and the voice of the map maker, from a future time and place, in gray.

Color and Emotion: Color is emotive: angry red, calm green, depressed gray, happy yellow. Pearce and Hermann use color hue to express shifting emotions from panel to panel on the Champlain map. Below, Champlain learns of an assassination plot against him, and the colors differentiate the different voices and shifting emotions of Champlain and the assassins.

There was a great dispute between our Indians and our impostor *[Nicholas de Vignau]*, who declared that there was no danger by the rapids and that we should go that way.
Our Indians said to him, 'You are tired of living'
and to me that I should not believe him and that **he was not speaking the truth.**

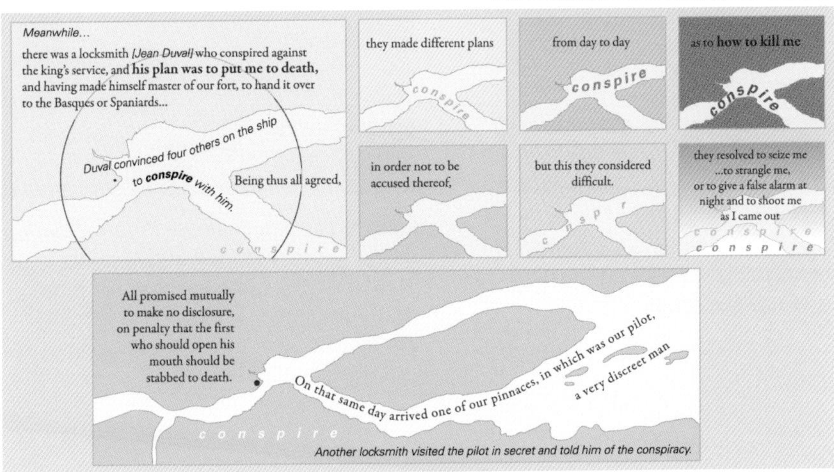

Meanwhile...
there was a locksmith *[Jean Duval]* who conspired against the king's service, and **his plan was to put me to death,** and having made himself master of our fort, to hand it over to the Basques or Spaniards...
Duval convinced four others on the ship to **conspire** with him.
Being thus all agreed,
conspire
they made different plans
conspire
from day to day
conspire
as to **how to kill me**
conspire
in order not to be accused thereof,
but this they considered difficult.
they resolved to seize me ...to strangle me, or to give a false alarm at night and to shoot me as I came out
conspire
All promised mutually to make no disclosure, on penalty that the first who should open his mouth should be stabbed to death.
On that same day arrived one of our pinnaces, in which was our pilot, a very discreet man
conspire
Another locksmith visited the pilot in secret and told him of the conspiracy.

Human Perception and Color

Our perceptual characteristics, abilities, and limitations can guide effective color use on maps.

Color Dimensions: Our eyes respond to blue, green, and red wavelengths of energy (with overlap) so we can sense the entire spectrum (red, orange, yellow, green, blue, indigo, violet). One way to think about how people perceive colors is in terms of three dimensions of color perception: hue, value (lightness), and intensity (saturation, chroma).

Hue is the name for our human experience of particular electromagnetic energy wavelengths. Hues are qualitatively different, thus good for showing qualitative data.

Value is the perceived lightness or darkness of a hue. Values are quantitatively different, thus good for showing quantitative data.

Intensity describes the purity of a hue (increasing to right). Intensity is subtle and OK for showing binary (yes, no), qualitative, and quantitative data.

Light Source: The colors on a map vary as the light source varies. The same colors will look different when viewing a map

Under daylight, incandescent, or LED lighting
As the intensity of the light varies
On a computer monitor, which emits light – thus the colors will be brighter and more saturated than on paper maps

When selecting colors for a map, consider the conditions under which your map will be viewed.

Low-intensity lighting: use more intense, saturated colors
High-intensity lighting: use less intense, less saturated colors
Computer monitor: use less intense, less saturated colors
Look critically at your map under lighting conditions similar to those of your map's audience, and adjust the colors to suit

Look at the colors above. Then move to a darker room and look at the colors again. They change. Choose colors for maps that work under appropriate lighting conditions.

Map Surface: colors on a map vary as the surface the map is displayed on varies.

Glossy paper: colors more intense and vibrant
Matte paper: colors less intense and dulled
Projectors, depending on the intensity of the bulb, may reproduce colors much more or less intense than you expect
Computer monitors will make colors vibrant, as the color on computer monitors is emitted rather than reflected (as on paper)
Look critically at your map on the medium the map will be presented on, and adjust the colors if necessary

Color Interactions: The appearance of any color on a map depends on surrounding colors. This optical illusion, called simultaneous contrast, makes the left gray dot (below) look slightly darker than the right gray dot (for most people).

Different colors can also look the same, depending on their background. Color subtraction makes the two small squares below look similar.

Yet, they are not.

Perceptual Differences: The appearance of color on a map varies, depending on the particular eye-brain system looking at it.

Older map viewers

Benefit from more saturated colors
Have particular difficulty in differentiating shades of blue
Benefit from increasing the type size a bit

Younger map viewers

Like brighter, saturated colors – but not too saturated
Dislike dull, gray, or mixed colors like brown
Perform tasks well with maps that use saturated and unsaturated colors
Understand quantitative, ordered data shown with color value by age 7 or 8

Color-blind viewers typically see red and green as the same. In the U.S., 3% of females and 8% of males are color-blind.

If reds and greens show important differences on your map, a significant number of viewers will not see these differences
Consider using reds and blues or greens and blues instead
Check internet resources for selecting color-blind safe colors

Creating Color on Maps

The specification and production of colors are often very different from the way in which we see them. Color specification systems are schemes that organize and help produce different colors. There are many different color specification systems, and map makers will encounter many of them. Three major categories of color specification systems are important: predefined color systems, perceptual color systems, and process color systems.

Predefined Color

Predefined color specification systems are like paint chips from the paint store. Thousands of predefined colors are specified by names or codes. Predefined colors ("spot" colors) are used by some commercial printers and are commonly used when mapping data with set color conventions (such as on geology maps).

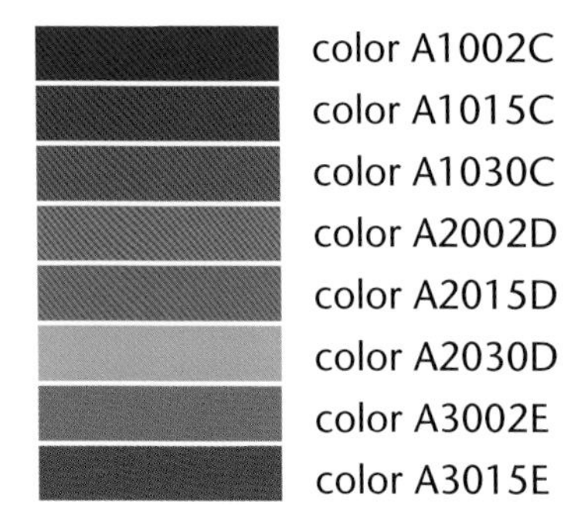

Predefined colors will be converted to another color system when printing on a computer printer or when using a commercial printer who uses process color.

Pantone is a common predefined color system used in mapping.

Perceptual Color

Perceptual color specification systems, such as Munsell, are based on human perceptual abilities. Perceptual tests have produced a set of colors that the average person can differentiate. Thus, no two colors in the Munsell system look exactly alike. The Munsell system consists of a series of color samples, each a single hue with varying value and intensity.

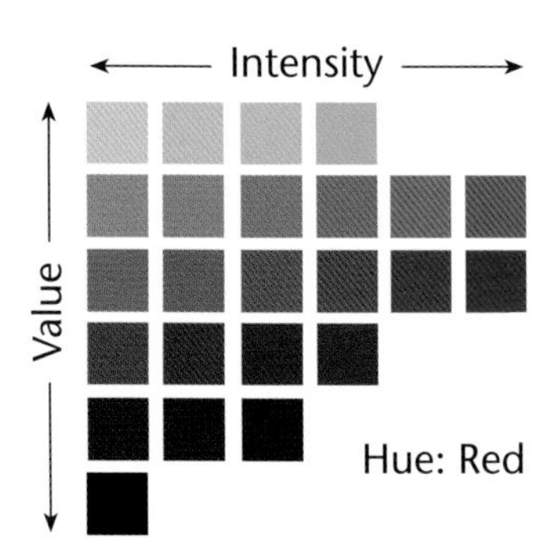

The Munsell system is excellent for selecting appropriate colors for your map, but it will be converted to another system in order to print.

Munsell colors are used as the basis of the ColorBrewer site (colorbrewer.org) created by Cindy Brewer and Mark Harrower. The site converts Munsell colors into other color specification systems, so you can easily use the colors in most mapping software.

Process Color: Printing

Process color specification systems use three or four colors to create all other colors. Printed colors typically use the subtractive primaries and rely on reflected light. When you combine cyan (C), magenta (M), and yellow (Y), you produce black – all light is absorbed (subtracted) from your vision. Thus, cyan, magenta, and yellow are the subtractive primaries.

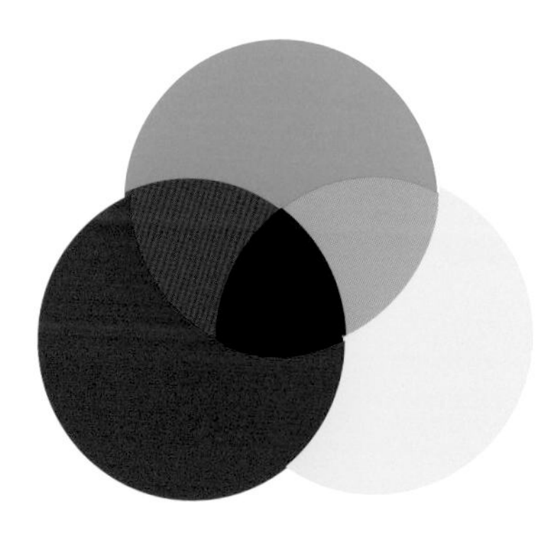

Black is added as a fourth "color" (K, thus CMYK) to avoid the muddy dark brown that is the result of combining cyan, magenta, and yellow.

Subtractive primaries are often used by commercial printers and are common on inkjet computer printers. Different amounts of CMY and K produce thousands of other colors. The CMYK color system should be used for most commercially printed maps.

Process Color: Monitors

Computer monitors also use three colors to create all other colors. Monitor colors typically create color with the additive primaries and rely on emitted light. Because the light is emitted, the colors are more intense. When you combine red (R), green (G), and blue (B), they add up to pure white. Thus red, green, and blue are the additive primaries.

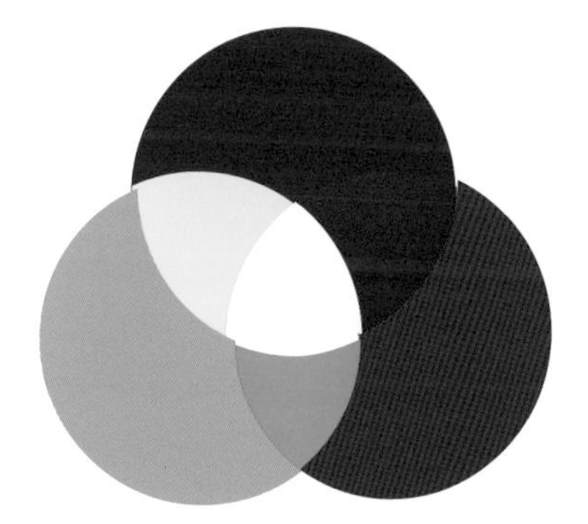

The hexadecimal color specifications used in HTML (HyperText Markup Language) are RGB. The first two digits are red, second two digits green, and third two digits blue. 00 is no color, and FF is maximum color.

The RGB color system should be used for maps printed with computer printers. RGB will have to be converted into CMYK or predefined color if you plan to print with a commercial printer.

Using Color on Maps

While color on maps is complex and wondrous, its application is hastened by a few basic guidelines that help match characteristics of your data to characteristics of color. Analogous vs. complementary colors can be used to suggest similarity and difference. Sequential color schemes suggest order in your data. Diverging color schemes also suggest data order, but away from a critical value. Qualitative color schemes suggest qualitative differences. Taken together, regardless of what kind of map you are making, these core ideas get you a long way towards effective use of color on maps.

Color Schemes

Analogous vs. Complementary Colors

Analogous colors are similar to each other and adjacent on a color wheel. Use them when mapped phenomena are similar. Complementary colors contrast strongly with each other and are opposites on a color wheel. Use them when mapped phenomena are different.

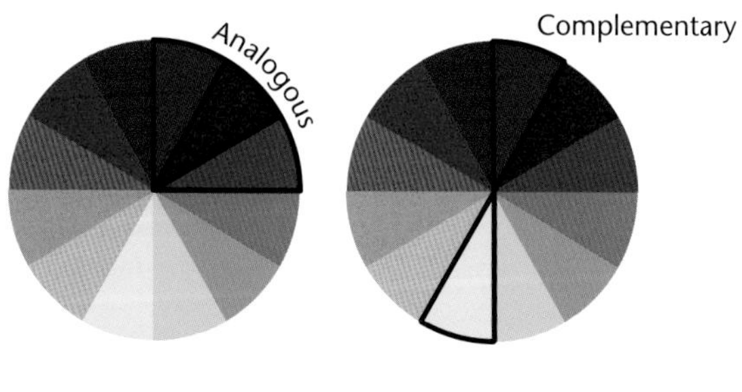

Sequential

Commonly used and easy to understand for ordered (quantitative) data ranging from low to high. Varying one hue from light to dark is common. A more complex mix of complementary hues with value creates schemes that seem more distinctive.

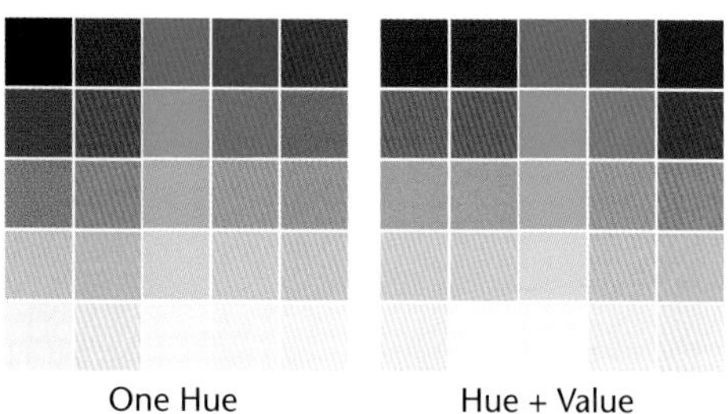

Diverging

Also called double-ended, these schemes use two hues and value differences to emphasize data variation away from a critical value (such as zero, when the data have positive and negative values) or class. These schemes tend to emphasize the extreme data classes, and can be more challenging to interpret.

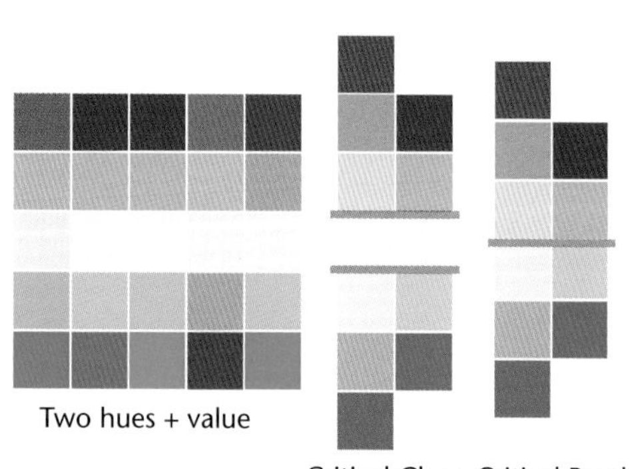

Qualitative

These schemes use different color hues to show qualitative variation in data. Color value and/or intensity are held constant. As with all color schemes, consider the symbolic meaning of the hues you choose and how they relate to your data.

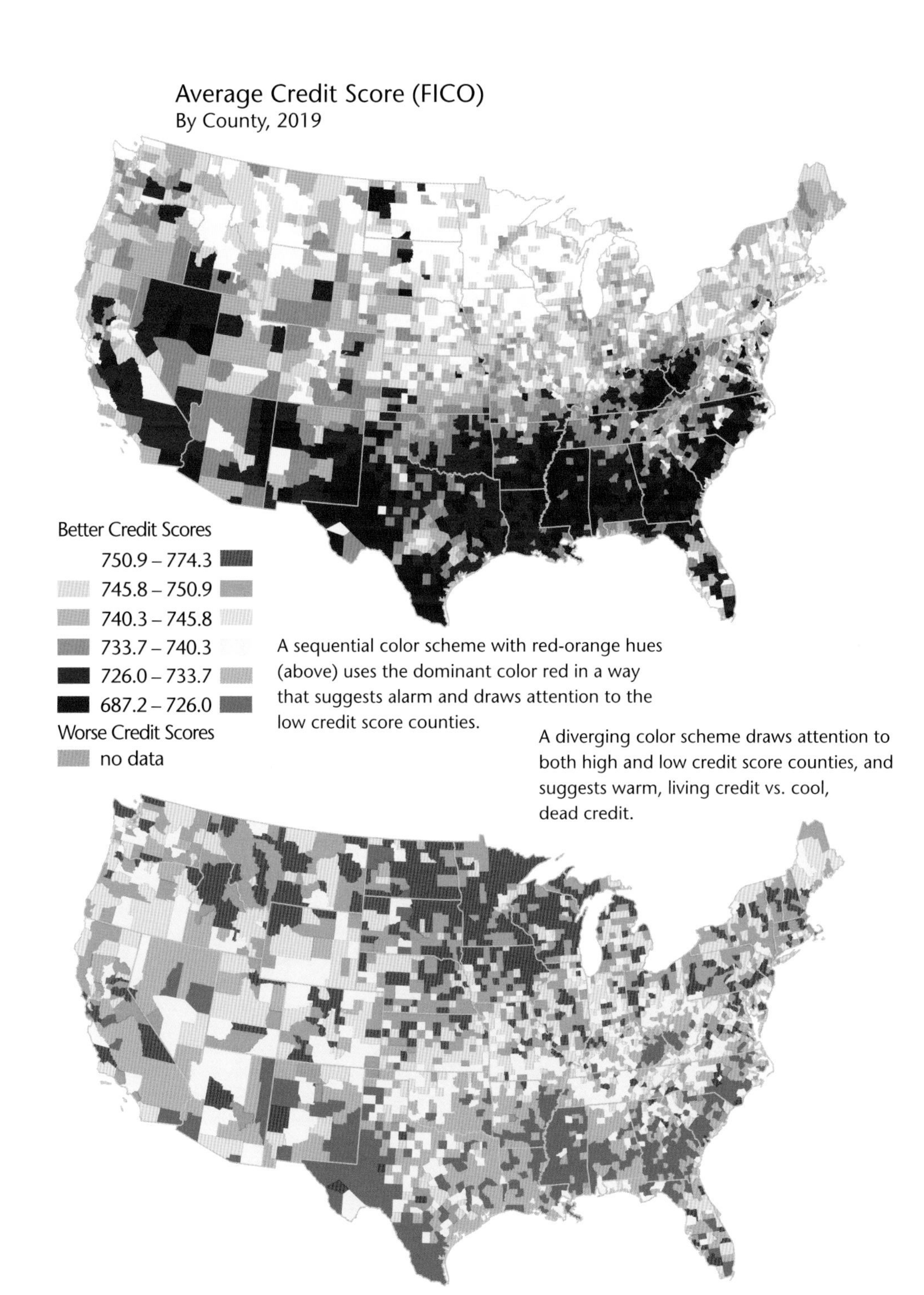

A sequential color scheme with red-orange hues (above) uses the dominant color red in a way that suggests alarm and draws attention to the low credit score counties.

A diverging color scheme draws attention to both high and low credit score counties, and suggests warm, living credit vs. cool, dead credit.

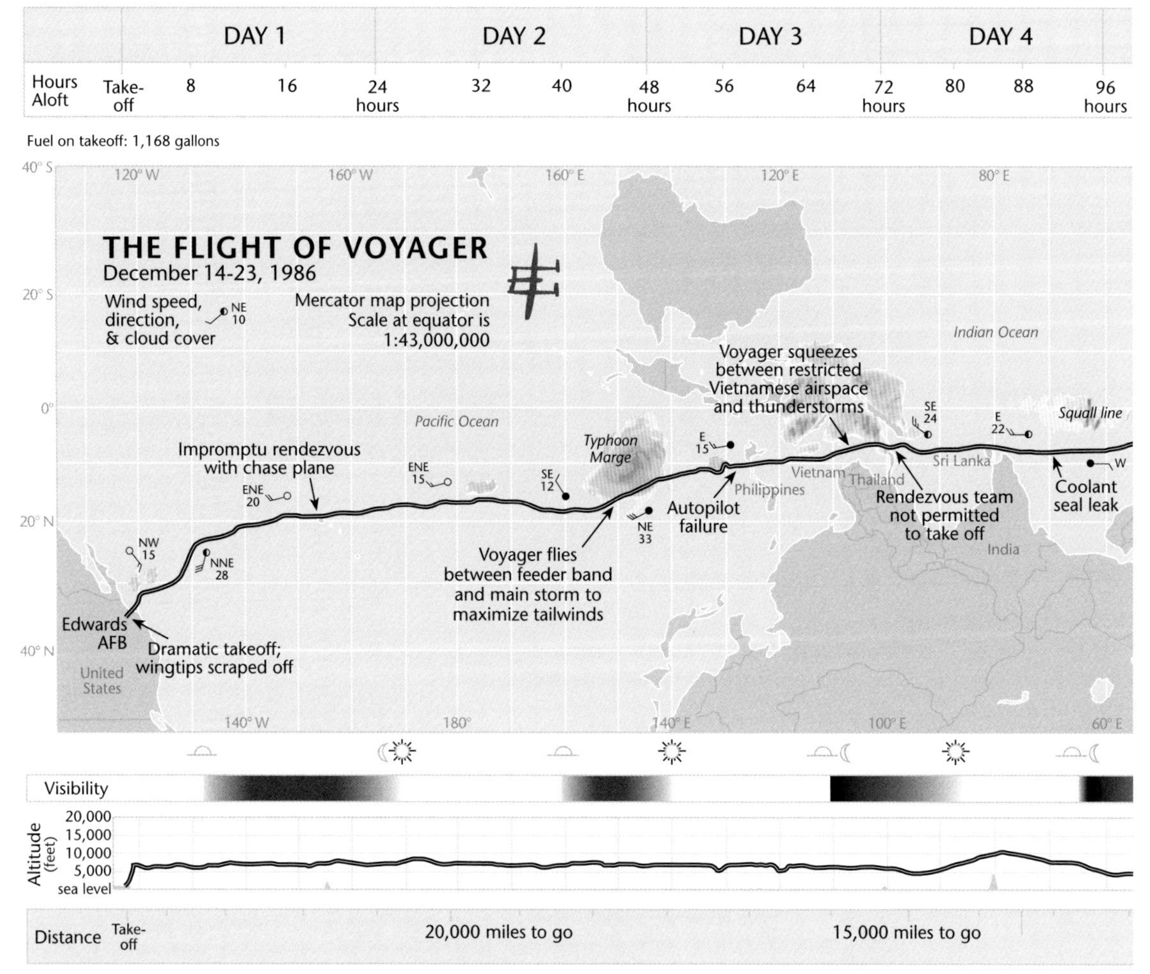

Flight data courtesy of Len Snellman and Larry Burch, Voyager meteorologists
Mapped by David DiBiase and John Krygier, Department of Geography, University of Wisconsin-Madison 1987

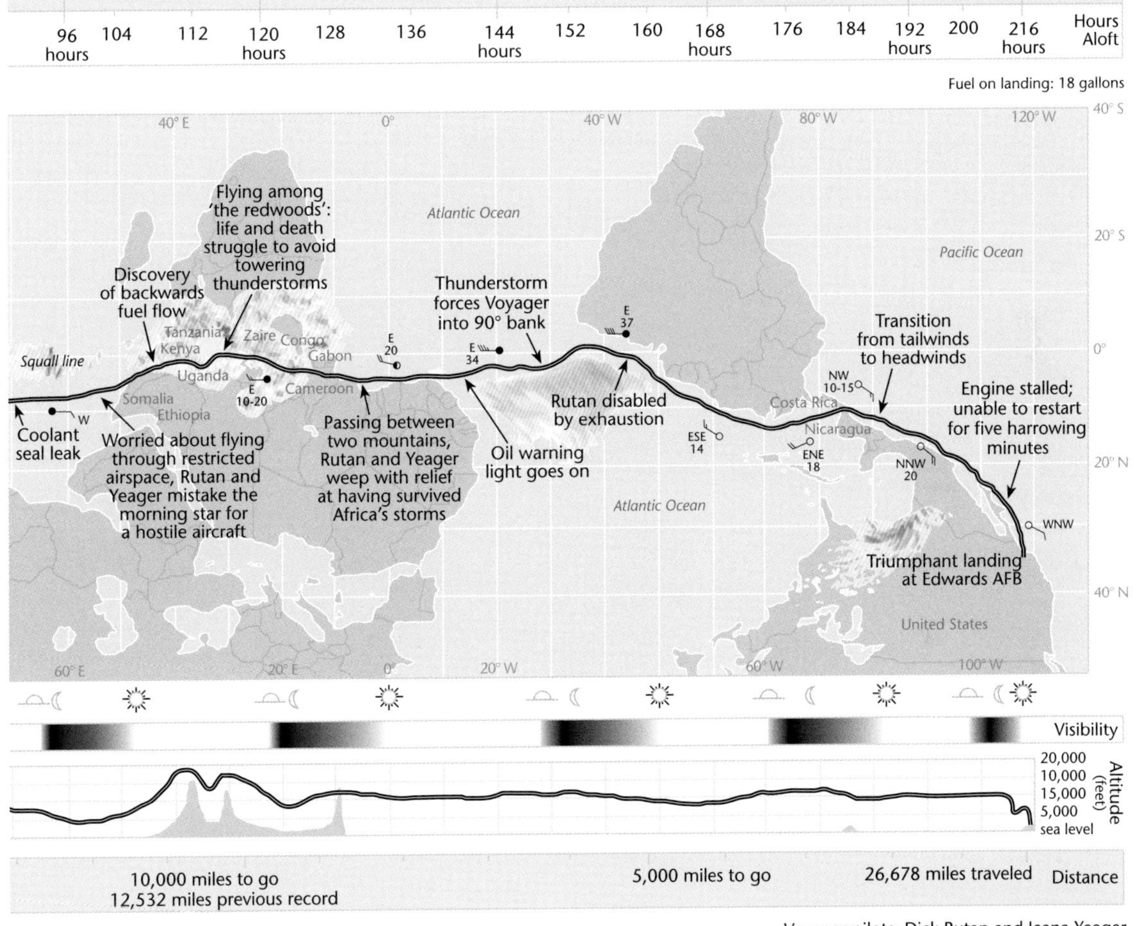
DAY 5
DAY 6
DAY 7
DAY 8
DAY 9
96 hours
104
112
120 hours
128
136
144 hours
152
160
168 hours
176
184
192 hours
200
216 hours
Hours Aloft
Fuel on landing: 18 gallons
Atlantic Ocean
Pacific Ocean
Flying among 'the redwoods': life and death struggle to avoid towering thunderstorms
Discovery of backwards fuel flow
Thunderstorm forces Voyager into 90° bank
Transition from tailwinds to headwinds
Engine stalled; unable to restart for five harrowing minutes
Squall line
Coolant seal leak
Worried about flying through restricted airspace, Rutan and Yeager mistake the morning star for a hostile aircraft
Passing between two mountains, Rutan and Yeager weep with relief at having survived Africa's storms
Oil warning light goes on
Rutan disabled by exhaustion
Triumphant landing at Edwards AFB
Tanzania
Kenya
Zaire
Congo
Gabon
Uganda
Cameroon
Somalia
Ethiopia
Costa Rica
Nicaragua
United States
Atlantic Ocean
Visibility
20,000
10,000
15,000
5,000
sea level
Altitude (feet)
10,000 miles to go
12,532 miles previous record
5,000 miles to go
26,678 miles traveled
Distance

Voyager pilots: Dick Rutan and Jeana Yeager
Voyager designer: Burt Rutan

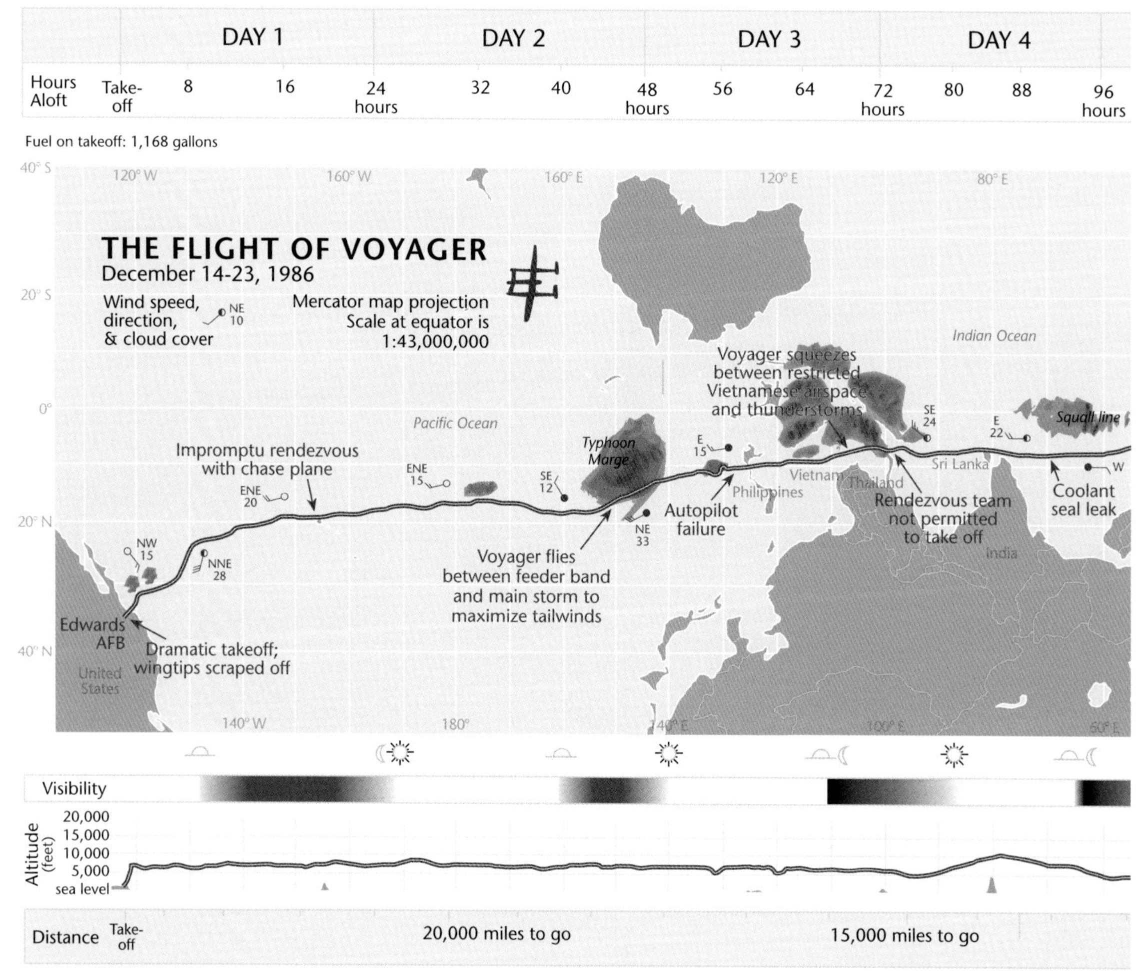

Flight data courtesy of Len Snellman and Larry Burch, Voyager meteorologists
Mapped by David DiBiase and John Krygier, Department of Geography, University of Wisconsin-Madison 1987

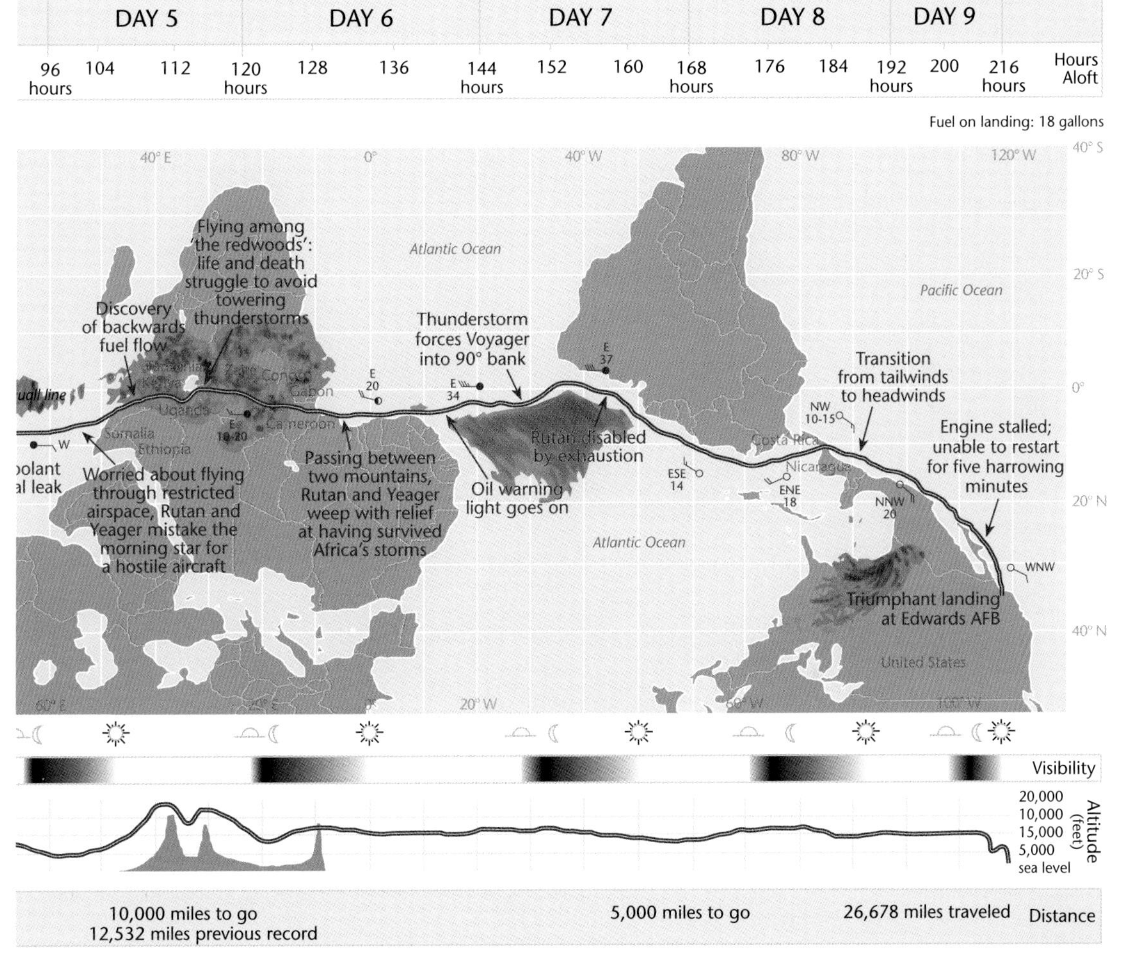

Voyager pilots: Dick Rutan and Jeana Yeager
Voyager designer: Burt Rutan

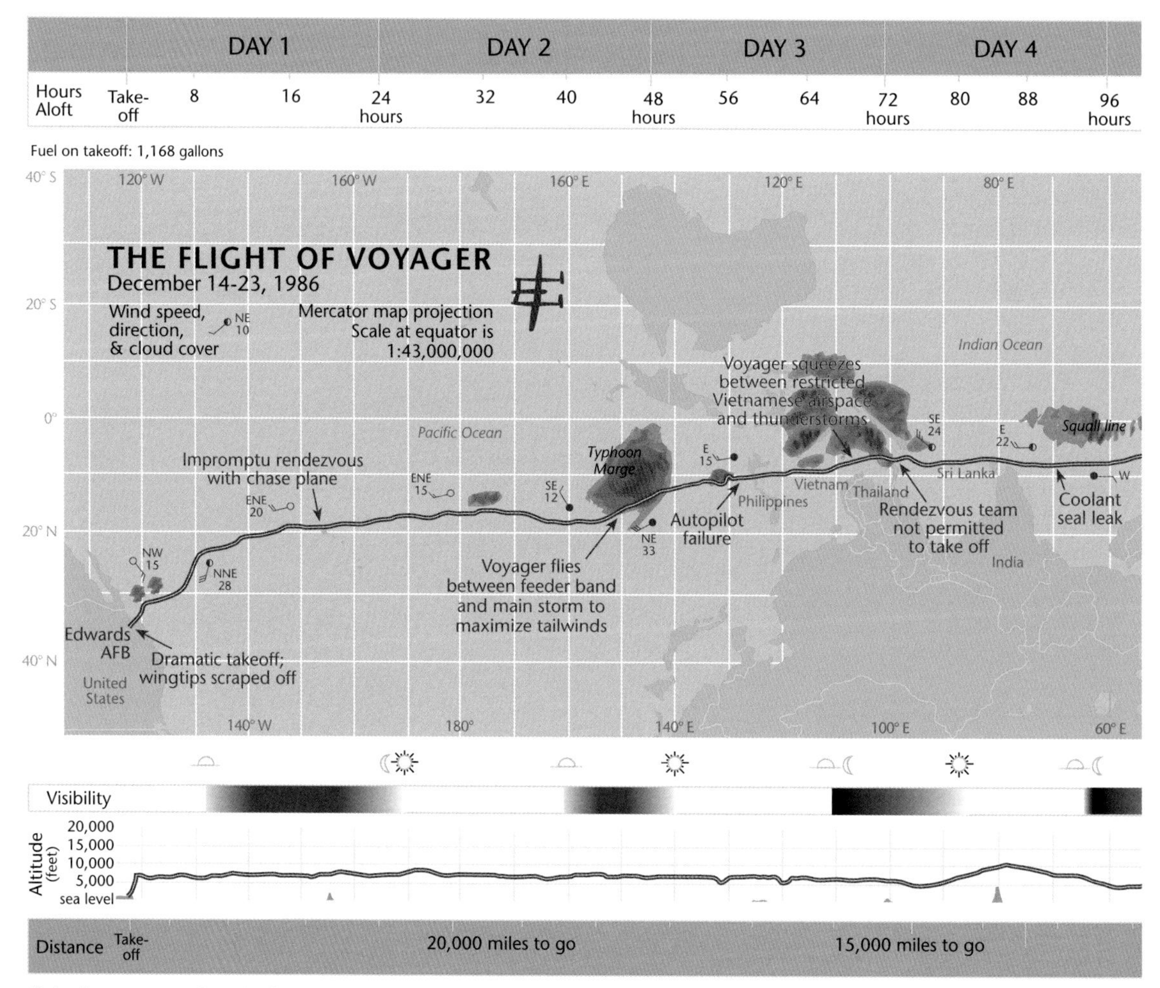

Flight data courtesy of Len Snellman and Larry Burch, Voyager meteorologists
Mapped by David DiBiase and John Krygier, Department of Geography, University of Wisconsin-Madison 1987

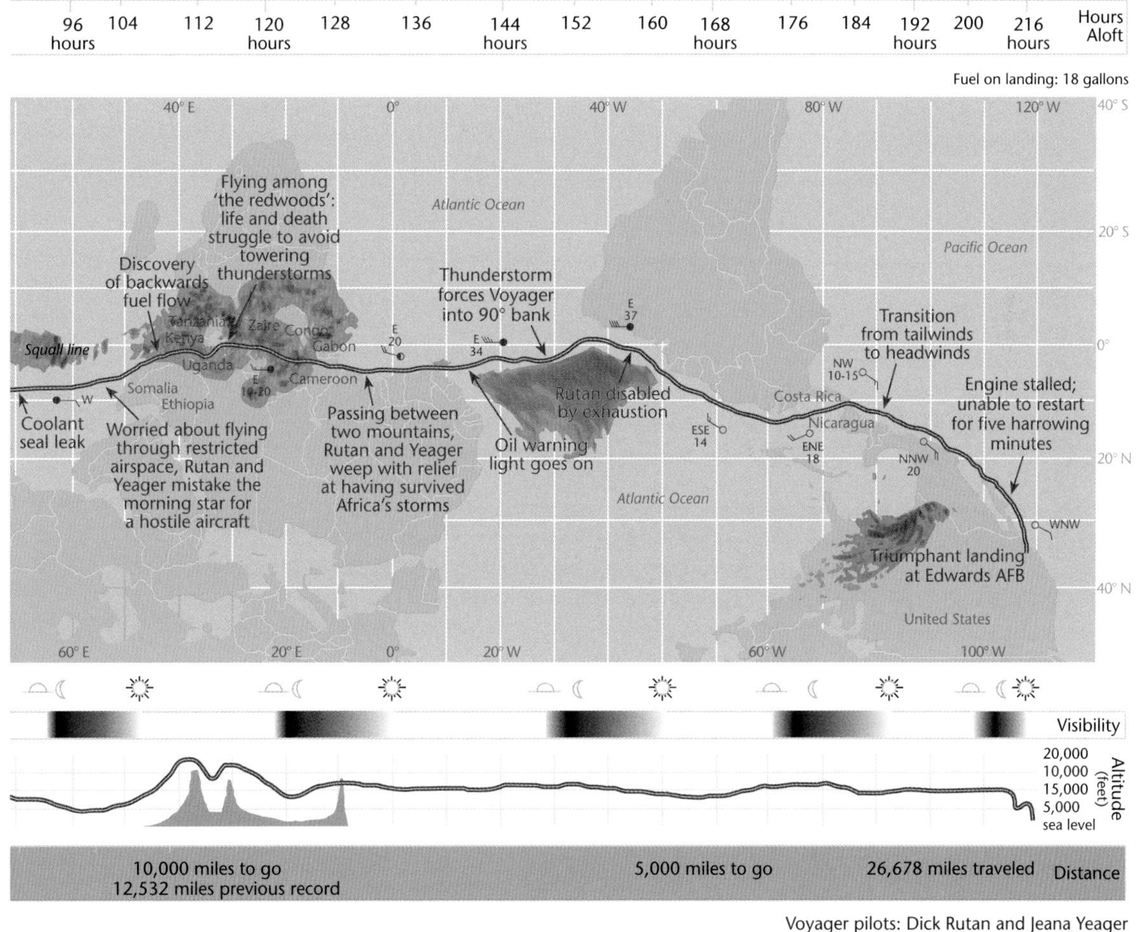

Voyager pilots: Dick Rutan and Jeana Yeager
Voyager designer: Burt Rutan

If we don't want to see the map of Central America covered in a sea of red, eventually lapping at our own borders, we must act now.

Ronald Reagan (1986)

The real object of coloring is to cover the errors and imperfections that are always numerous in maps that are highly colored; to color a well-engraved map is a positive blemish...

Appleton & Co.'s Monthly Bulletin of New Publications (1870)

I do not advance that the face of our country would change if the maps which Philadelphia sends forth all over the Union were more decently colored, but certainly it would indicate that the Graces were more frequently at home on the banks of our lovely rivers, if the engravers were able to sell their maps less boisterously painted and not as they are now, each county of each state in flaming red, bright yellow, or a flagrant orange dye, arrayed like the cover produced by the united efforts of a quilting match. When I once complained of this barbarous offensive coloring of maps, the geographer assured me that he would not sell them unless bedaubed in this way; "for," said he, "the greatest number of the large maps are not sold for any purpose of utility, but to ornament the walls of barrooms. My agents write continually to me to color high." This reason was given me by one of the first geographers of the United States, who has himself a perfectly correct idea of the tasteful coloring of maps.

Francis Lieber, "On Hipponomastics: A Letter to Pierce M. Butler," *Southern Literary Messenger*, 3:5 (1837).
Thanks to Penny Richards for this quotation.

...the fairness of the colour is the art of the map...

The School of Arts (1820)

More...

Cindy Brewer's research on color for maps has been integrated into the very useful colorbrewer.org website. It is a great way to select effective color for maps. Her book *Designing Better Maps* (2024) includes much useful guidance for color use on maps.

A great article on natural color maps is Tom Patterson and Nathaniel Vaughn Kelso's "Hal Shelton Revisited: Designing and Producing Natural-Color Maps with Satellite Land Cover Data," available with a bunch of other cool stuff at the shadedrelief.com website.

Color Oracle is a very useful free software application that simulates three types of color blindness on your computer screen (colororacle.org).

Edward Tufte engages color in all of his books, including a whole chapter on "Color and Information" in *Envisioning Information* (1990).

For some solid background on the history and theory of color: John Gage, *Color and Culture: Practice and Meaning from Antiquity to Abstraction* (1999) and *Color and Meaning: Art, Science, and Symbolism* (2000); also see Charles A. Riley II, *Color Codes: Modern Theories of Color in Philosophy, Painting and Architecture, Literature, Music, and Psychology* (1995). An edited book on maps and color with a historical bent is *Maps and Colours: A Complex Relationship* by Diana Lange and Benjamin Van Der Linde (2024).

Sources: The State College night map was redrawn from the original created in the Deasy GeoGraphics Lab (now the Gould Center) at Penn State. The *Ancient Courses, Mississippi River Meander Belt* maps are available in digital form from the *Lower and Middle Mississippi Valley Engineering Geology Mapping Program* (lmvmapping.erdc.usace.army.mil). Excerpts of Pearce and Hermann's map "They Would Not Take Me There: People, Places and Stories from Champlain's Travels in Canada, 1603-1616" are used by permission. See Margaret Pearce and Michael Hermann, "Mapping Champlain's Travels: Restorative Techniques for Historical Cartography," *Cartographica* 45:1 (2010). The map is available from the Canadian-American Center at the University of Maine (www.umaine.edu/canam).

Hello Jaki, I'm Brandon Henry, we met at the community center...
Sure, I remember you.
I was impressed by how your concern for your treelawn --
I love my treelawn --
-- ended up as a justice sensitivity metric for the whole area.
Yeah, it surprised us too. We had no idea what we were doing when we started.

Well, I was wondering if you know even now.
What do you mean?
It's not about the connector, it's about Biopharma.
Biopharma?
Yeah, the huge plant they're building north of the city. I've been tracking them for months.
They're a big deal...

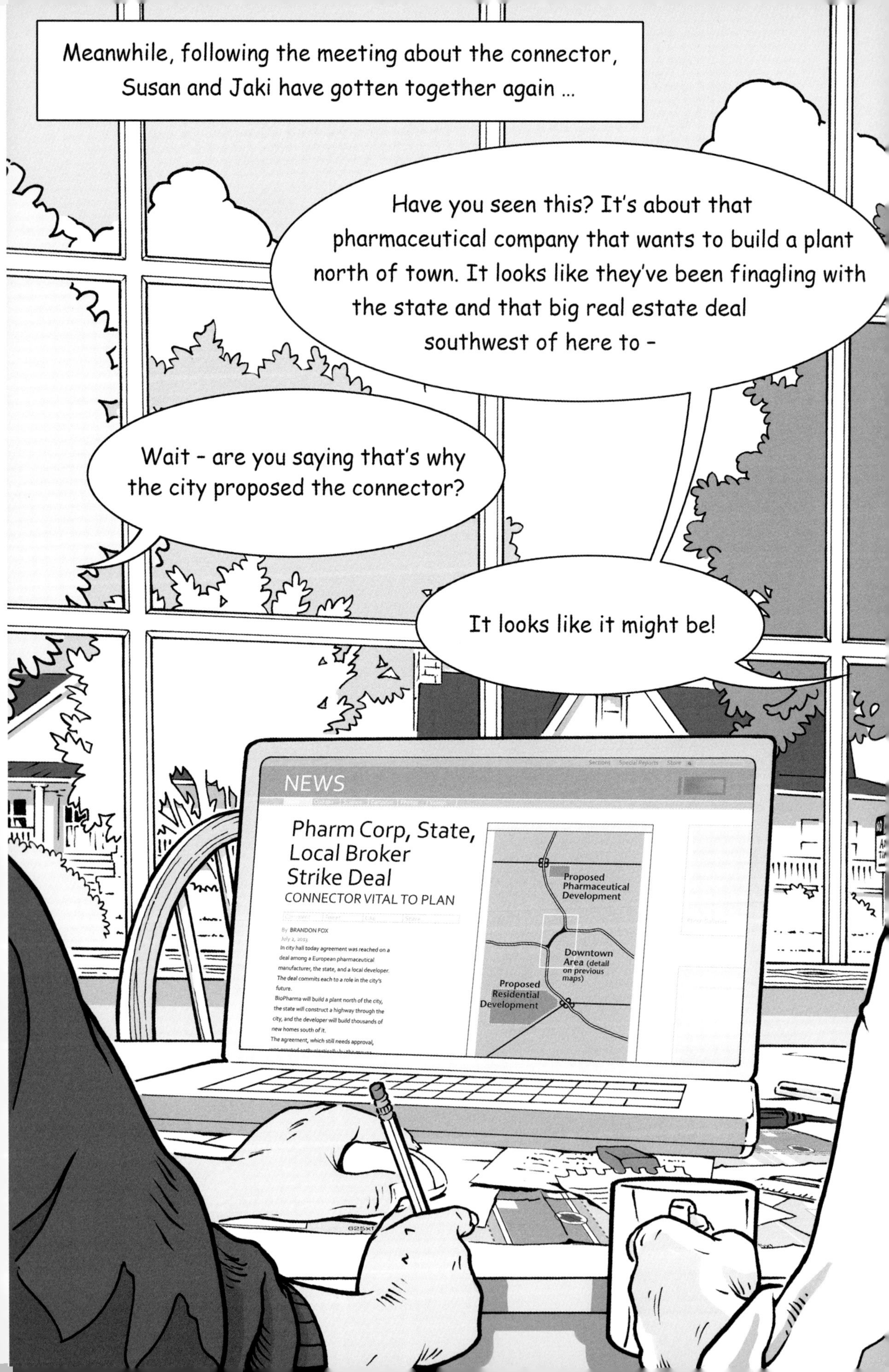
Meanwhile, following the meeting about the connector, Susan and Jaki have gotten together again ...
Have you seen this? It's about that pharmaceutical company that wants to build a plant north of town. It looks like they've been finagling with the state and that big real estate deal southwest of here to -
Wait - are you saying that's why the city proposed the connector?
It looks like it might be!
Sections
Special Reports
Store
NEWS
Pharm Corp, State, Local Broker Strike Deal
CONNECTOR VITAL TO PLAN
By BRANDON FOX
July 2, 2013
In city hall today agreement was reached on a deal among a European pharmaceutical manufacturer, the state, and a local developer. The deal commits each to a role in the city's future.
BioPharma will build a plant north of the city, the state will construct a highway through the city, and the developer will build thousands of new homes south of it.
The agreement, which still needs approval,
Proposed Pharmaceutical Development
Downtown Area (detail on previous maps)
Proposed Residential Development

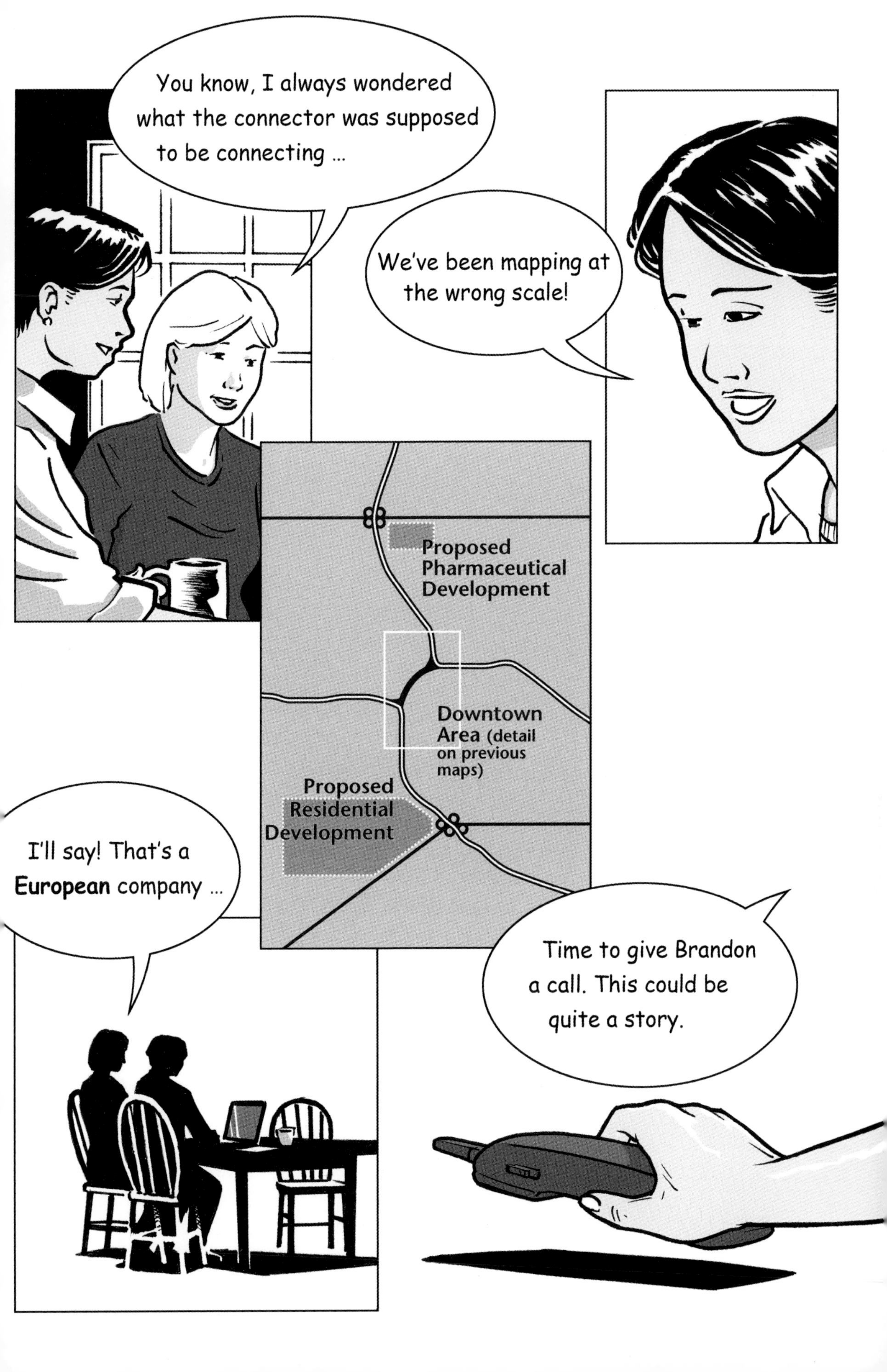
You know, I always wondered what the connector was supposed to be connecting ...
We've been mapping at the wrong scale!
Proposed Pharmaceutical Development
Downtown Area (detail on previous maps)
Proposed Residential Development
I'll say! That's a **European** company ...
Time to give Brandon a call. This could be quite a story.

Brandon did a great job. What a horrible company!

NEWS

Sections Special Reports

News | Opinion | Science | Cartoons | Photos | Videos

Pharmaceutical Plant Has Bad Track Record

ENVIRONMENTAL, HUMAN RIGHTS VIOLATIONS

By BRANDON FOX

July 2, 2013

Comment | Tweet | Like | Share

The pharmaceutical plant poised to land in a field to the northwest of town has run up a terrifying record of environmental violations and human rights abuses around the world.

In Mexico and Peru the company exceeded emissions standards for a variety of toxic chemicals. In Thailand they regularly employee children under the age of 16. In Russia their new plants has been tied to organized crime. In Canada they've been accused of resisting Canadian government price controls by removing their drugs from the market. And in Alabama they've been accused of union busting activities at two new plants.

Repeated efforts to get company officials on record have been met with silence.

"We make and sell drugs," a comp ... news stories."

Canada: Removed popular drugs from market because of price controls

Russia: Inv in new plant ti organized

Alabama (U.S.): Union busting at two new plants

Mexico: Toxic emissions exceed Mexican standards

Peru: Toxins from plant found in local water sources

Thailand: Childre under 16 years ol routinely employe

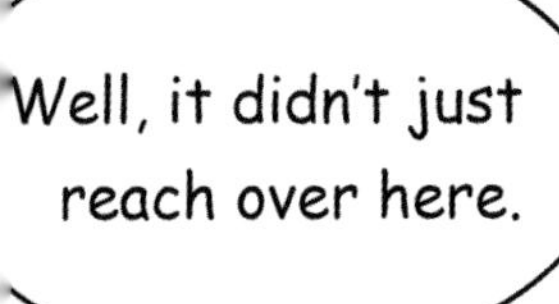
Well, it didn't just reach over here.
First it worked with the state's business recruitment division. They worked with movers and shakers here in town.

That got it to the mayor's office and the city planners. And they're the ones who want to widen the road.

I think it's time to make a whole bunch of maps for the hearing on the road the city council's planning.
At a whole bunch of different scales –

From the company's effect on the world all the way down to the tree they want to cut down in your yard – it's all connected!

ATLANTIC OCEAN
Equator
Caribbean Sea
Trinidad
Paria
Ciudad Bolívar
VENEZUELA
Llanos
Caracas
Valencia
Maracaibo
Lago de Maracaibo
La Goajira
Barinas
Río Apure
Orinoco
Río Meta
Georgetown
BRITISH GUIANA
Paramaribo
SURINAM
DUTCH GUIANA
FRENCH GUIANA
Cayenne
GUIANA HIGHLAND
Sierra Parima
Cerro Duida
Boa Vista
Río Branco
Río Negro
Manáos
AMAZON LOWLAND
Selvas
Río Amazon
Solimoes
Río Madeira
Río Purus
Río Jurua
Tabatinga
Iquitos
Río Javary
Río Napo
Río Putumayo
Río Yapura
Río Apaporis
Río Guaviare
COLOMBIA
Bogota
Medellín
Cali
Pasto
Cartagena
Barranquilla
Puerto Colombia
Santa Marta
Magdalena Plain
Cord. del Choco
Gulf of Darien
Panamá
Colón
Isthmus
Pacific Ocean
Buenaventura
Quito
Chimborazo
Ambato
Riobamba
Cuenca
ECUADOR
ORIENTE
Guayaquil
Esmeraldas
Equator
Talara
Paita
Salaverry
Lima
Callao
Pisco
Chincha
Cerro de Pasco
Oroya
Huancayo
Cuzco
Vilcanota Knot
Puno
L. Titicaca
El Misti
Arequipa
Mollendo
Arica
Iquique
Atacama Desert
Peruvian Current
Tropic of Capricorn
PAC
Río Ucayali
Río Tigre
Río Marañón
Río Huallaga
Pongo de Manseriche
Montaña
Cordillera Oriental
Cordillera Central
Cord. Blanca
Cord. Negra
Andes de Carabaya
Cord. de Cochabamba
Altiplano
Maritime
Andes
ANDES
La Paz
Oruro
L. Poopó
Cordillera Real
Salar de Uyuni
Sucre
Potosí
Cochabamba
Santa Cruz
Yungas
BOLIVIA
LLANOS DE MOJOS
Río Mamoré
Río Grande
Río Beni
Río Guaporé
Serra dos Parecis
CHIQUITOS PLATEAU
Matto Grosso
MATTO GROSSO
Campos
Cuyabá
PANTANAL
Corumbá
Chuquicamata
Río Roosevelt
Río Juruena
Río Manoel
BRAZILIAN PLATEAU
Río Tapajóz
Río Xingú
Río Tocantins
Río Araguaya
Ilha do Bananal
Serra do Roncador
Planalto de Matto Grosso
Goyaz
GOYAZ MASSIF
Federal District
S. do São Domingos
Ilha de Marajo
Río Pará
Pará
Maranhão
Río Parnahyba
Serra Grande
Chapada do Apoay
Chapada do Araripe
Borborema Plateau
Paulo Affonso Falls
Cabo de S. Roque
Natal
Parahyba
Recife
Bahia Embayment
Río São Francisco
Chapada Diamantina
OLD LAND
HIGHLAND

How can the words mean more than they say?

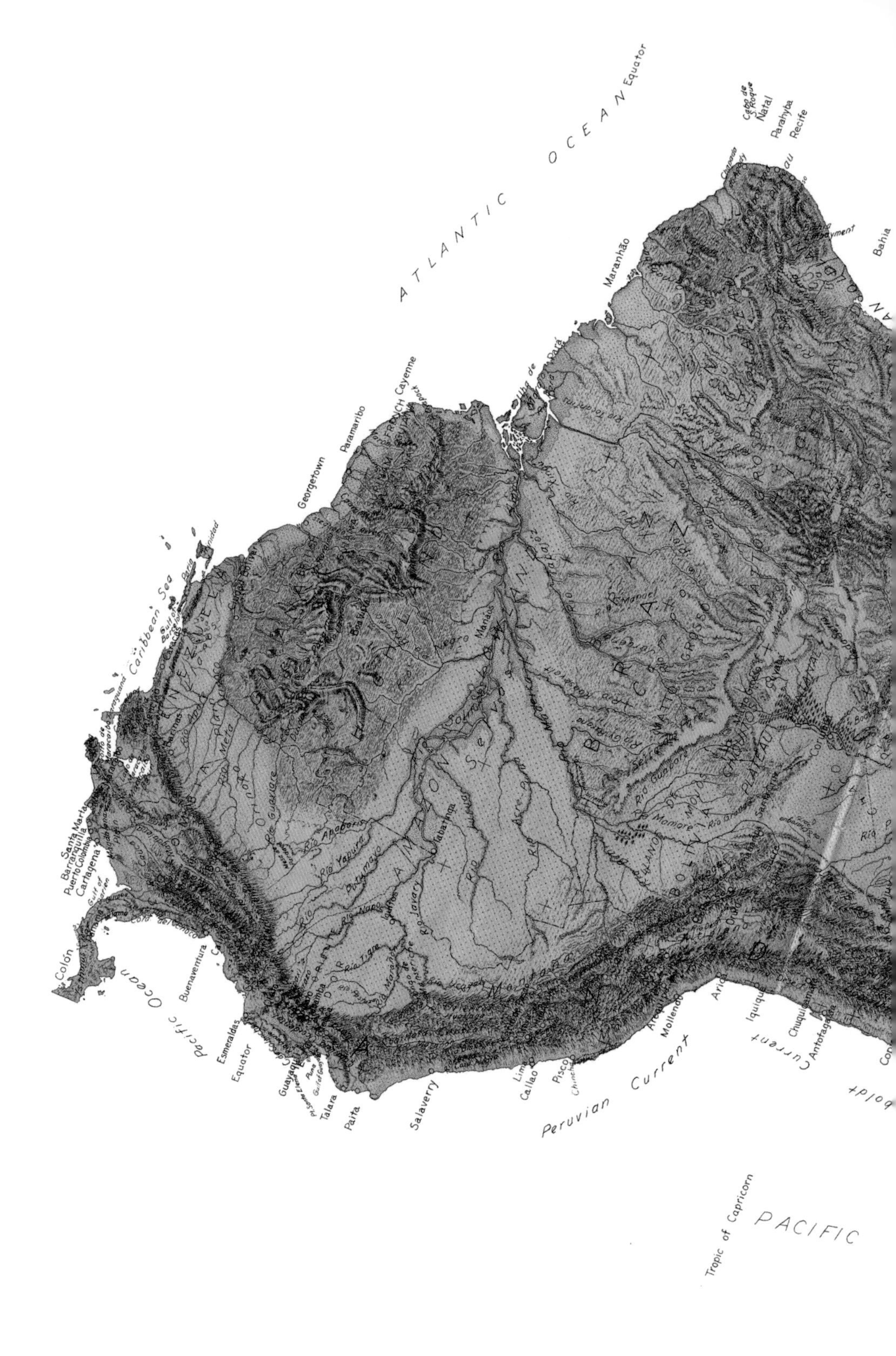
ATLANTIC OCEAN
Equator
Cabo de S. Roque
Natal
Parahyba
Recife
Bahia
Maranhão
Cayenne
Paramaribo
Georgetown
Caribbean Sea
Santa Marta
Barranquilla
Puerto Colombia
Cartagena
Colón
Buenaventura
Pacific Ocean
Esmeraldas
Equator
Guayaquil
Talara
Paita
Salaverry
Lima
Callao
Pisco
Mollendo
Arequipa
Iquique
Chuquicamata
Antofagasta
Peruvian Current
Tropic of Capricorn
PACIFIC
Rio Negro
Manáos
Rio Madeira
Rio Javary
Rio Napo
Rio Putumayo
Rio Yapura
Rio Tigre
Rio Guaporé
Rio Mamoré
Rio Meta
Rio Tocantins

CHAPTER 9 Words on Maps

The beauty of words on maps is often not evident – embedded, as they are, in an array of other symbols. A "word map" of South America, shown on the previous two pages, accompanies Guy-Harold Smith's *Physiographic Diagram of South America* (1935) as a reference; it easily suggests the geography of South America despite being nothing but words. Words themselves contain meaning on maps, but they also lend invaluable structure and coherence to maps.

What Words Mean

It's names that make a map a map. Without names, maps fade into pictures or photographs. They lose their great power, which is to christen and claim. Named and renamed, no part of the world has escaped the map's heavy hand. Every inflection of the land, every body of water, has been categorized and christened. Every path, byway, roadway, and street, every hamlet, village, town, and city has been tagged.

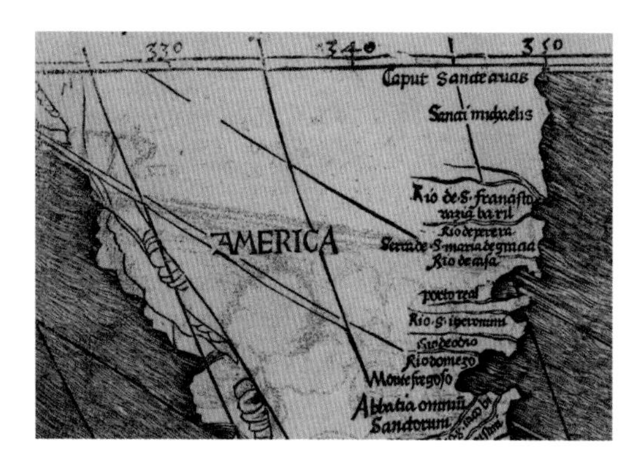

Martin Waldseemüller's 1507 map named the continent west of the Atlantic "America." Americans have been Americans ever since.

RED
RED RED
RED REDgreen
REDgreenOCHREpurple
buff green orange
olive olive mottled green
greenOCHRE purplegreenpurple green
olive orange yellow yellow
olive REDgreen yellow purple
olive yellow greenish yellow greenmottled yellow
oliveGOLDgreenGOLD REDgreenGOLDorange
mauvegreen greenish yellow greenolive green
buff green purpleGOLDmauve purple lemon
olive mottled greenmottled oliveREDruby purplegreen
greenish yellow olive yellow marigold GOLD
olive green greenOCHRE purple greenish yellow orange purple
olive purple buff green RED mauvepurple tangerine mottled
GOLDgreen dappled GOLDolive yellow olive green green
purple green purple mottled yellow clotted green
greenish yellow greenish yellow
bruise REDgreenish yellow mauve green greenpurple
RED green yellow GOLD mottled green mauvemarigold purple
orange green chartreusepurple green GOLDgreenish yellow RED dappled RED green
mauve mauve marigold yellow greenish yellow mottled green green
green clotted green orange greenish yellow dandelion marigold green
green marigold purple mauve green greenish yellowOCHRE clotted chartreuse
tangerine green OCHRE RED purple greenish yellow OCHRE chartreuse mauve greenish yellow green
yellow OCHRE REDpurple dappled mauve green olive
chartreuse clotted olive green clotted yellow GOLD ruby OCHRE
green orange mottled RED green purple greenRED
olive purple purplegreen yellow RED clotted purple
yellow mottled greenish yellow OCHRE GOLDdappled mottled clotted
GOLDgreen clotted buff green purple mauve
yellow green GOLDgreen OCHRE OCHRE
greenish yellowpurpleyellowgreenish yellow clotted mauve bruise
OCHRE green mauvegreen dappled mottled
chartreuse greengreenish yellow olive
purpleGOLD mauve chartreuse olive
marigold yellow
yellow green purple
greenish yellow mottled olive
dappled green green

A map of autumn leaf color in the Boylan Heights neighborhood in Raleigh, NC, made entirely of color names. Even by themselves, names can do a lot of work.

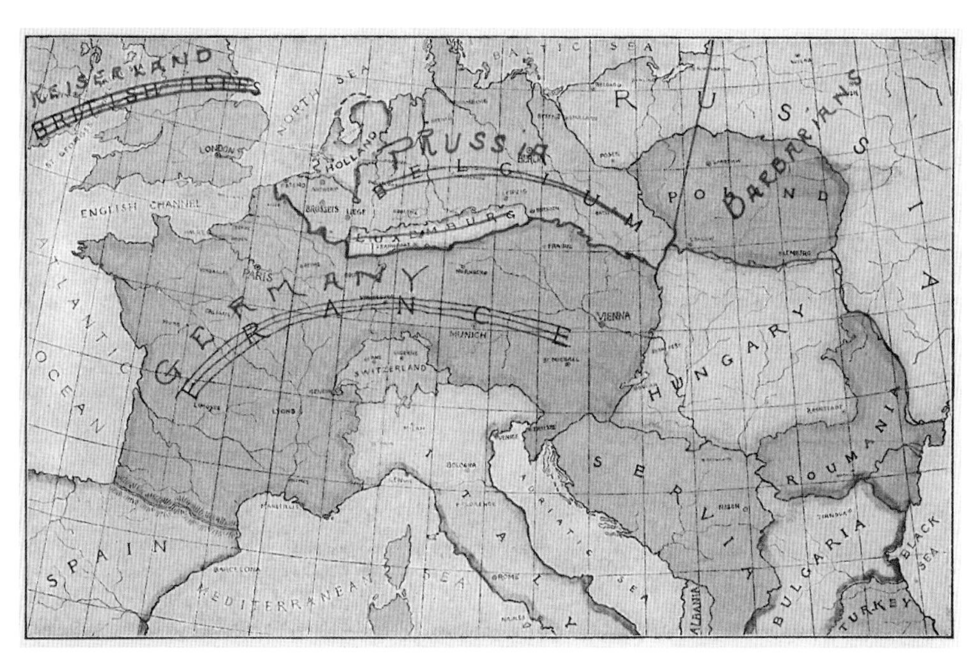

Assigning names can be contentious. In 1914 a map of Europe was published in *Life* magazine. "An offended reader" of the magazine "corrected" the map and returned it to *Life*, where it was republished.

When the public owner of London's famous tube map demanded all spoofs of the tube map be removed from an artist's website, threatening legal action, another map began to circulate on the web.

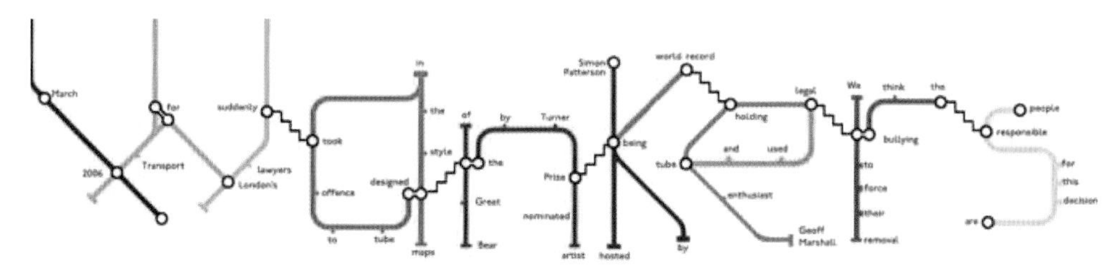

What Words Look Like: Type Anatomy

Map type is as diverse as any other symbol. Variations in type anatomy shape its look: its overall form (typeface, font), the way the strokes on letters finish (serifs), the size of the main body of lowercase letters (x-height), the parts that stick up above and below the main body (ascenders, descenders), and the size of the type.

Typeface | Font is a set of letters and numbers with a unique design. A font is a subset of a typeface, including all letters and numbers of a specific size. Font is often used to mean typeface. This particular typeface is Times Roman:

Making

Maps

x-height is the height of the most compact letters in a typeface, such as an a, o, or e. Type with a greater x-height is typically easier to read.

Point size 48 point type here, where 72 points equals one inch. Type size is determined by the height of the original lead foundry block, and is not the same as the height of the letter; thus 8 point Times Roman is smaller than 8 point Helvetica

12 point Helvetica

14 point Helvetica

18 point Helvetica

24 point Helvetica

48 point

Serifs are finishing strokes added to the ends of letters. Helvetica has no serifs (sans serif); Times Roman has them.

Typefaces | Fonts vary immensely and for mapping uses are typically divided into two broad classes – serif fonts, such as Times Roman, and sans serif fonts, such as Helvetica:

Making

Maps

Sans serif

x-height

Ascender is the portion of certain letters that rises above the x-height, such as in the letters k or f.

Descender is the portion of certain letters that falls below the x-height, such as in the letters p or g.

Typefaces (fonts) are often classified according to their historical development, from early typefaces based on the style of hand lettering to later, much more abstract and geometric, styles.

Sabon is a humanist or old style typeface, designed to emulate the look of calligraphy

Baskerville is a transitional typeface, sharper and with more contrast than humanist typefaces

Bodoni is a modern typeface, abstract with strong thick/thin contrast

Clarendon is an Egyptian or slab serif typeface, with heavy serifs, designed to be used in ads

Gill Sans is a humanist sans serif typeface, calligraphic like humanist typefaces but lacking the serifs

Helvetica is a transitional sans serif typeface, like transitional typefaces lacking the serifs

Futura is a geometric sans serif typeface based on basic geometric forms

A type family is a collection of variations on a single typeface (font) appropriate for different uses. Type families may be small, or have dozens of variants.

This is Bodoni regular

This is Bodoni light

This is Bodoni light italic

This is Bodoni regular italic

This is Bodoni bold

This is Bodoni bold italic

This is Bodoni extra bold

This is Bodoni estra bold italic

Type variations on your map should mean something. Convey information with type style, size, weight, and form.

What Words Look Like: Type as Map Symbol

Map type can be used as a map symbol to differentiate qualities (typeface, type color hue, italics) and quantities and order (type size, weight, type color value) or both (type spacing, case). Make sure that variations in type on your map reflect variations in your data.

Typeface (Font): Qualities

Typeface has a significant visual impact on your map. Different typefaces can be used to suggest qualitative aspects of data and to shape the overall feel of your map.

Serif type, such as Sabon, implies tradition, dignity, and solidity.

Sans Serif type, such as Stone Sans, implies newness, precision, and authority.

Typeface (Font) Considerations

- Avoid combining more than two typeface families on a map unless designed to be used together (e.g., Stone Sans and Stone Serif)
- With care, combine more than two typeface families on a map for large, complex maps with lots of different kinds of labels.
- Evaluate compatibility if different typefaces are combined
- Avoid combining two serif or two sans serif type styles on one map
- Serif type is easier to read in blocks of text
- Decorative type styles such as Bradley Hand are difficult to read and can look goofy. But they are fun you say! Bah!
- Typeface (Font) choices are subtle but vital and deserve a good review by someone with a good eye.

Type Size: Order

Type size variations imply quantitative, ordered differences. Larger sizes imply more importance or greater quantity, smaller sizes less importance or less quantity.

More important

Less important

Type Size Considerations

- Type less than 6 points in size is hard to read on paper, 10 points on computer screen
- Use a 2 point difference in small type sizes and 3 point difference in medium and larger type sizes if you want a noticeable difference
- Most people have difficulty distinguishing more than five to seven categories of data symbolized by type size on a map
- Adjust for your final medium. Increase type size for computer display or posters, decrease it for the print medium

Type Weight: Order

Type weight variations imply ordered (quantitative) differences. Bold type implies more importance or greater quantity, standard or light type less importance or less quantity.

More important

Less important

Type Weight Considerations

Bold implies significance and power, yet may look **pudgy**
Some fonts have different bold options: **semibold** vs. **bold** for example
Bold makes **gray type** more legible
Don't underline, use **bold** or *italic* instead
Ordered categories using size and **bold:**
8 pt. standard
8 pt. bold
10 pt. standard
10 pt. bold
12 pt. standard
12 pt. bold

Type Form: Qualities, Order

Type form variations can suggest qualities (italics, color hue), order (color value), and both (spacing, case).

Water feature

BIG WATER FEATURE

Type Form Considerations

Italics is conventional for qualitative data differences, water, and other natural features
Variations in the color hue (red, green, blue) of type imply qualitative data differences
Variations in the color value (light red, dark red) of type imply ordered data differences
Carefully evaluate light values of type for legibility on your final medium
Do not tint your type if the final product is printed with a dot-screen (books, newspapers) unless the font size is large and the tint relatively dark (50%)
E x t e n d e d or spaced type is OK for extended area features, but don't go o v e r b o a r d
Condensed type may look squished
UPPER CASE implies greater quantity or importance but is more difficult to read than a mix of upper- and lower-case letters and seems a bit shouty

Masking creates a halo around type to enhance it's readability over a tinted background
Masking creates a halo around type to enhance it's readability over a tinted background
Compare **positive type** to **reversed type** to enhance readability over a tinted background
Compare **positive type** to **reversed type** to enhance readability over a tinted background

The Geo-Smiley Terror Spree map and this book use Stone Sans typeface (font). While designed for low-resolution laser printers, Stone works well in any medium. It is harder to screw up on the map than with fussier typefaces.
Type size clearly distinguishes the title from other map type, as well as the story text from state abbreviations. What's important is larger and more noticeable.
Geo-Smiley Terror Spree
Luke Helder, a university student from Minnesota, went on a 2-week spree of bombings throughout the Midwestern U.S. in 2002 in an attempt to create a giant smiley face across the country.
MN
Home of Mr. Helder:
Smiley Terror Spree Begins
May 3, 2002
Eye 2
NE
IA
Eye 1
IL
NV
Capture of
Mr. Helder
May 18, 2002
CO
Last Bomb
Incomplete
Smile
TX
Value (type form) suggests order and provides depth for the map. White type reversed out of the states is small in size but readable, falling onto a different visual layer than the black type.
Type weight distinguishes the main components of the bombing spree face – eyes and incomplete smile – from other elements of the story. The smiley face is the story.

Geo-Smiley Terror Spree

Luke Helder, a university student from Minnesota, went on a 2-week spree of bombings throughout the Midwestern U.S. in 2002 in an attempt to create a giant smiley face across the country.

Sabon, an **old style** serifed **typeface** (font) designed to look a bit like calligraphy, is too elegant and traditional a typeface choice for this goofy modern story.

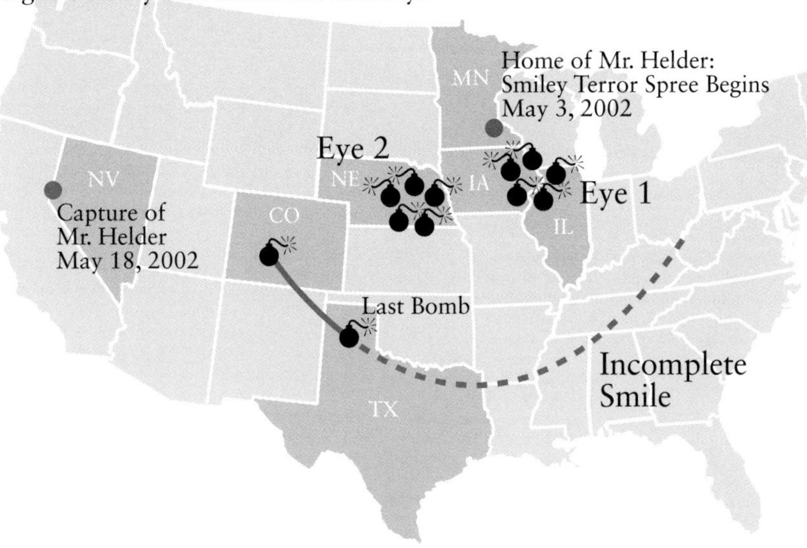

Geo-Smiley Terror Spree

Luke Helder, a university student from Minnesota, went on a 2-week spree of bombings throughout the Midwestern U.S. in 2002 in an attempt to create a giant smiley face across the country.

Futura, a modern, geometric **sans serif typeface** (font), is better than Sabon but is slightly distracting: the look of the typeface is somewhat more noticeable than it should be.

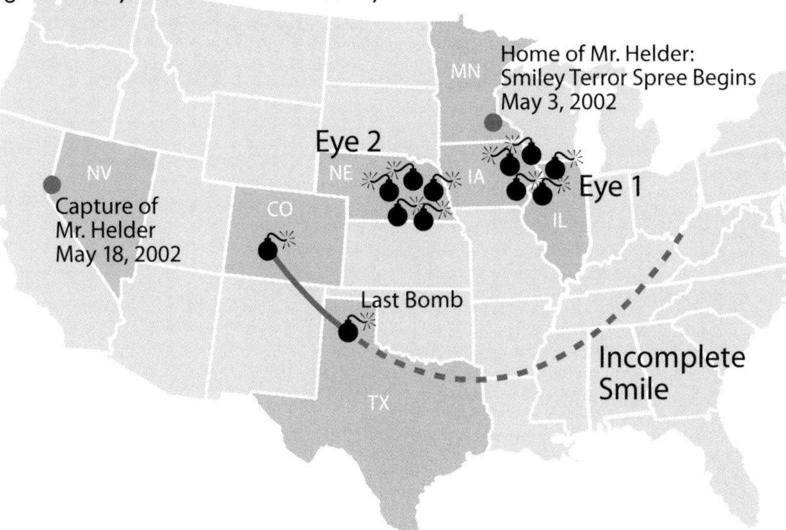

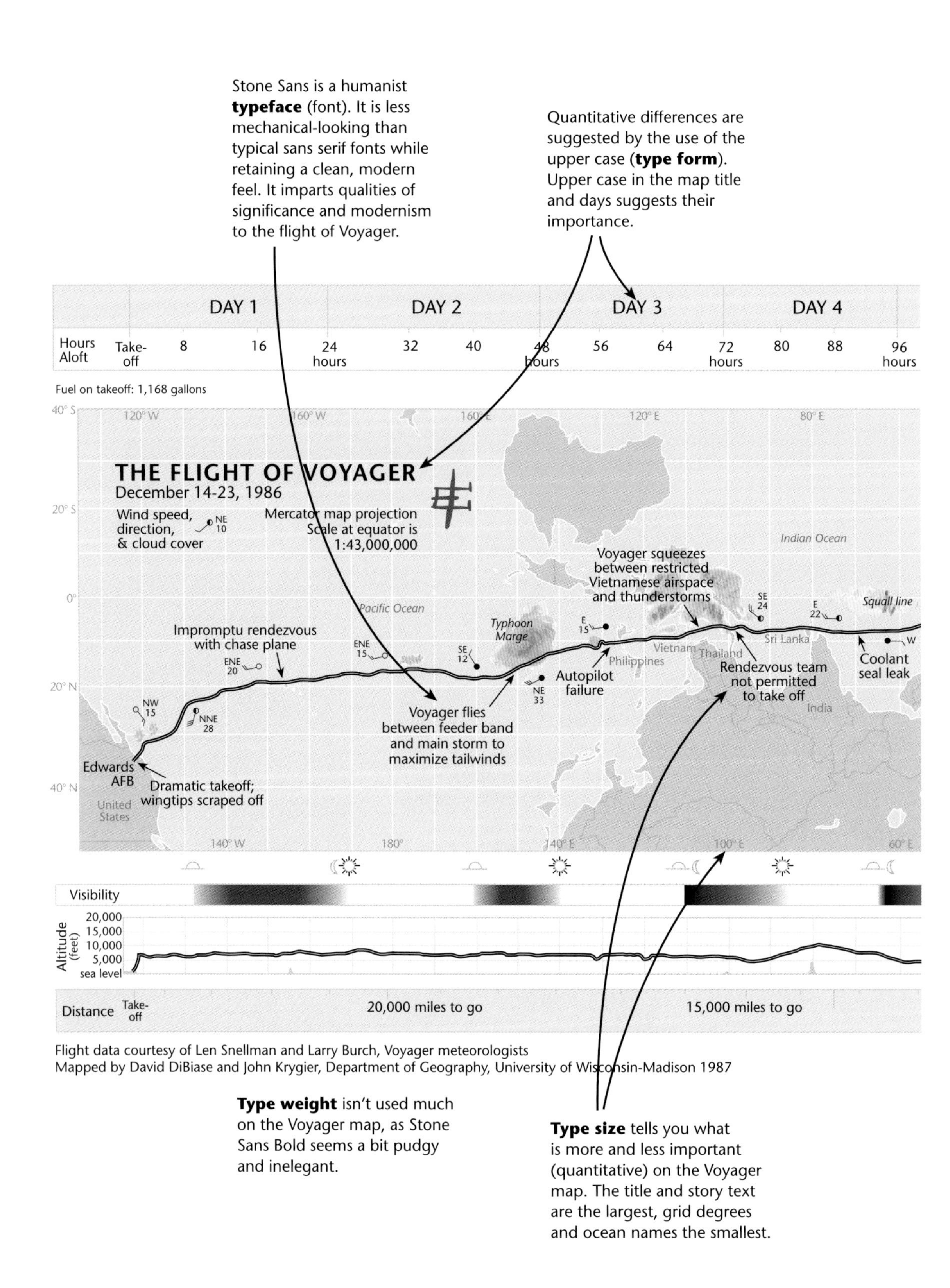
Stone Sans is a humanist **typeface** (font). It is less mechanical-looking than typical sans serif fonts while retaining a clean, modern feel. It imparts qualities of significance and modernism to the flight of Voyager.
Quantitative differences are suggested by the use of the upper case (**type form**). Upper case in the map title and days suggests their importance.
DAY 1
DAY 2
DAY 3
DAY 4
Hours Aloft
Take-off
8
16
24 hours
32
40
48 hours
56
64
72 hours
80
88
96 hours
Fuel on takeoff: 1,168 gallons
40° S
120° W
160° W
160° E
120° E
80° E
THE FLIGHT OF VOYAGER
December 14-23, 1986
20° S
Wind speed, direction, & cloud cover
NE 10
Mercator map projection
Scale at equator is 1:43,000,000
Indian Ocean
Voyager squeezes between restricted Vietnamese airspace and thunderstorms
0°
Pacific Ocean
Typhoon Marge
SE 24
E 22
Squall line
E 15
Impromptu rendezvous with chase plane
ENE 15
SE 12
Vietnam
Thailand
Sri Lanka
W
ENE 20
Philippines
Coolant seal leak
Rendezvous team not permitted to take off
20° N
Autopilot failure
NE 33
India
NW 15
NNE 28
Voyager flies between feeder band and main storm to maximize tailwinds
Edwards AFB
Dramatic takeoff; wingtips scraped off
40° N
United States
140° W
180°
140° E
100° E
60° E
Visibility
Altitude (feet)
20,000
15,000
10,000
5,000
sea level
Distance
Take-off
20,000 miles to go
15,000 miles to go
Flight data courtesy of Len Snellman and Larry Burch, Voyager meteorologists
Mapped by David DiBiase and John Krygier, Department of Geography, University of Wisconsin-Madison 1987
Type weight isn't used much on the Voyager map, as Stone Sans Bold seems a bit pudgy and inelegant.
Type size tells you what is more and less important (quantitative) on the Voyager map. The title and story text are the largest, grid degrees and ocean names the smallest.

Quantitative differences are suggested by the use of value (**type form**). Gray type is used for less important information, black type for more important information.

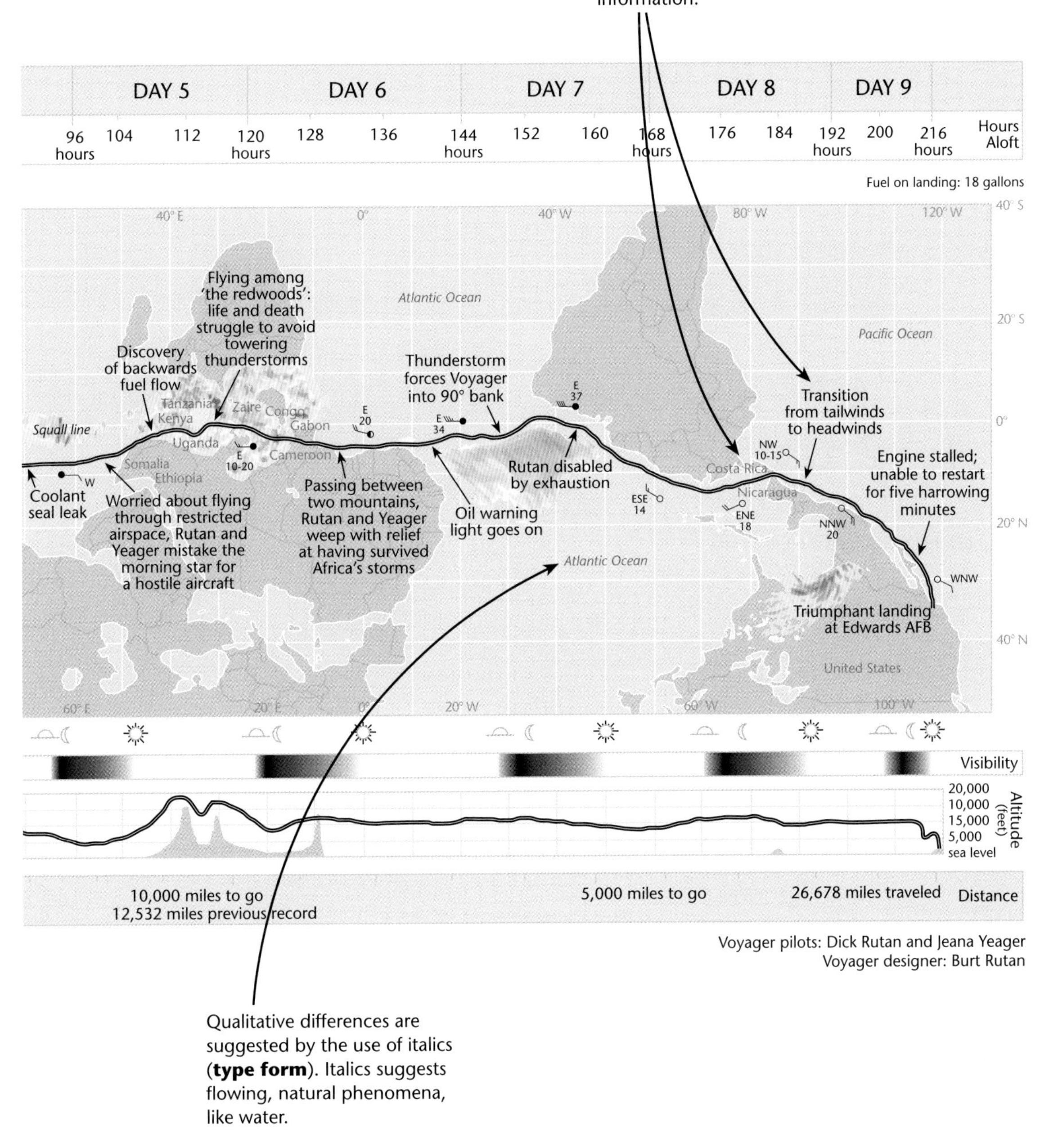

Qualitative differences are suggested by the use of italics (**type form**). Italics suggests flowing, natural phenomena, like water.

Arranging Type on Maps

Effective type placement clarifies the relationship between a label and the symbol (point, line, area) to which it refers. These guidelines are flexible and a starting point. GIS applications may offer automated feature labeling, which normally follows traditional type placement rules. Always evaluate any automated type placement for clarity and legibility.

Labeling Point Data

When labeling point symbols on a map, start in the densest part of the map and work outward. For each symbol, follow these priorities, where 1 is best, 8 is worst.

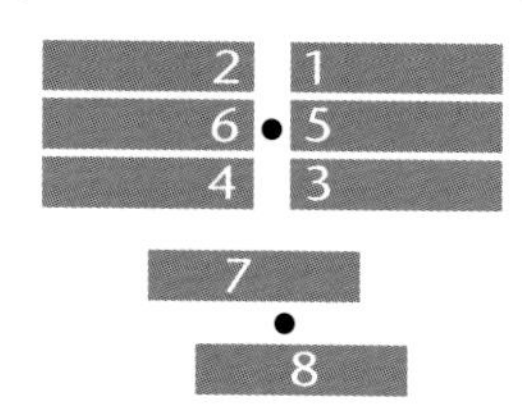

Show characteristics of the labeled location with type placement.

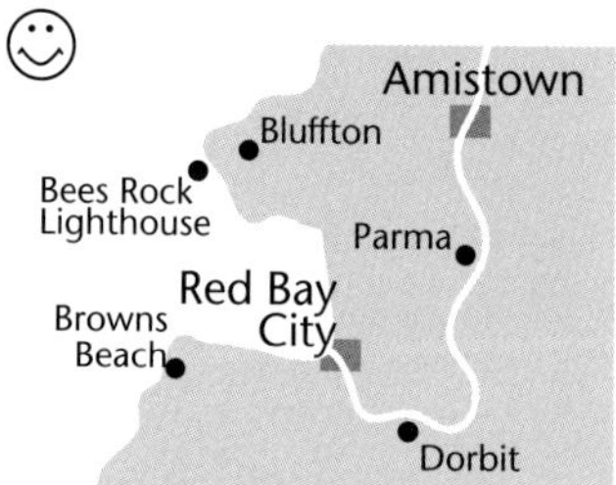

Label ports and harbor towns on the sea
Label inland towns on the land
Label land features on land, water features on water
Label towns on the side of the river on which they are located
Align type to grid (latitude) if grid is included

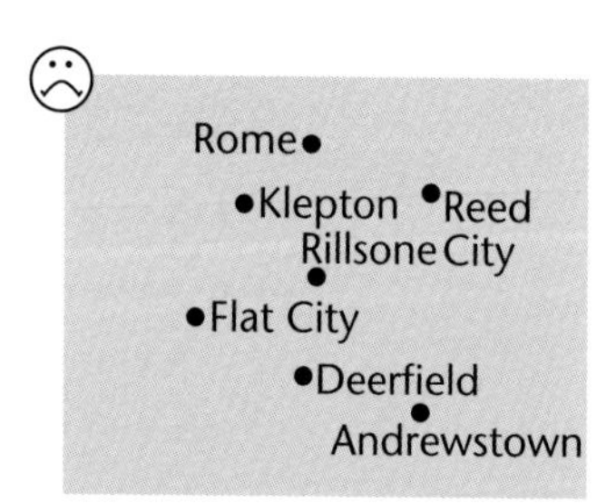

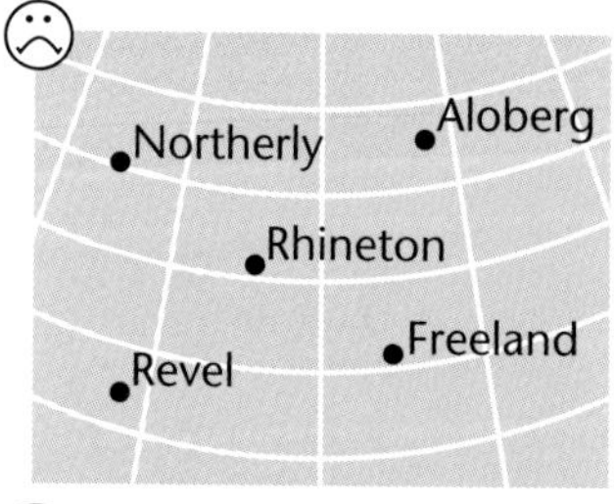

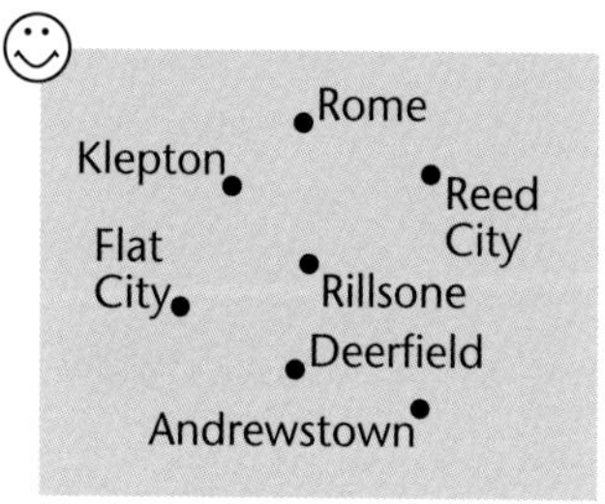

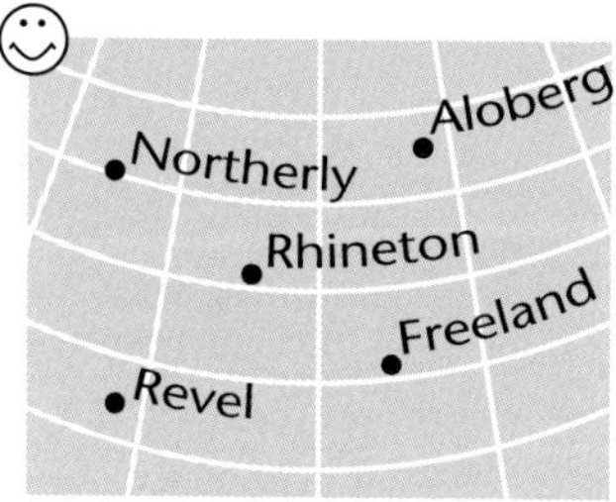

Labeling Line Data

Curve or slant type, following symbol
Keep type above symbol if possible
Keep type as horizontal as possible for ease of reading
Never place type upside-down
With vertical type, place the first letter of the label at the bottom
Repeat rather than stretch type

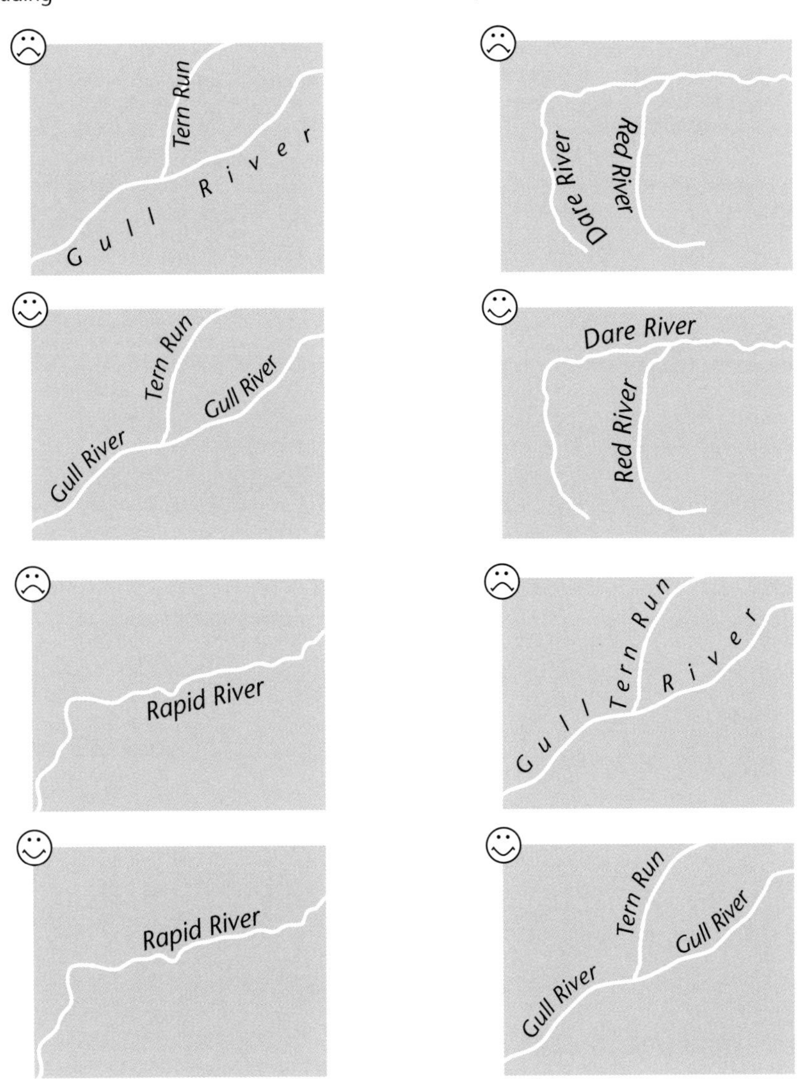

Labeling Area Data

Curve and space type to fit areas. Ensure that the area and the label are clearly associated.

Entire area label should follow a gentle and smooth curve
Keep area labels as horizontal as possible (they are easier to read)
Avoid vertical and upside-down labels (they are harder to read)
Keep labels away from area edges
Avoid hyphenating or breaking up area labels

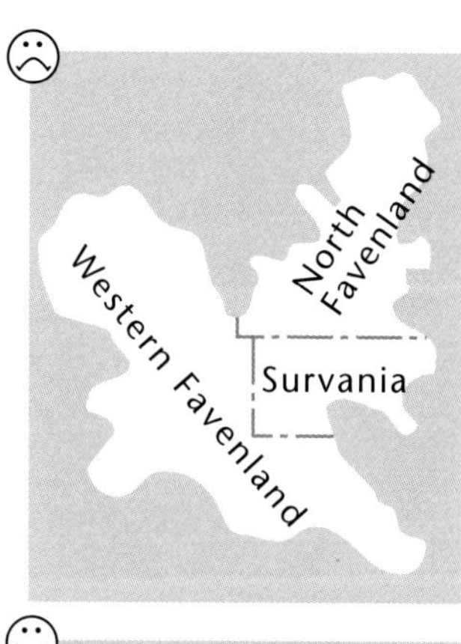

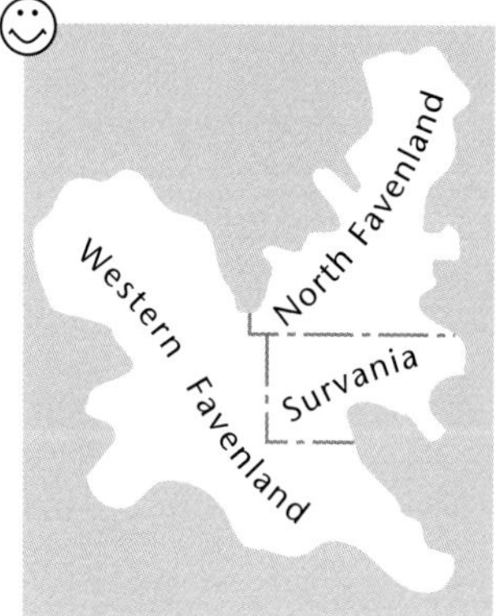

Distinguish overlapping areas by varying type size, weight, form
Label linear areas like line symbols

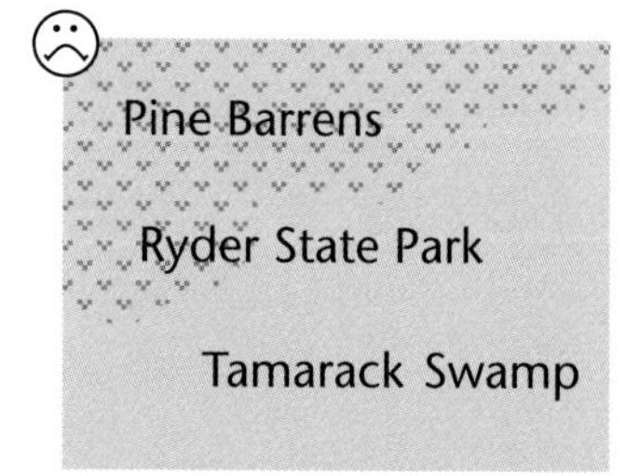

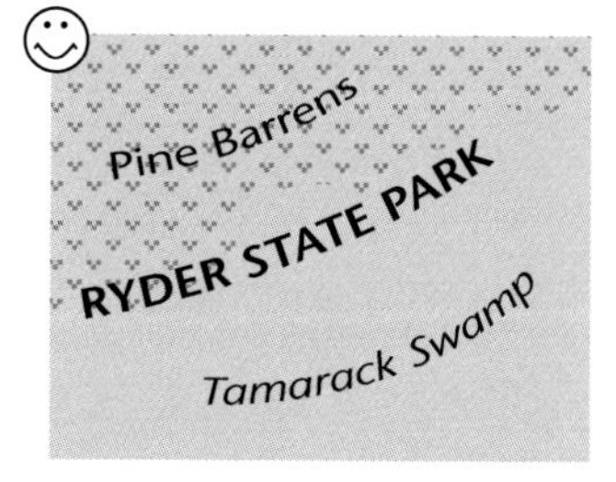

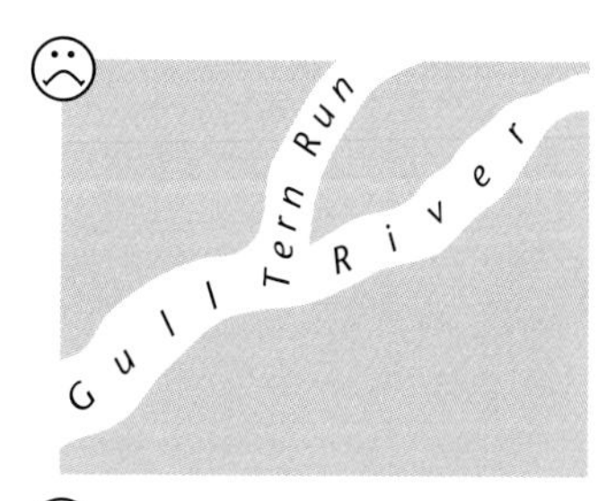

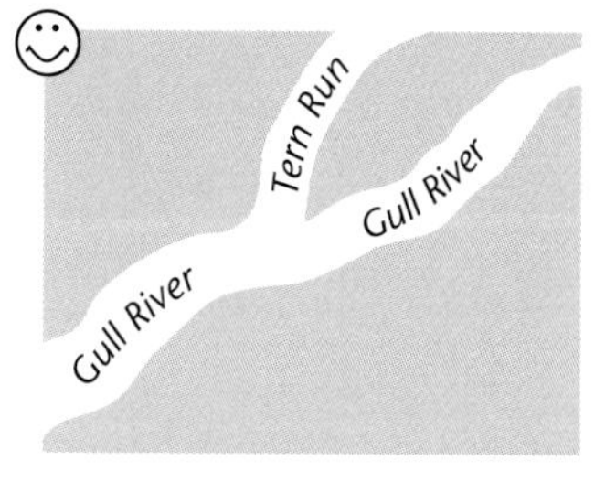

Typographic Minutiae and Maps

Effective type on and around maps requires understanding the basics of typography. Of particular relevance are kerning, letterspacing, line spacing, and alignment. Careful design of type will make your map more functional and beautiful.

Kerning

The combination of Ta and Ve below, when not kerned, look like there is too much space between them. Kerning adjusts the spacing between particular pairs of letters to make them look uniform and less distracting.

T alc V enus
Talc Venus

Kerning is automatically adjusted with most digital type placement algorithms in mapping software, but can also be set by hand.

Evaluate type on or around your map that may need kerning
Kerning is more important for larger type sizes
Increase kerning for reversed and curved type

Letterspacing

Letterspacing, or tracking, changes the spacing between all letters. Normal, positive, and negative letterspacing are shown below.

Talc Venus
T a l c V e n u s
TalcVenus

Avoid negative letterspacing, or take care to avoid hard-to-read, scrunched-up type
Increase letterspacing slightly for a more open, airy feel in a block of text
Increase letterspacing by using upper-case letters to label area features

Line Spacing

Line spacing, or leading, adjusts spacing between lines of text.

This paragraph of text is in Stone Sans Serif 9 point type with 11 point spacing. Normal spacing is set to 120% of the point size of the type.

This paragraph of text is in Stone Sans Serif 9 point type with 9 point spacing. Blocks of text "set solid" look cramped and are harder to read.

This paragraph of text is in Stone Sans Serif 9 point type with 14 point spacing. Too much line spacing causes text blocks to break into what looks like separate lines of text.

Evaluate type on or around your map that may need line spacing
Maintain a consistent line spacing for similar features labeled on a map
Avoid "set solid" line spacing; instead try a smaller type size
Avoid too much line spacing, as such labels may be misinterpreted as multiple feature labels

Alignment

Avoid left-right justification if it causes distracting spacing problems
Ragged right alignment is the norm, but too much ragged is distracting. Use hyphenation sparingly in text blocks, and avoid it on map labels
Ragged left is more difficult to read in blocks, but use it for map labels referring to symbols to the right of the label

Bruce: So, who are you?
North: I'm North.
Bruce: I see your name on maps, very impressive.

North (1994)

"I can't read the names on this [map] because they are all in English."

Christian realized he would have to show his friend how to read a map. "The top is north," he said. "The little circles are towns and villages. Blue means rivers and lakes. The thin lines are roads and the thick ones railways."

"There's nothing at all here," said Big Tiger, pointing to one of the many white patches.

"That means it's just desert," Christian explained. "You have to go into the desert to know what it looks like."

Fritz Muhlenweg, *Big Tiger and Christian* (1952)

...words on maps have a meaning in the original language, but when translated this is not the case...

D.W. Nash, *"On the Geology of Egypt"* (1837)

Rock which does not cover,
Coral reef, detached,
Wreck always partially submerged.

A number of sunken wrecks,
Obstruction of any kind,
Limiting danger line.

Foul ground, discolored water,
Position doubtful,
Existence doubtful.

John Krygier, from map symbol descriptions, Section O of Chart #1, *Nautical Chart Symbols and Abbreviations* (2007)

More...

Most cartography texts devote some space to type on maps. Borden Dent's type chapter in *Cartography: Thematic Map Design* (2008) is probably the best. For useful thinking about non-Roman typefaces, try the first 60 pages of Ruben Pater's *The Politics of Design* (2016).

A very useful book on type is Ellen Lupton, *Thinking with Type: A Critical Guide for Designers, Writers, Editors, and Students* (2010). Why not also peruse Warren Chappell and Robert Bringhurst, *A Short History of the Printed Word* (2000), and Robert Bringhurst, *Elements of Typographic Style* (2004); Edward Catich's *The Origin of the Serif* (1991) is a thought-provoking book about letters. Want more on the anatomy of type? *The Anatomy of Type: A Graphic Guide to 100 Typefaces,* by Stephen Coles (2012).

You're probably not going to be using graffiti lettering, but *Art in the Streets* (2021) remains the best introduction. Who knows? You may pick up some ideas.

Sources: Guy-Harold Smith's *Physiographic Diagram of South America* (1935) is reproduced courtesy of the Geographical Press. The Waldseemüller map is from Wikipedia (wikipedia.org). The Boylan Heights leaf map is from Denis Wood, *Everything Sings: Maps for a Narrative Atlas* (Siglio Press, 2010). The scribbled-on map of Europe was published in *Life* magazine, December 17, 1914. The Wankers tube map was posted on Geoff Marshall's sillymaps web page; the map was made by an anonymous contributor. The type placement pairs are redrawn from Eduard Imhof, "Positioning Names on Maps," *The American Cartographer* 2:2, 1975.

Arctic
Atlantic
Amazon
Pacific

Less is more?

Arctic
Atlantic
Amazon
Pacific

CHAPTER 10 Map Generalization and Classification

Fewer data can be better. Bill Bunge made a map he called *The Continents and Islands of Mankind* to make a point: as land and water barriers are about equally effective nowadays, there is no reason to keep mapping the continents and oceans when we are concerned with human affairs. This, then, is a map of places where there are more than 30 people per square mile. The shapes of the continents are obvious enough without being drawn. More importantly, they are beside the point. So leave them off the map! What matters here is where people are. Fewer data more effectively make the point.

Map Generalization

Our human and natural environments are complex and full of detail. Maps work by strategically reducing detail and grouping phenomena together. Driven by your intent, maps emphasize and enhance a few aspects of our world and de-emphasize everything else. In contrast, Tufte's dictum "to clarify, add detail" serves as a critical check. Ruthlessly cut away the superfluous – "reality junk" – so you can add more levels of information to clarify and achieve your goals for the map.

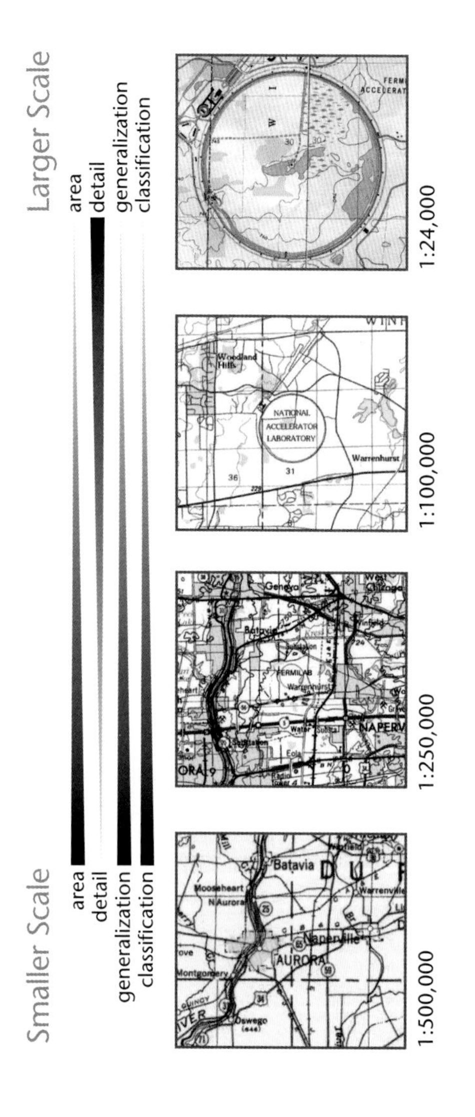

Larger-Scale Maps

Less area
More detail
Less generalization
Less classification

Transformation from large to small scale requires generalization and classification.

City changes from area to point

Diversity of city sizes combined into a few categories (small, medium, large)

Minor streets and roads removed

Different types of streets and roads combined into a few categories

Houses, then major buildings, removed

Small streams removed

Detail removed from rivers and roads

Change areas on statistical maps from local to regional to national

Smaller-Scale Maps

More area
Less detail
More generalization
More classification

Selection

Maps select a few (and don't select most) features from the human or natural environment. Selected features are vital to the intent of your map. Unnecessary features should not be selected. Selection reduces clutter and enhances the reader's ability to focus on what is most important on a map.

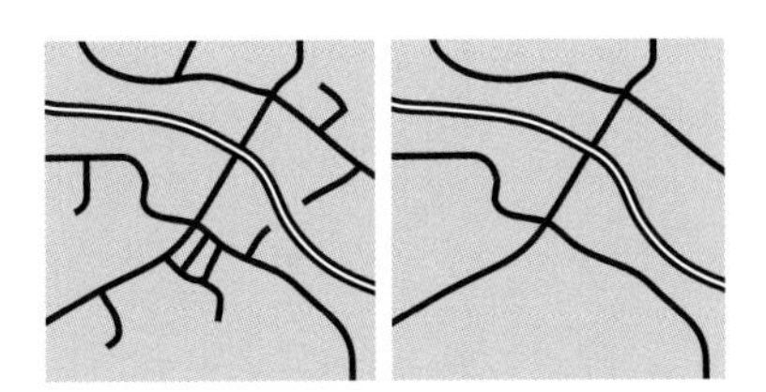

Selection is the responsibility of the map maker. Selection is automated with digital mapping. For example, at a larger scale (more detail, less area) all cities over 1000 in population are shown. At a smaller scale (less detail, more area) only cities over 100,000 population are shown.

Questions to ask when selecting map features, keeping in mind your goals for the map:

Is the feature necessary to make your point?
Will removing the feature make the map harder to understand?
If less important features are removed, do more important features stand out more clearly?
Does removal of less important features lead to a less cluttered map?

Dimension Change

A dimension change in a feature is often necessary when changing scale and useful for removing unnecessary detail from a map. A city changes from an area to a point, a river from an area to a line. Conversely, a group of points may be transformed into an area, or a group of small areas into one larger area.

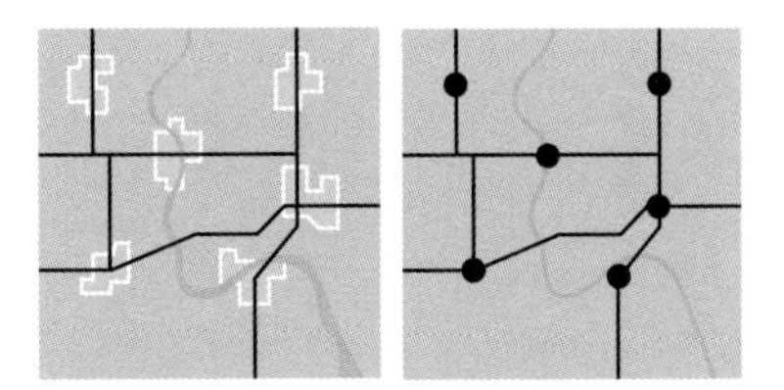

Scale change is the most common reason for dimension change, but map makers also change dimensions of features to remove clutter from a map in order to support its point. Computer mapping software can change feature dimensions based on the scale of the map.

Questions to ask that guide dimension change, keeping in mind your goals for the map:

Would changing dimensions of a feature remove unnecessary detail?
Does changing the dimensions of a feature in any way affect how it is understood by the map reader?
Does changing the dimensions of map features help make the map less cluttered?

Simplification

Features on maps are often simplified. Simplification can enhance visibility, reduce clutter, and, with digital data, reduce the size of the digital map file. Smaller-scale maps (showing more area) tend to have more simplified features than larger-scale maps. Eliminate detail that is not necessary for the map to make your point.

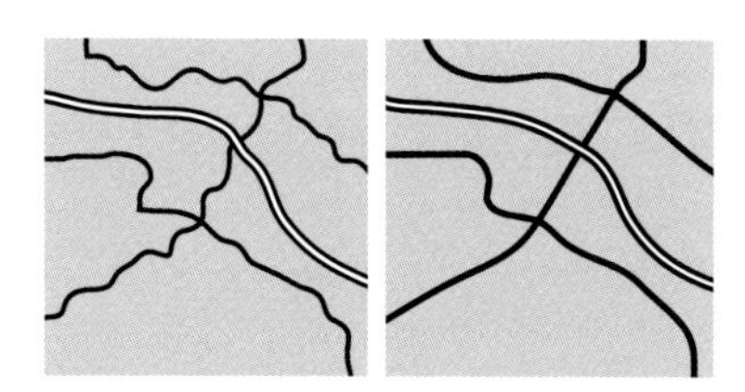

Simplification of features is done to the point where they are less complex, yet still recognizable. Identifying characteristics should be retained. Mapping software simplifies features by methodically removing detail. A line is simplified by removing every other point, for example.

Questions to ask when simplifying features, keeping in mind your goals for the map:

How simplified can a feature be and still be recognized?
Does the removal of detail remove any vital information?
Does the simplification of a feature make it more noticeable?
Does the simplification of a feature make the map less cluttered looking?

Smoothing

Smoothing map features reduces their angularity. Smoothing is related to simplification but focuses on adjustments in the location of a feature or possibly the addition of detail. Smoothing affects the qualitative look of features. Lots of smoothing needed? Maybe you need better (more detailed) data.

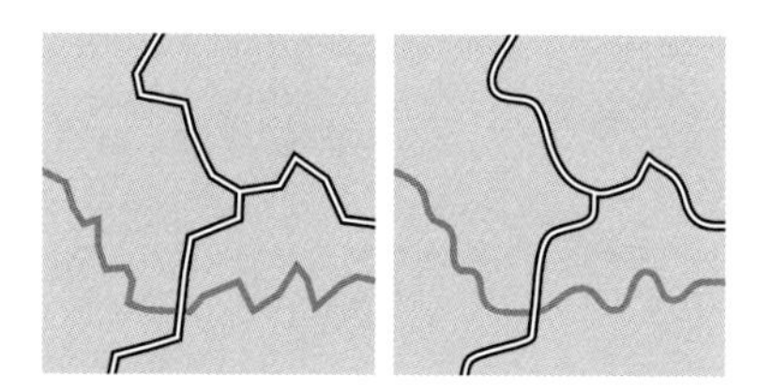

Map features that are naturally sinuous (often natural features) are more heavily smoothed. Features that are not smooth in the real world are lightly smoothed or not smoothed at all. Computers smooth features by rounding angles that exceed a set limit.

Questions to ask when smoothing features, keeping in mind your goals for the map:

How much can you smooth a feature without losing its character?
Does smoothing a feature make it more difficult to recognize?
Does smoothing a feature make it easier to recognize?
Does the smoothing of a feature make the map less cluttered looking?
Do I need more detailed data?

Displacement

Displacement moves vital map features apart that visually interfere with one another. Moving features away from their actual location may sound like a terrible idea, but it makes the features easier to distinguish and understand.

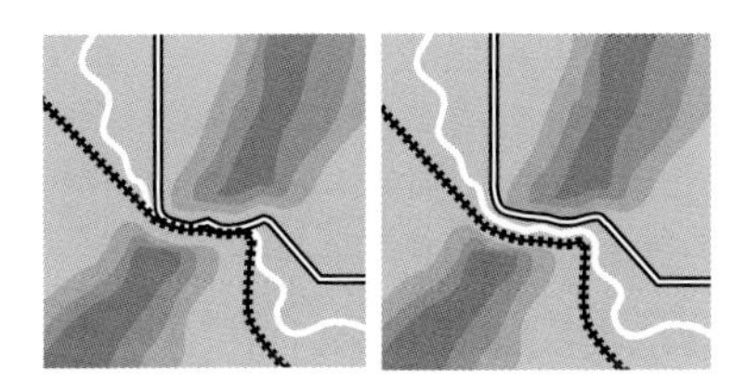

Scale change is a common reason for displacement. Displacement of map features sacrifices location accuracy for visual clarity. Displacement of point and line features is common on maps where such features are crowded together in certain areas.

Questions to ask that guide displacement, keeping in mind your goals for the map:

Are important map features interfering with one another?
Will the slight movement of a map feature make it and neighboring features easier to distinguish?
Will the slight movement of a map feature lead to confusion because the feature has been moved?

Enhancement

Enhancement of map features occurs when the map maker knows enough about the feature being mapped to add details that aid in understanding. A few bumps, for example, are added to a road the map maker knows is winding and angular so that the map reader understands the actual character of the road.

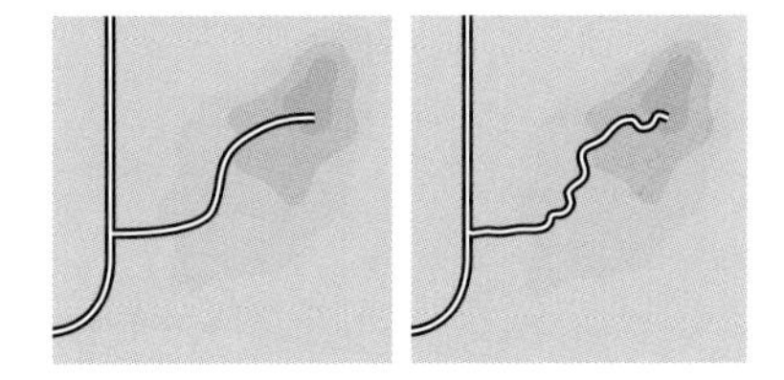

Enhancement adds detail, as opposed to removing detail, which most generalization procedures do. Enhancement must be used with care. The enhancement should not be deceptive, but instead should help the map reader understand important features on the map.

Questions to ask that guide enhancement, keeping in mind your goals for the map:

Do you know enough about a feature to enhance it?
Will enhancement help the map reader to better understand the feature and the map?
Does enhancing a feature make it easier to recognize?
Could enhancing a feature possibly lead to misunderstanding by the map reader?

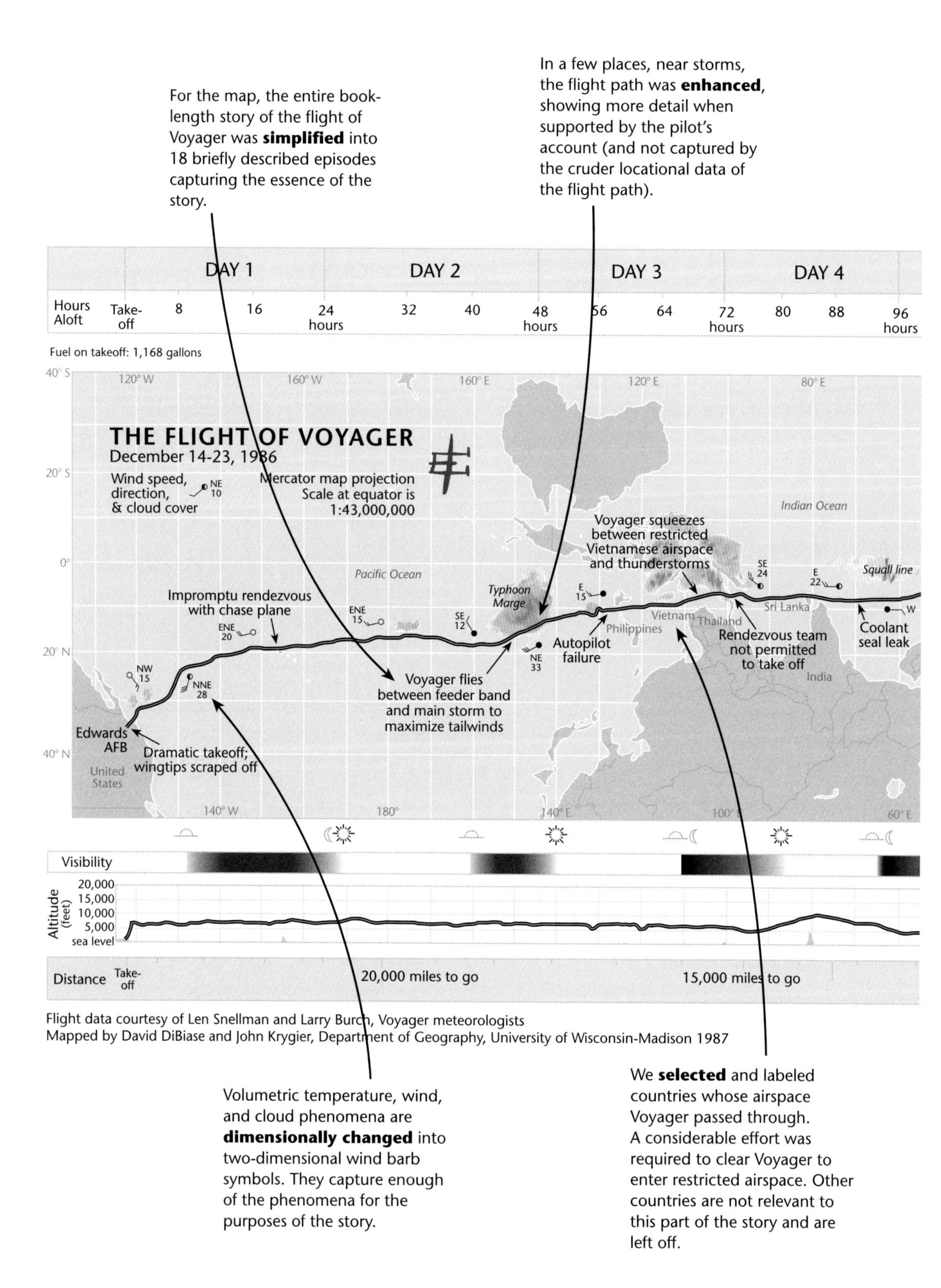
For the map, the entire book-length story of the flight of Voyager was **simplified** into 18 briefly described episodes capturing the essence of the story.
In a few places, near storms, the flight path was **enhanced**, showing more detail when supported by the pilot's account (and not captured by the cruder locational data of the flight path).
DAY 1
DAY 2
DAY 3
DAY 4
Hours Aloft
Take-off
8
16
24 hours
32
40
48 hours
56
64
72 hours
80
88
96 hours
Fuel on takeoff: 1,168 gallons
THE FLIGHT OF VOYAGER
December 14-23, 1986
Wind speed, direction, & cloud cover
NE 10
Mercator map projection
Scale at equator is 1:43,000,000
Indian Ocean
Pacific Ocean
Voyager squeezes between restricted Vietnamese airspace and thunderstorms
Squall line
Typhoon Marge
Impromptu rendezvous with chase plane
Autopilot failure
Rendezvous team not permitted to take off
Coolant seal leak
Philippines
Vietnam
Thailand
Sri Lanka
India
Voyager flies between feeder band and main storm to maximize tailwinds
Edwards AFB
Dramatic takeoff; wingtips scraped off
United States
Visibility
Altitude (feet)
20,000
15,000
10,000
5,000
sea level
Distance
Take-off
20,000 miles to go
15,000 miles to go
Flight data courtesy of Len Snellman and Larry Burch, Voyager meteorologists
Mapped by David DiBiase and John Krygier, Department of Geography, University of Wisconsin-Madison 1987
Volumetric temperature, wind, and cloud phenomena are **dimensionally changed** into two-dimensional wind barb symbols. They capture enough of the phenomena for the purposes of the story.
We **selected** and labeled countries whose airspace Voyager passed through. A considerable effort was required to clear Voyager to enter restricted airspace. Other countries are not relevant to this part of the story and are left off.

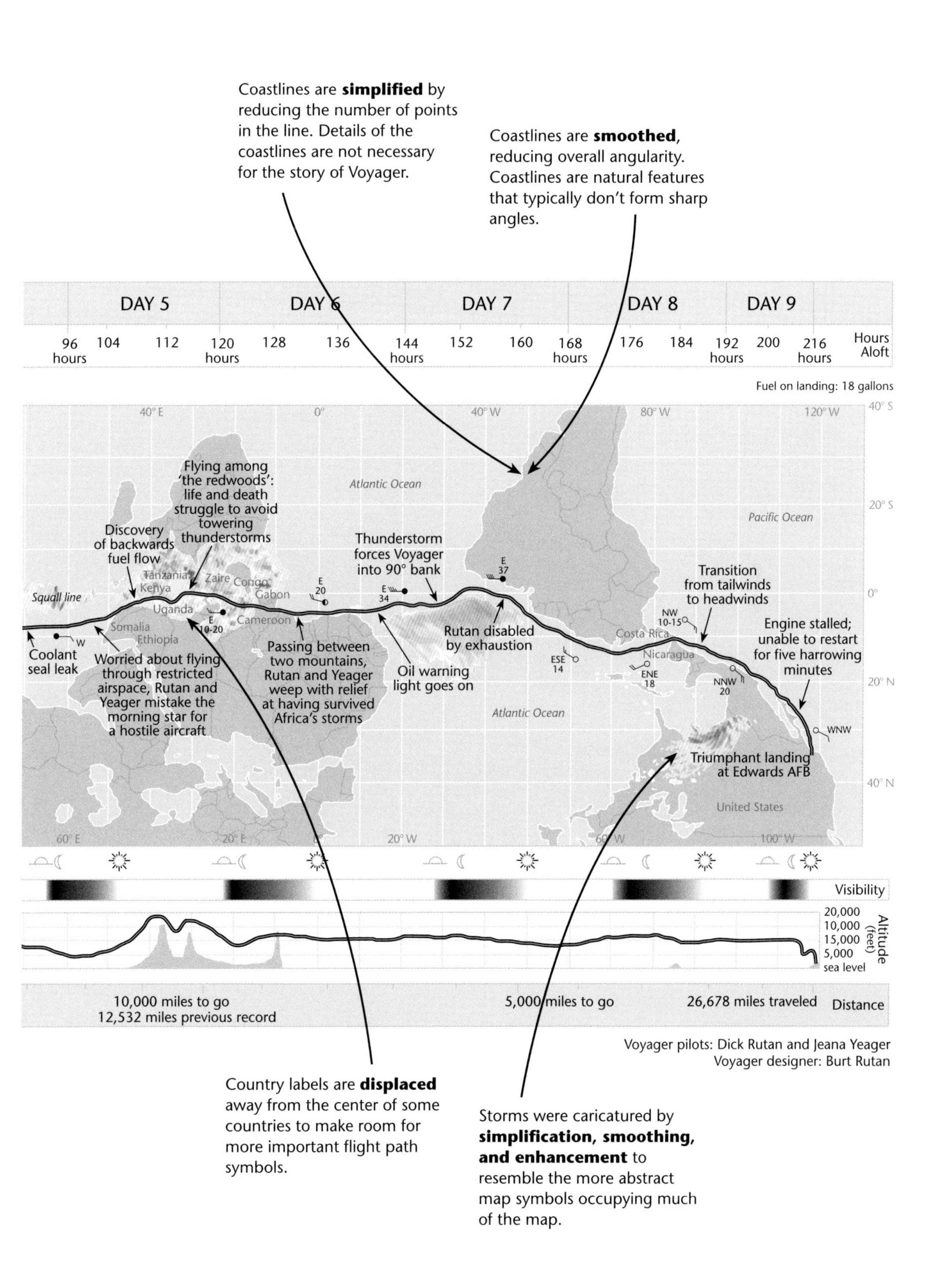

Coastlines are **simplified** by reducing the number of points in the line. Details of the coastlines are not necessary for the story of Voyager.
Coastlines are **smoothed**, reducing overall angularity. Coastlines are natural features that typically don't form sharp angles.
DAY 5
DAY 6
DAY 7
DAY 8
DAY 9
96 hours
104
112
120 hours
128
136
144 hours
152
160
168 hours
176
184
192 hours
200
216 hours
Hours Aloft
Fuel on landing: 18 gallons
40° E
0°
40° W
80° W
120° W
40° S
20° S
0°
20° N
40° N
60° E
20° E
0°
20° W
60° W
100° W
Atlantic Ocean
Pacific Ocean
Atlantic Ocean
United States
Flying among 'the redwoods': life and death struggle to avoid towering thunderstorms
Discovery of backwards fuel flow
Thunderstorm forces Voyager into 90° bank
Transition from tailwinds to headwinds
Engine stalled; unable to restart for five harrowing minutes
Squall line
Tanzania
Kenya
Zaire
Congo
Gabon
Uganda
Somalia
Ethiopia
Cameroon
Costa Rica
Nicaragua
Rutan disabled by exhaustion
Coolant seal leak
Worried about flying through restricted airspace, Rutan and Yeager mistake the morning star for a hostile aircraft
Passing between two mountains, Rutan and Yeager weep with relief at having survived Africa's storms
Oil warning light goes on
Triumphant landing at Edwards AFB
W
E 10-20
E 20
E 34
E 37
NW 10-15
ESE 14
ENE 18
NNW 20
WNW
Visibility
20,000
10,000
15,000
5,000
sea level
Altitude (feet)
10,000 miles to go
12,532 miles previous record
5,000 miles to go
26,678 miles traveled
Distance
Voyager pilots: Dick Rutan and Jeana Yeager
Voyager designer: Burt Rutan
Country labels are **displaced** away from the center of some countries to make room for more important flight path symbols.
Storms were caricatured by **simplification, smoothing, and enhancement** to resemble the more abstract map symbols occupying much of the map.

Data Classification

Every map is generalized. It doesn't matter if it's a detailed topographic survey sheet showing trees, buildings, etc., or a map of populations. Classification operates in every map as well, distinguishing trees from buildings on the topographic map and breaking populations into classes such as small, medium, and large. Data classification is shaped by your goals for the map. In general, features in the same class should be more similar than dissimilar; features in different classes should be more dissimilar than similar.

Qualitative Point Data

A student polls community members on social issues. The first, unclassified, map is not very revealing. The classification on the second map is OK, but the third reveals more about the political landscape. Include the unclassified data so map viewers can decide if your classification is justified.

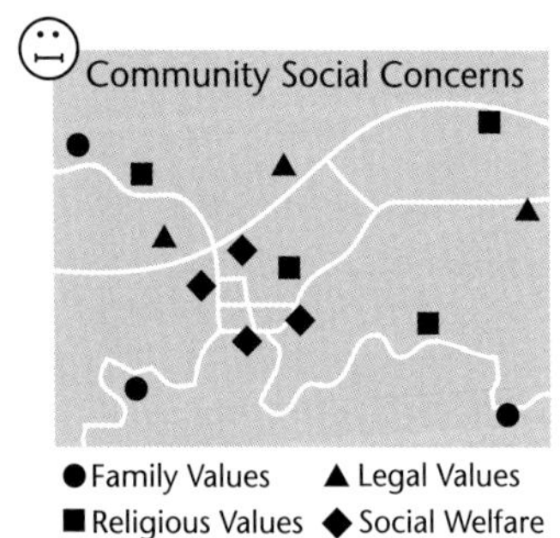

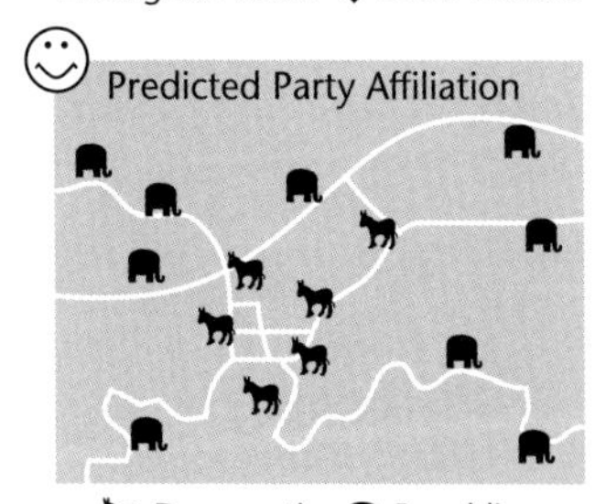

Qualitative Line Data

Roads are often classified in terms of who builds and maintains them (federal, state, local). However, this classification is not the best if your map is for tourists. Your goal for the map (tourism) should shape your data classification. Choose tourism-based classes for the roads.

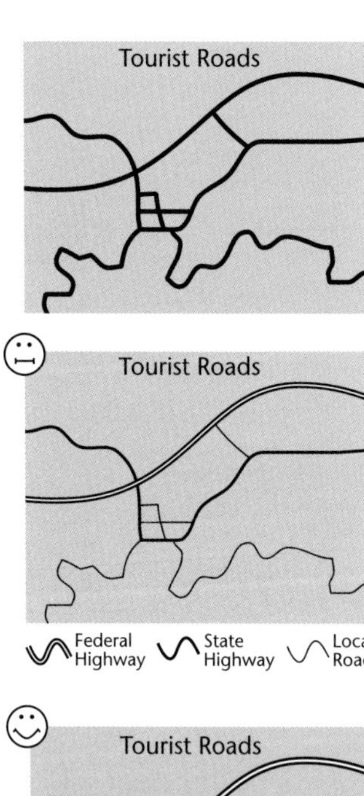

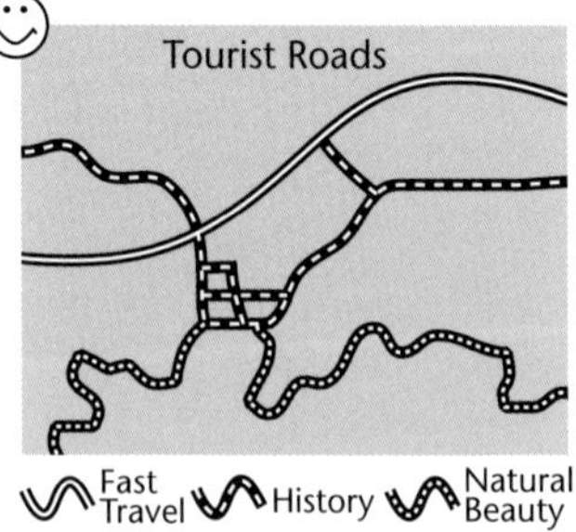

Qualitative Area Data

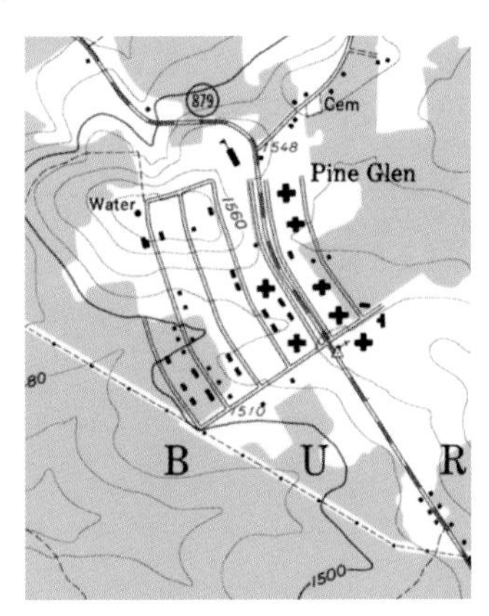

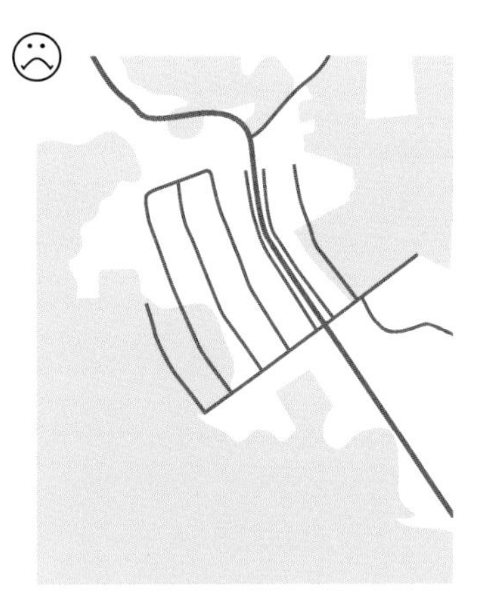

Be aware that qualitative data on maps has been classified. Determine the criteria for classification. A map may classify data based upon criteria suitable for one purpose but not necessarily for others.

For ecology projects, U.S. Geological Survey (USGS) topographic maps (above left and right) are often consulted. They classify vegetation into two categories: vegetation (gray areas) or no vegetation. One might assume that the classification is based on ecological criteria.

Alas, vegetation areas are actually classified based on military criteria. Vegetation areas are those with tree cover at least 6 feet tall that can hide military troops. It is a classification for army guys, not ecologists!

Instead, find a vegetation map (right) with a classification based on ecological criteria.

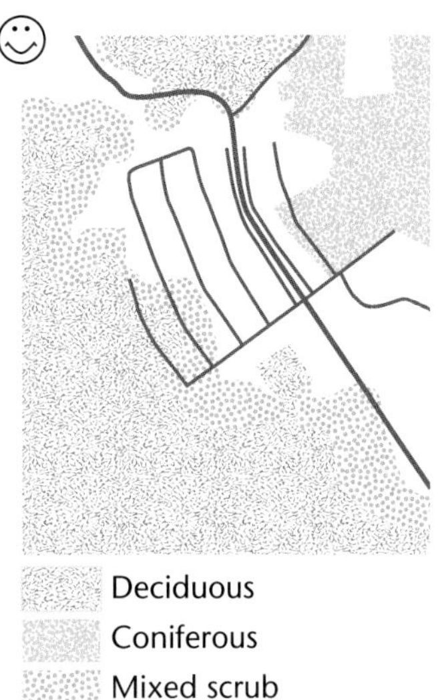

Quantitative Point Data

A map created for a community meeting about well test results should clearly show what is most important: whether the well water is safe for humans or not. The third map, below, is best for that goal.

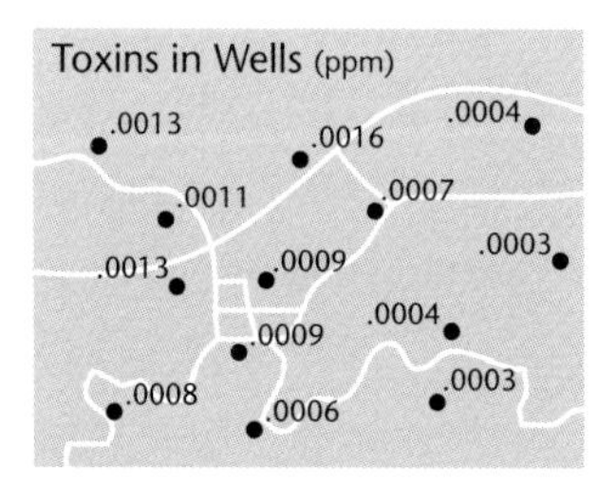

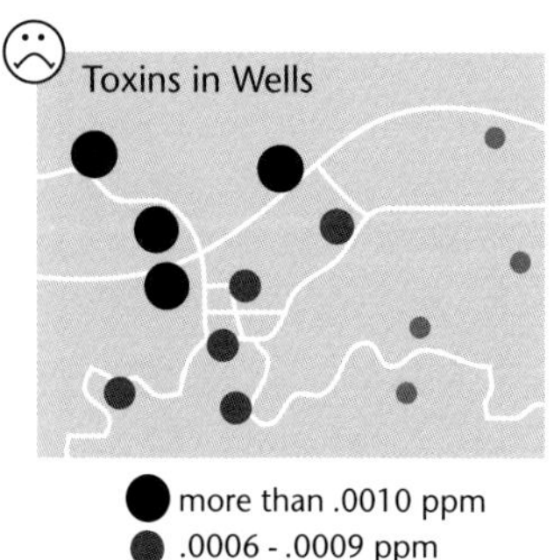

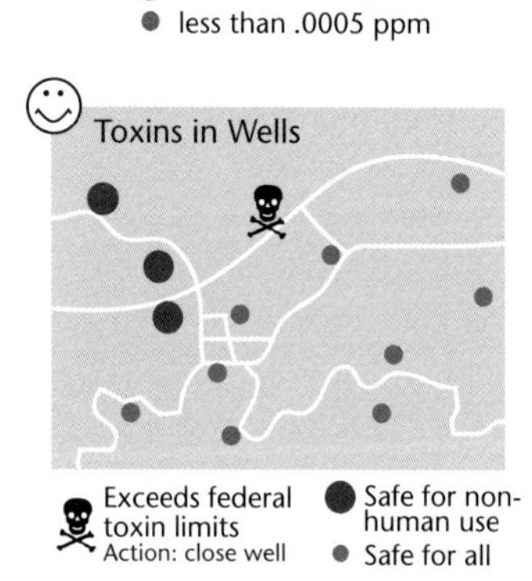

Quantitative Line Data

A map intended to help guide the restructuring of police patrol routes should classify data, in this case average vehicle speeds, in categories appropriate to the task: increase, maintain, or decrease patrols.

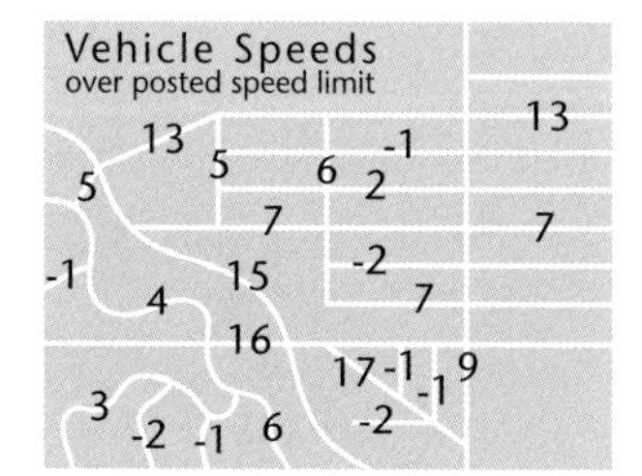

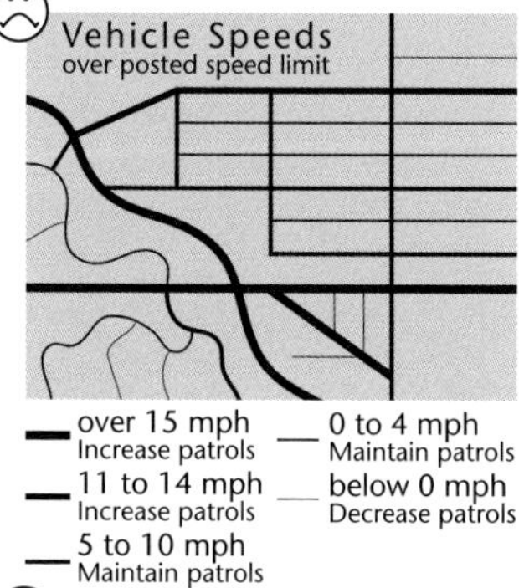

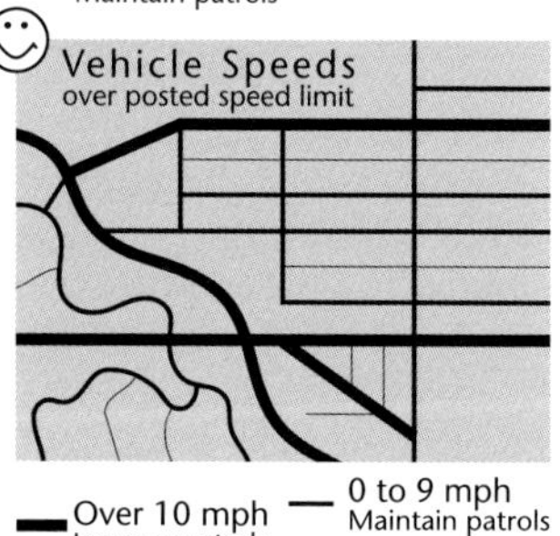

Quantitative Area Data

With quantitative data aggregated in areas, first decide the number of classes. Fewer classes often result in distinct patterns; more classes often result in complex patterns. Which option is best depends on why you are making the map. This map shows the density of mobile homes (dark equals higher density).

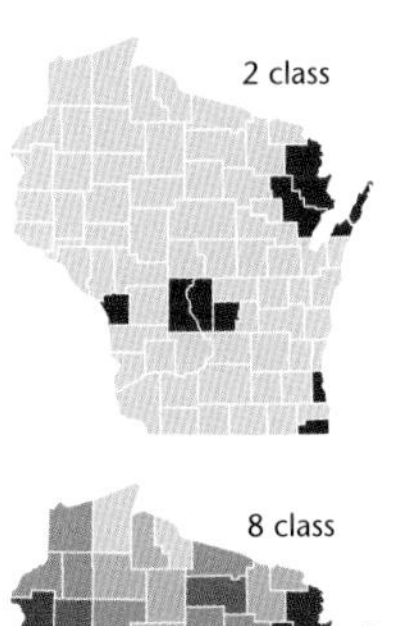

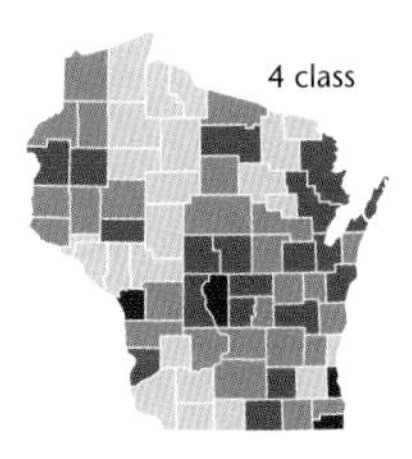

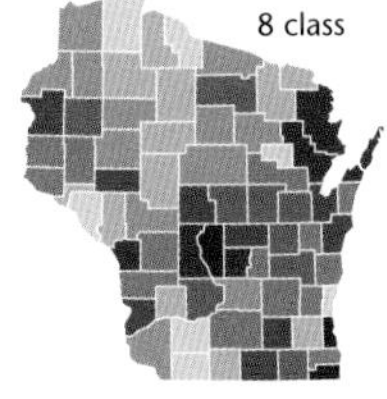

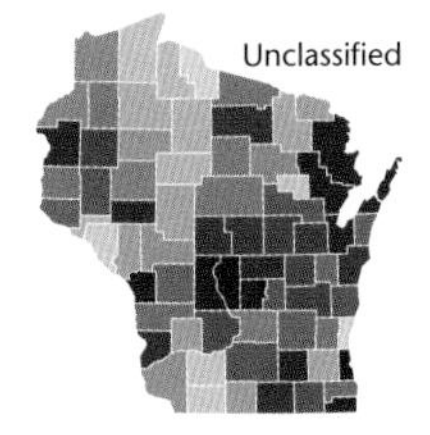

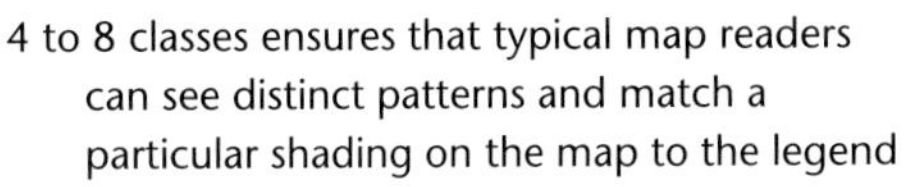

A 2-class map is good for binary (yes|no) data or data with negative and positive values

4 to 8 classes ensures that typical map readers can see distinct patterns and match a particular shading on the map to the legend

Over 8 classes produces more complex patterns, but map readers may not be able to match a shading on the map to the legend

Unclassified data (each area has a unique shading corresponding to its unique value) produce the most complex patterns

Reclassify your data with different numbers of classes and look at how the patterns change. Think about your data and goals for the map, and make an intelligent decision

In addition to choosing the number of classes, you must decide where to place boundaries between the classes. Classification schemes set these boundaries. The maps below show the density of mobile homes (dark equals higher density) unclassified and as a 5-class map using four different classification schemes.

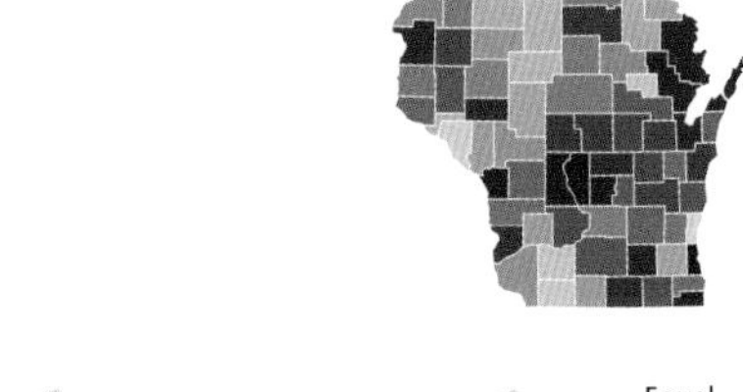

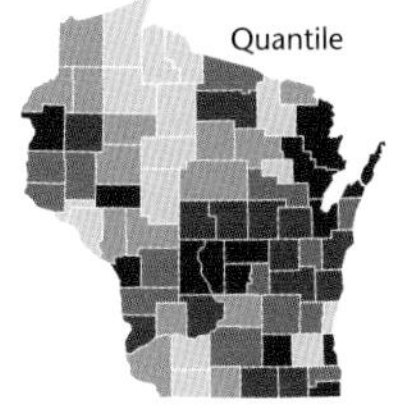

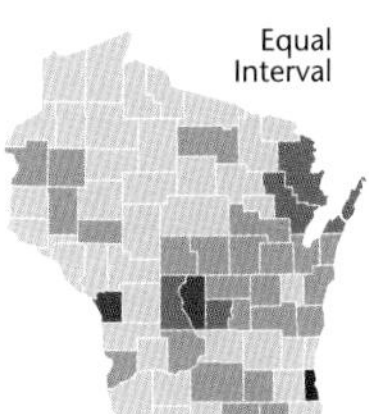

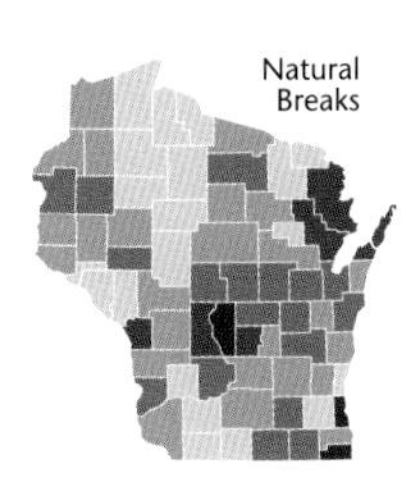

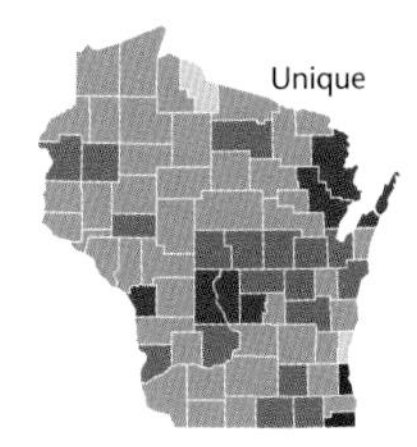

Thinking Drives Classification

Poverty is a contentious issue. Debates rage over defining poverty, why it exists, and how to address it. The U.S. Census Bureau provides official data on poverty in the U.S., and various classifications of Census 2021 poverty data follow.

It is easy to calculate the percent of people in each county in the U.S. who live in a state of official poverty. But choosing how to map the data is not as easy. Common (and equally valid) data classification schemes – methods for placing boundaries between the classes on a map – are easy to generate but difficult to choose from. Understanding the benefits and problems with each classification scheme is vital, as is clarifying why you are making the map. Together, these guide the thinking behind choosing the most appropriate classification scheme for your data.

Graphing Data

Selecting a classification scheme without examining your data as a graph is a bad idea. As examples in this section reveal, classification schemes can mask important characteristics of your data and even undermine the goal of your map.

Decent mapping software will generate a histogram for you while you are classifying your data. Or, make your own. The x-axis is your data variable (from low to high) and the y-axis the number of occurrences of each value.

The poverty data have a cluster of counties near the lower to mid-end of the graph, with a smaller number of counties skewed out to 43.9%. On a histogram you can see where a classification scheme places class boundaries, which values are grouped together, and which values are in different groups.

If a particular classification scheme seems to violate the basic classification rule (features in the same class should be more similar than dissimilar; features in different classes should be more dissimilar than similar), then consider a different scheme. Consider placing the graph on your final map, so map users can see how the data are classified.

Unclassified Scheme

To create an unclassified scheme, assign a unique visual shade to every unique data value. In essence, each unique data value is in its own class. Unclassified schemes make complex and subtle patterns by minimizing the amount of generalization.

This map (due in large part to the concentration of counties near the low end of the range of values) suggests that poverty is not a significant issue in most places, that the number of people living in poverty is somewhat similar across the U.S., and that there are relatively few places with very high poverty.

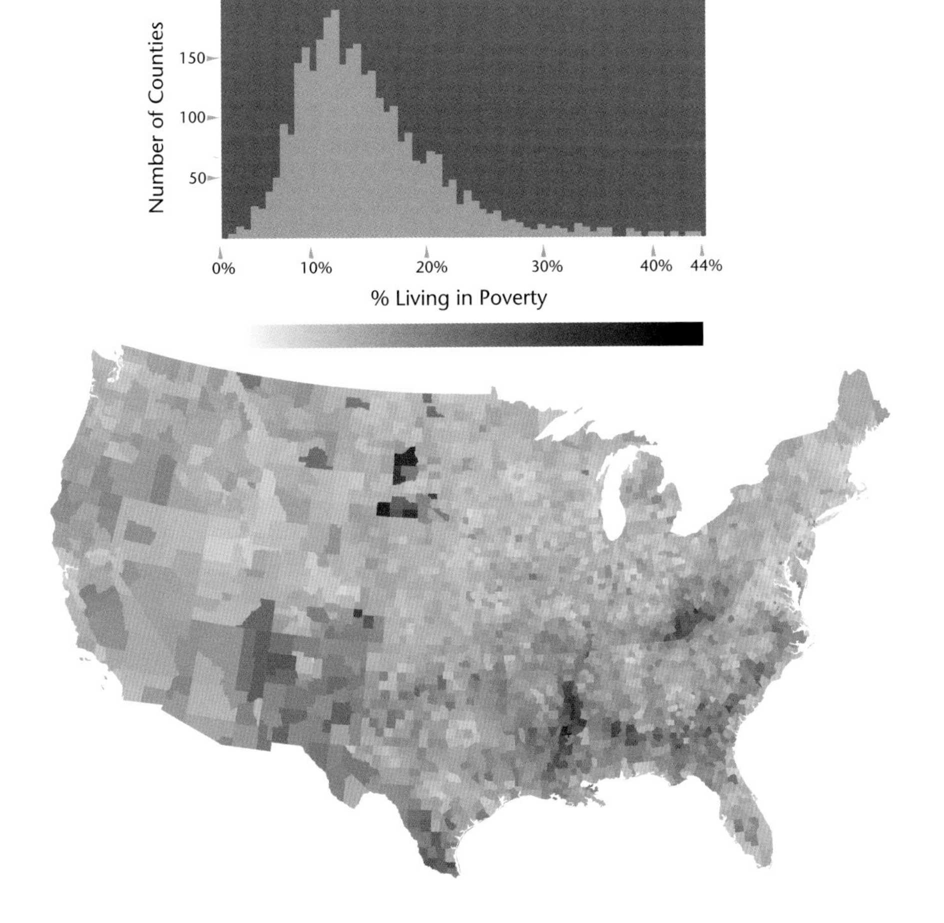

Quantile Scheme

Quantile schemes place the same number of data values in each class. Quantile schemes are appealing in that they always produce distinct map patterns. A quantile classification will never have empty classes or classes with only a few or many values. Quantile schemes look great.

The problem with quantile schemes is that they often place similar values in different classes or very different values in the same class. The map suggests that poverty is a significant issue in many counties, and the numerous counties in the top (darkest) classes impart a rather ominous view of poverty in the United States.

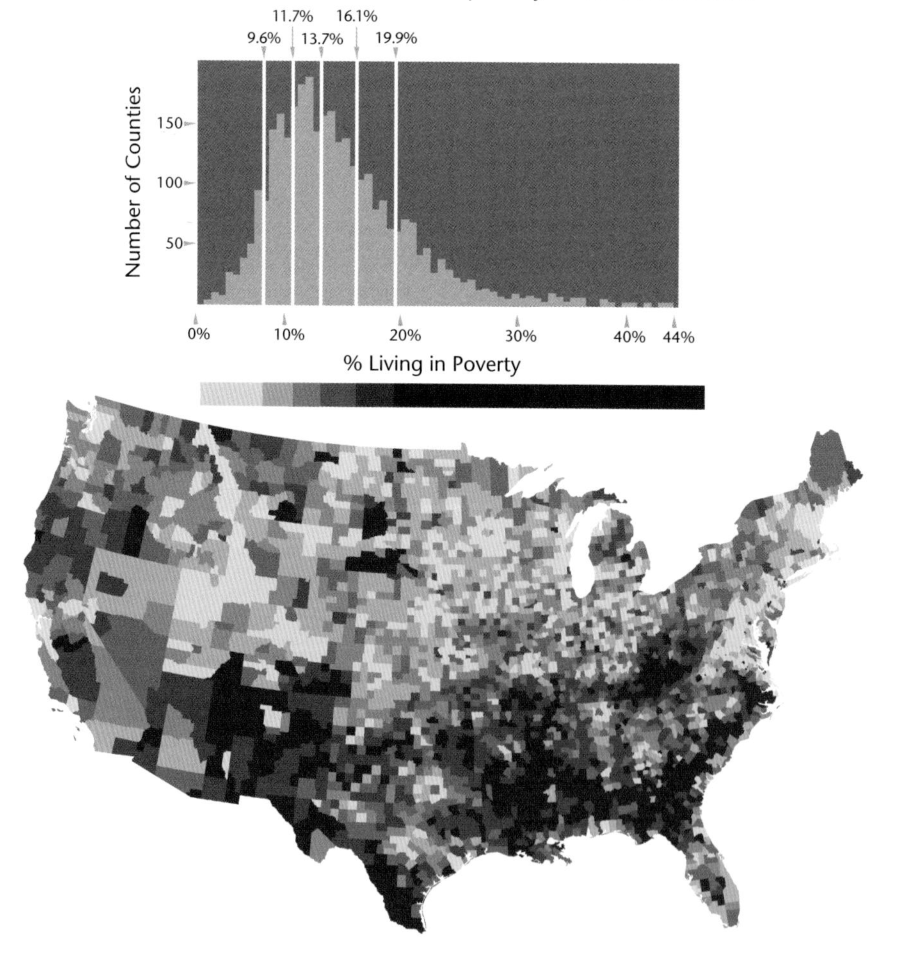

Equal-Interval Scheme

Equal-interval schemes place boundaries between classes at regular (equal) intervals. Equal-interval schemes are easily interpreted by map readers and are particularly useful for comparing a series of maps (which necessitates a common classification scheme).

Equal-interval schemes do not account for data distribution and may result in most data values falling into one or two classes, or classes with no values. The map suggests that poverty is not an issue in most places, as there are relatively few counties in the highest three classes.

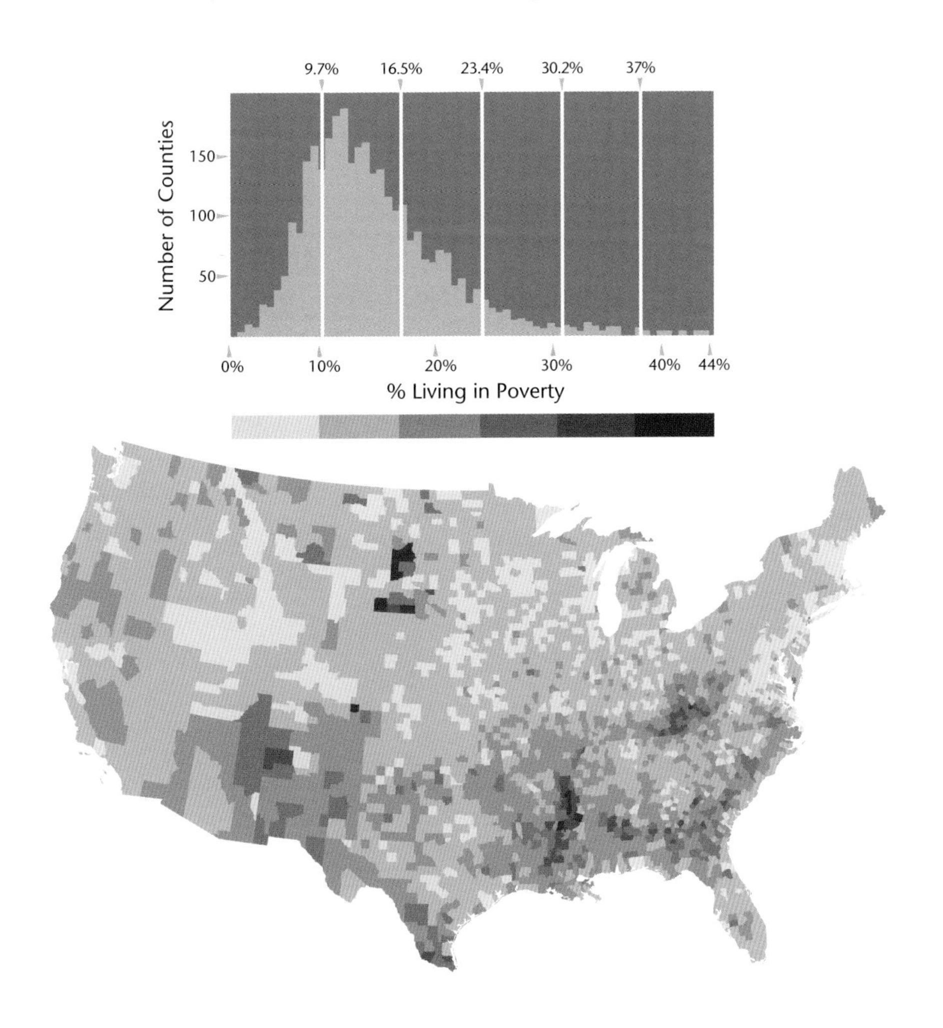

Natural-Breaks Scheme

Natural-breaks schemes minimize differences between values within classes and maximize differences between values in different classes. Class boundaries are determined by algorithms in mapping software that seek statistically significant groupings in a set of data.

Natural-breaks schemes can serve as a default classification scheme, as they take into account characteristics of the data distribution. This map makes poverty seem more significant than the equal-interval map does, but it is not quite as ominous as the quantile scheme.

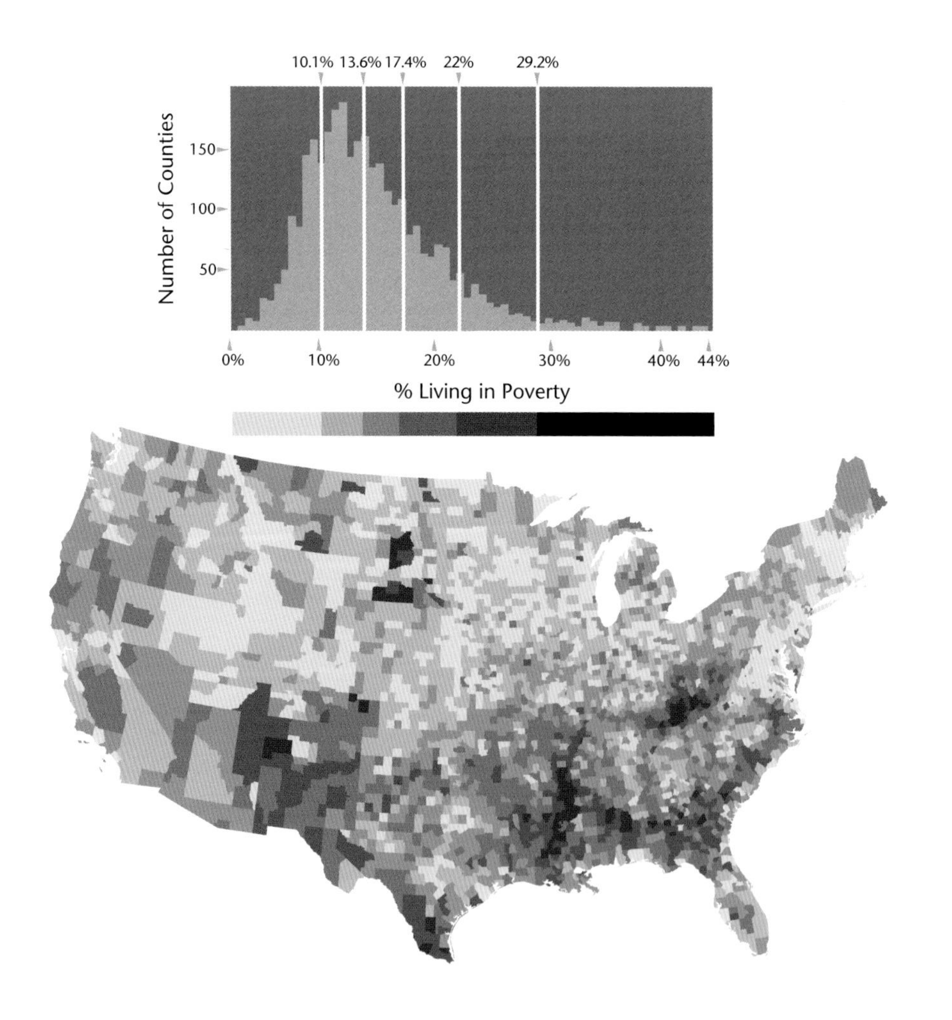

Unique Scheme

Class boundaries can be set by external criteria. A government program offers special funds to counties with over 25% poverty. A two-class map would show which counties qualify (and which don't). The map below is classified for a study of counties with very high poverty.

The researchers are not interested in counties below a 25% poverty rate. The remaining data are divided into roughly equal interval classes. While excellent for the study, this scheme may not be good in general, as it suggests poverty is isolated and rare.

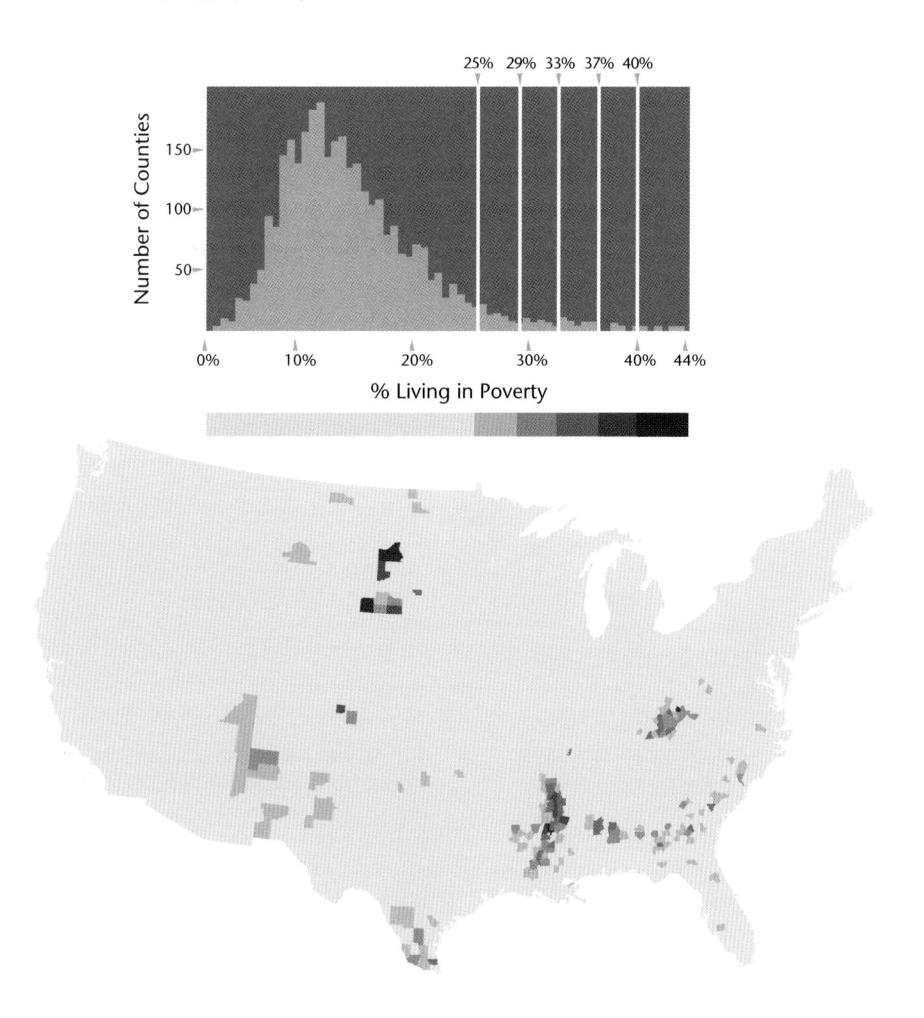

Think!

Looking at a number of classification schemes brings forth geographic facts, both the facts that are variously emphasized and those that are preserved through every variation.

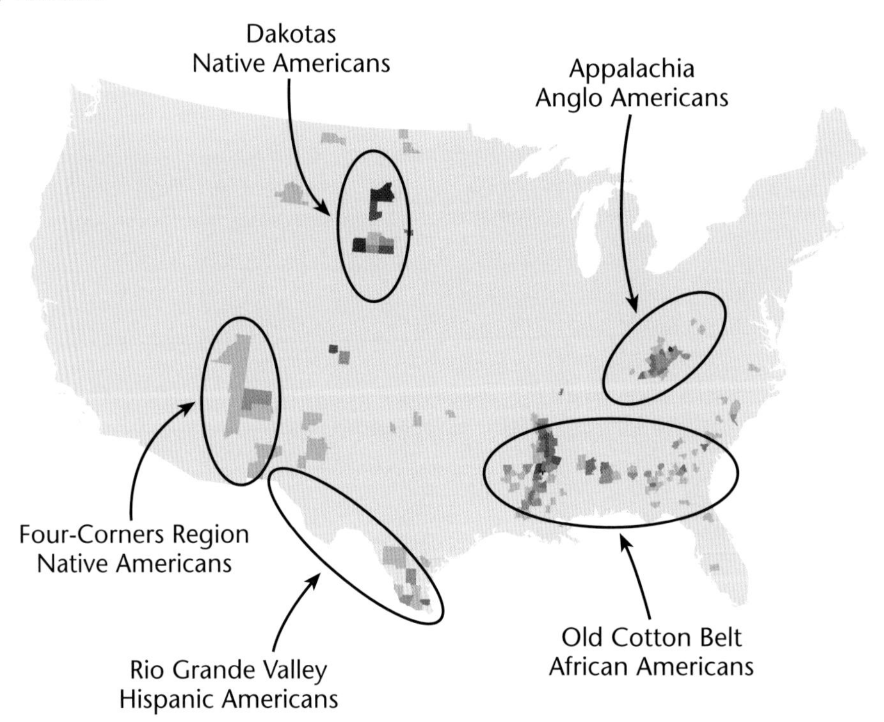

The unique scheme displays counties with the highest rates of poverty. While this pattern can be seen in every scheme, the unique scheme isolates – and so draws attention to – regions of the country where poverty is a long-standing reality. The unique scheme picked out the regions of significant social injustice.

Unless you looked at a lot of maps, you might not have identified these regions of injustice as anything other than those with high levels of poverty.

It takes many *different maps* to begin to make sense of the world.

Different maps... Racial Terrorism and Poverty in the American South

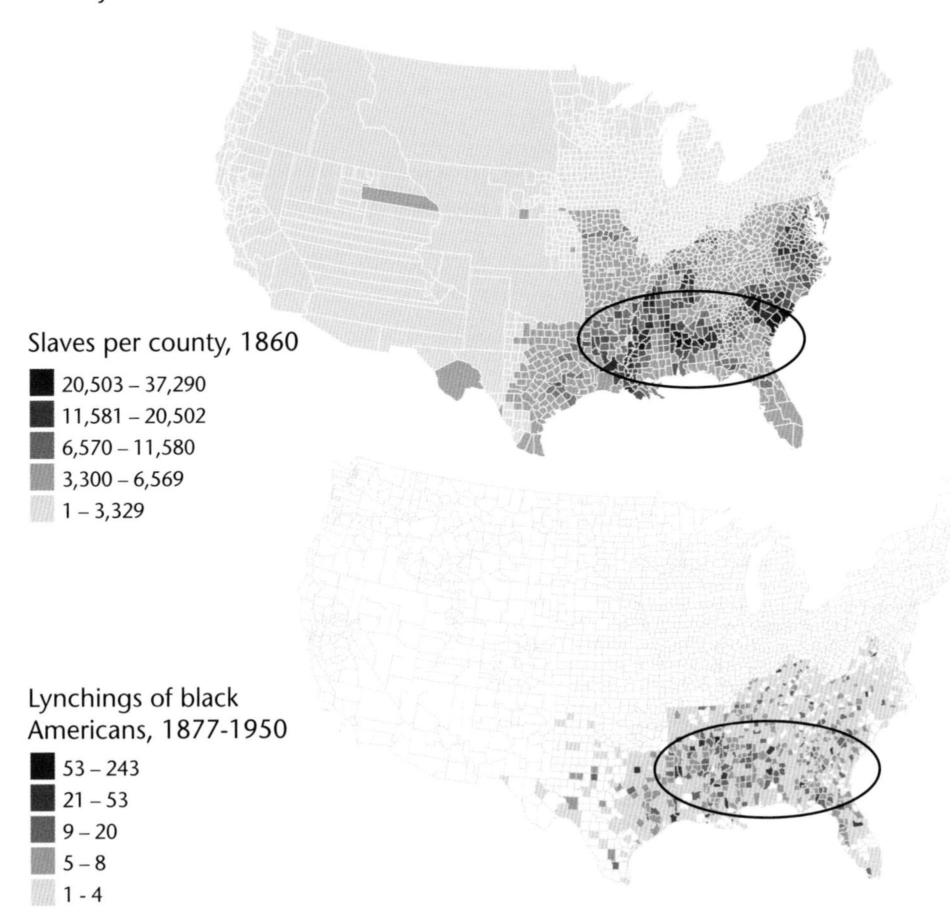

For five million pounds, I'd want a map that showed me looking at the map I'd just bought.

Rich Hall, *QI* (2005)

A recently patented "notion" for a map consists of having those portions intended to represent the rivers, lakes, and oceans filled with actual water. This is done by attaching the map to a back of wood of sufficient thickness. The rivers, etc., are dug out, filled with water, and glazed. Such maps may be hung upon the wall in the usual manner.

Boston Journal of Chemistry (1875)

Airey's Railway Map is almost unique in its way, devoting itself to its subject with a singleness of purpose which is really almost sublime, and absolutely ignoring all such minor features of the country it portrays as hills, roads, streets, churches, public buildings, and so forth. It is rather startling at first to find the Metropolitan Railway pursuing its course through a country as absolutely devoid of feature as was the "Great Sahara" in the good old African maps...

Charles Dickens, *London Guide* (1879)

I have witnessed the massacre
I am a victim of the map.

Mahmoud Darwish, *I Have Witnessed the Massacre* (1977)

More...

Mark Monmonier includes a succinct discussion of map generalization in *How to Lie with Maps* (1996). A comprehensive source for data classification is Terry Slocum et al., *Thematic Cartography and Geovisualization* (2023). The section "Visual and Statistical Thinking: Displays of Evidence for Making Decisions" from Edward Tufte's *Visual Explanations* (1997) dramatizes the power of thinking visually about statistics.

For very deep thoughts about classification in general, see Geoffrey Bowker and Susan Leigh Star, *Sorting Things Out: Classification and Its Consequences* (2000). Also see *Standards and Their Stories: How Quantifying, Classifying, and Formalizing Practices Shape Everyday Life* by Martha Lampland and Susan Leigh Star (2008).

A curious tale about what happens when you don't generalize is Jorge Luis Borges's short story "Funes the Memorius" from *Labyrinths: Selected Stories and Other Writings* (1964).

Sources: Bill Bunge's "Continents and Islands of Mankind" is re-created from his *Field Notes: Discussion Paper No. 2* (self-published, no date). The 2013 poverty data for the U.S. is from the U.S. Census, *Small Area Income and Poverty Estimates.* The slave data are from the *Teaching American History Project: Historic GIS Data.* The lynching data are from the *Equal Justice Initiative* (eji.org). Parts of this chapter are modified from Borden Dent et al., *Cartography: Thematic Map Design* (McGraw-Hill, 2008), and Terry Slocum et al., *Thematic Cartography and Geovisualization* (Prentice Hall, 2008).

How do you make data into visual marks?

WORLD MAP OF THE M
Equator
MALARIA
YELLOW FEVER
DENGUE
TYPHUS
PLAGUE
CHOLERA

CHAPTER 11 Map Symbolization

Map symbols allow us to put just about anything on a map. Real stuff, like tropical diseases, shown with grotesque hands, faces, feet, rats, worms, and insects on Boris Artzybasheff's creepy, WTF map from the 1950s. Maps also show statistical abstractions – like population per square mile, crime density, or suicide rates in states, provinces, or countries. Mostly real stuff is covered in this chapter, mostly abstract stuff in the next.

Common real things symbolized on maps, according to George Andre's The *Draughtman's Handbook of Plan and Map Drawing* (1891).

Over 100 years later, some of these categories and symbols seem archaic — others remain conventional — in GIS and digital mapping.

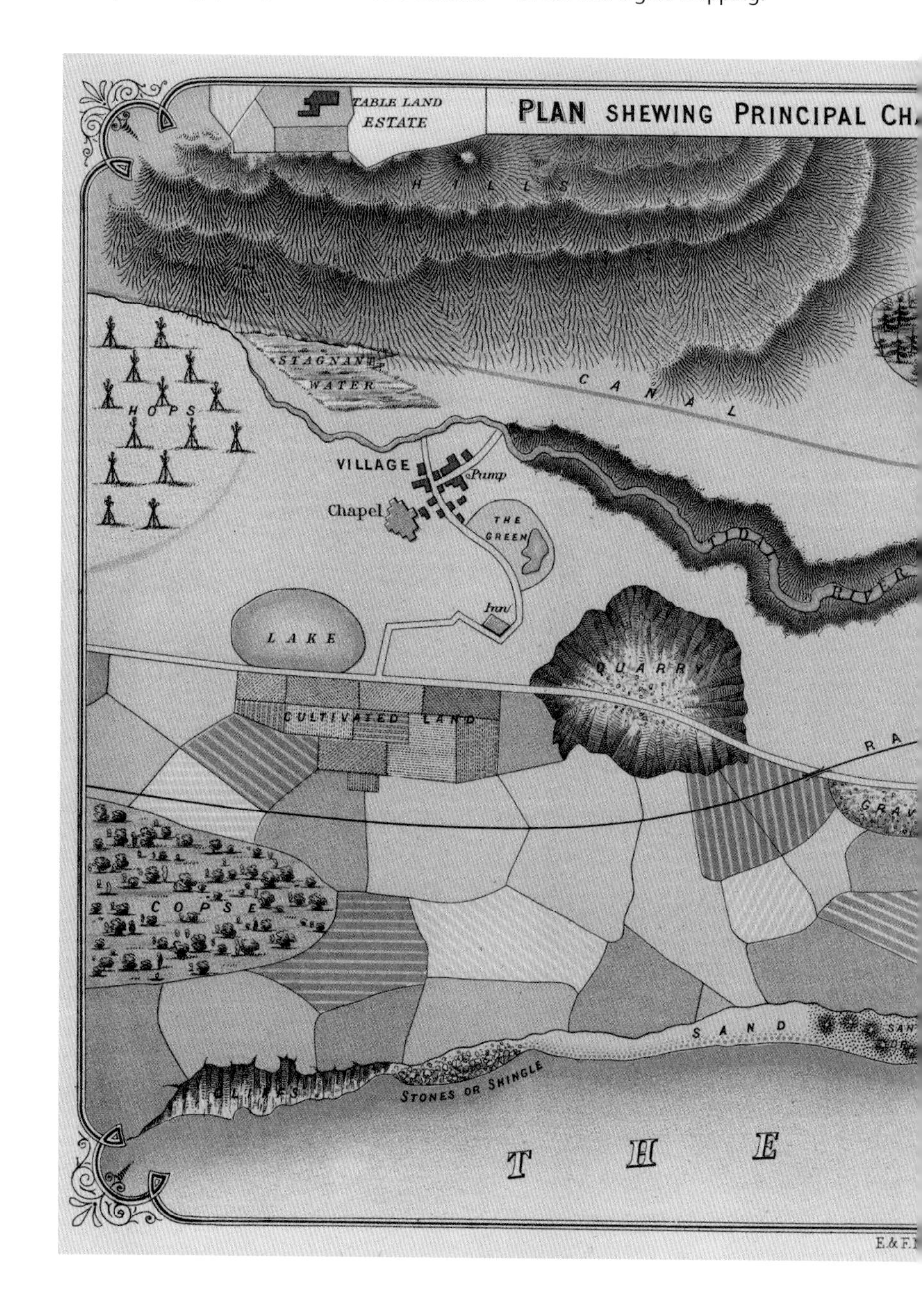

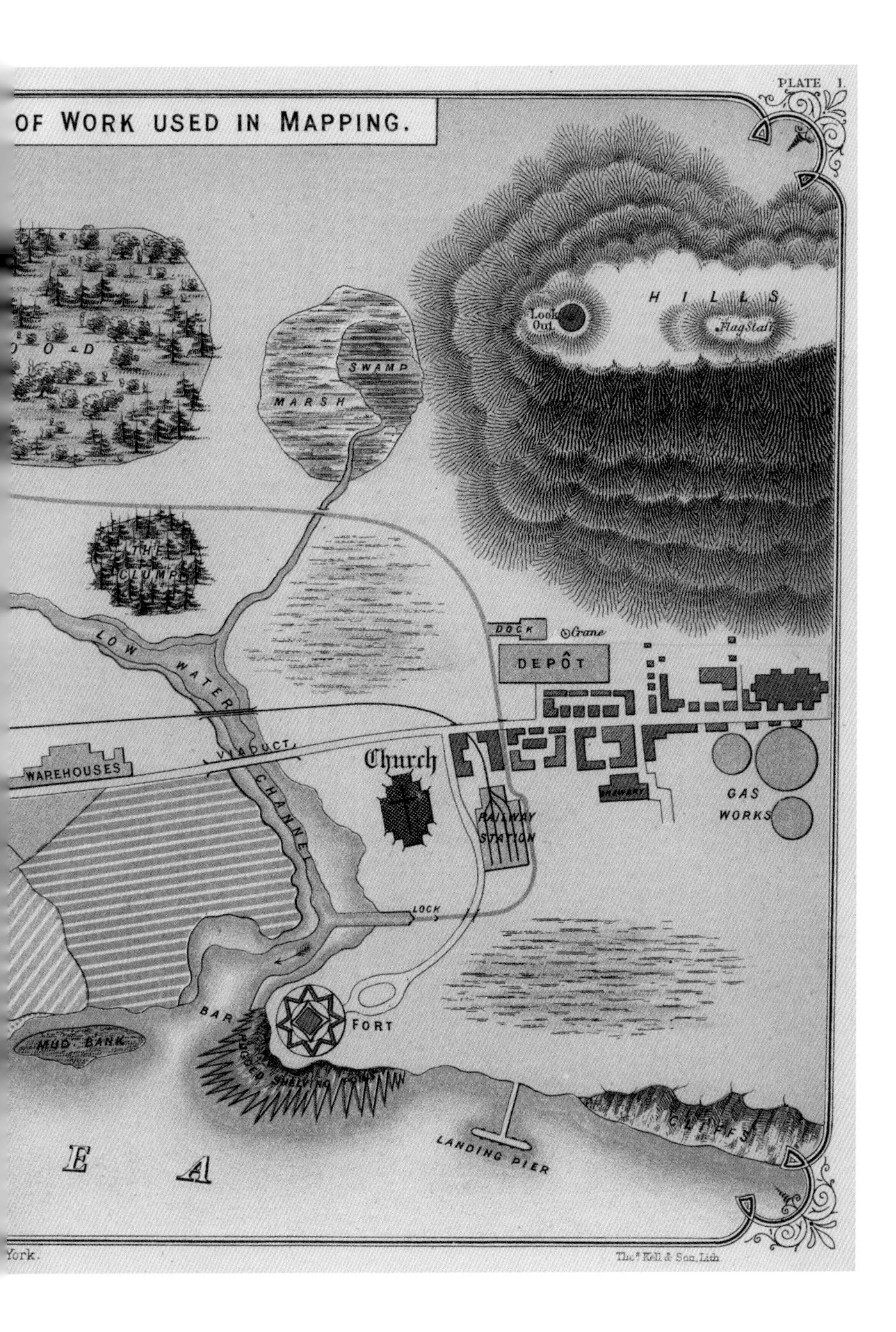

On old maps or new, map symbols work in particular ways that have persisted throughout the history of map making.

We can think about and more effectively make map symbols more systematically...

Ways to Think about Map Symbols

Everything on a map is a symbol. Map symbols, or signs, have two parts. The first is conceptual: an earthquake epicenter, a cold front, a sphere of influence. The second is a graphic mark. The mark is connected to the concept by a code or convention. For example, a cold front is often, though not always, shown as a blue line with regularly spaced triangles pointing in the direction of the front's movement:

Resemblance

Some map symbols look like particular data or concepts. A map showing the location of airports uses an airplane symbol. Airplanes make us think of airports.

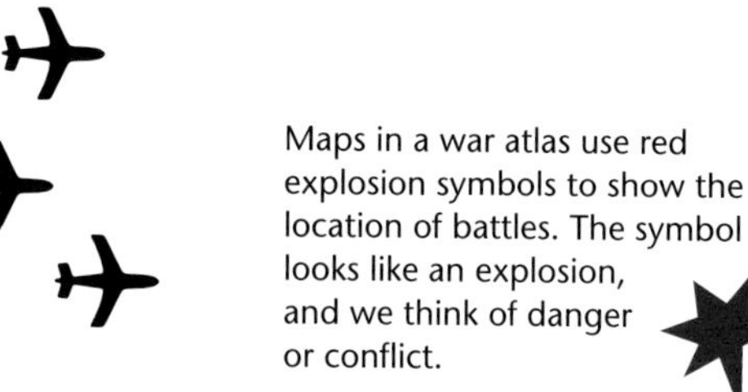

Maps in a war atlas use red explosion symbols to show the location of battles. The symbol looks like an explosion, and we think of danger or conflict.

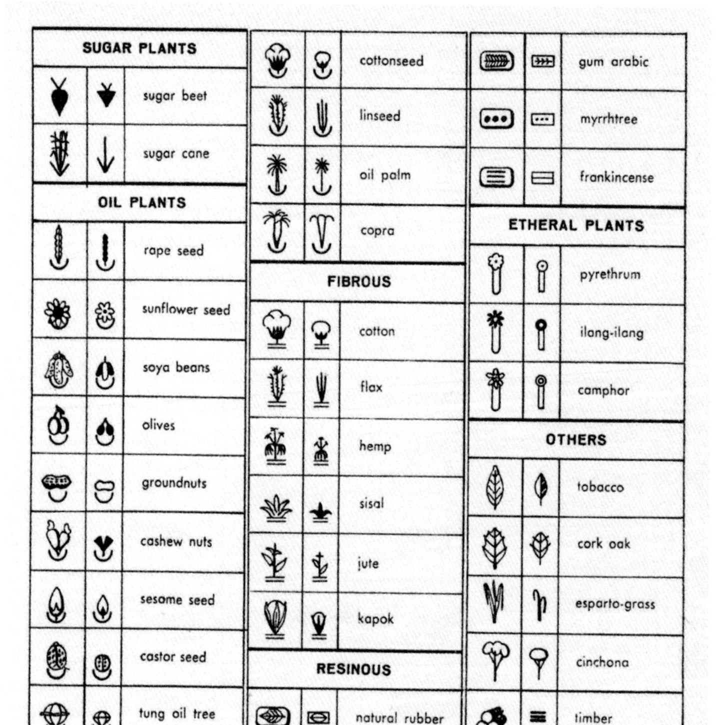

SUGAR PLANTS
sugar beet
sugar cane
OIL PLANTS
rape seed
sunflower seed
soya beans
olives
groundnuts
cashew nuts
sesame seed
castor seed
tung oil tree

(oil plants, continued)
cottonseed
linseed
oil palm
copra
FIBROUS
cotton
flax
hemp
sisal
jute
kapok
RESINOUS
natural rubber

(resinous, continued)
gum arabic
myrrhtree
frankincense
ETHERAL PLANTS
pyrethrum
ilong-ilong
camphor
OTHERS
tobacco
cork oak
esparto-grass
cinchona
timber

Fig. 11 – The signs for technical crops

Polish cartographer Lech Ratajski developed thousands of symbols for maps in the 1960s. In many cases these symbols attempted to capture the essence of resemblance to phenomena – here at two different levels of abstraction.

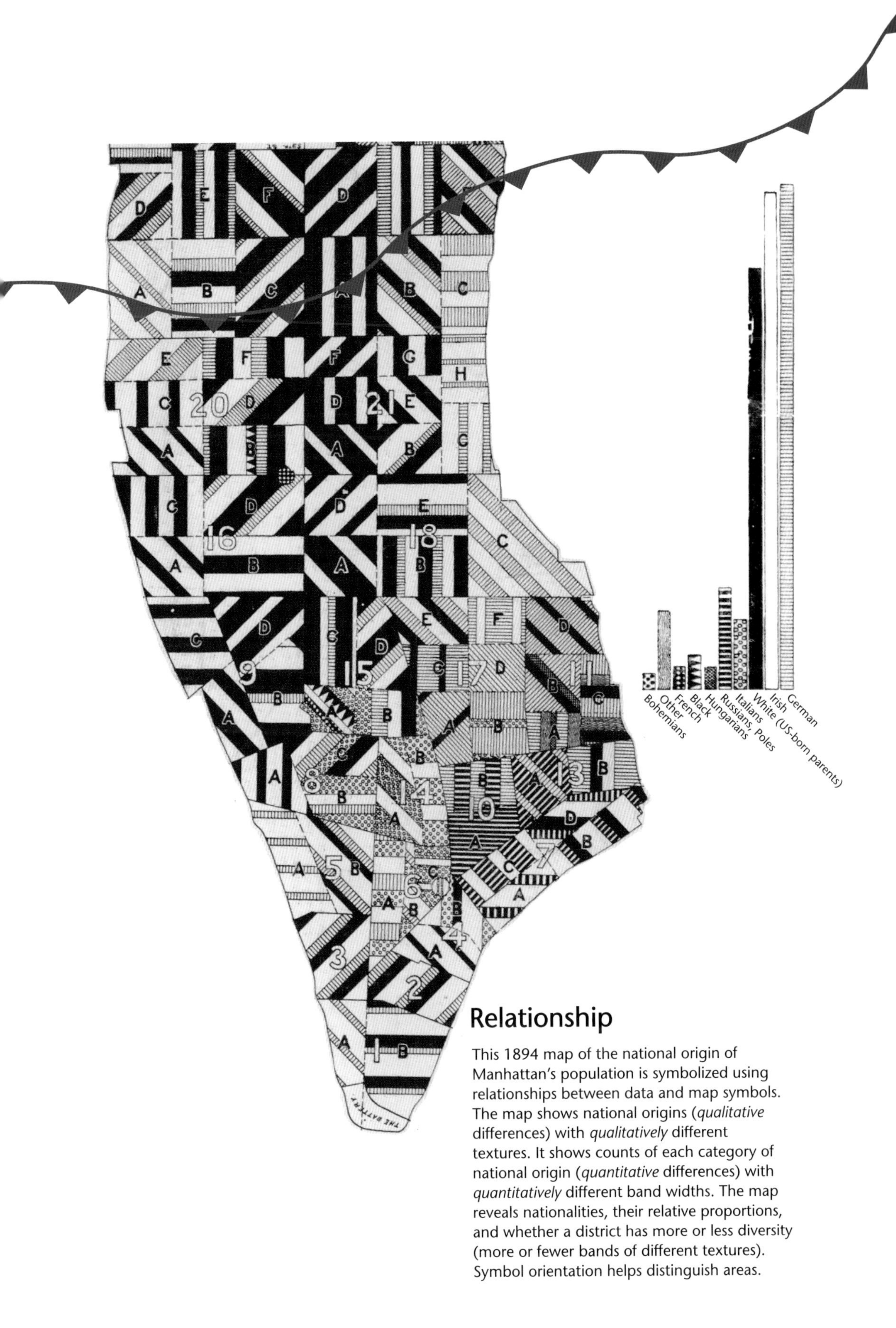

Relationship

This 1894 map of the national origin of Manhattan's population is symbolized using relationships between data and map symbols. The map shows national origins (*qualitative* differences) with *qualitatively* different textures. It shows counts of each category of national origin (*quantitative* differences) with *quantitatively* different band widths. The map reveals nationalities, their relative proportions, and whether a district has more or less diversity (more or fewer bands of different textures). Symbol orientation helps distinguish areas.

Convention

Of course, all map symbols are symbols by convention. But this is really obvious when symbols reveal cultural bias or don't resemble what they symbolize. Water, for example, is colorless, but it's blue on most maps because on a sunny day it reflects the color of the sky ... which sometimes is blue. But not usually. The blue of water is a convention, as in the map (below) where the intensity of the blue is used to indicate depth.

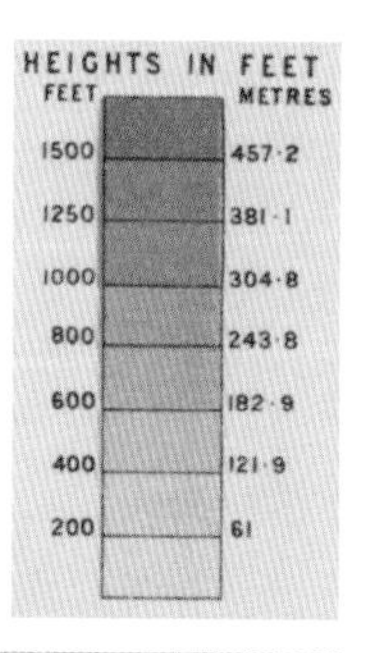

Similarly, the intensity of the earth tone is used to indicate elevation, as spelled out in the legend (left). This is known as layer tinting, which has long been used by map makers to characterize terrain.

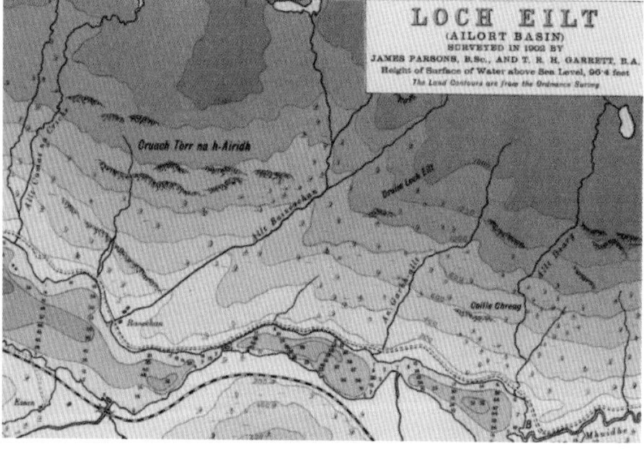

Below are more schemes for indicating elevations, lowest at the bottom. All are conventions you must learn in order to understand the maps that use them.

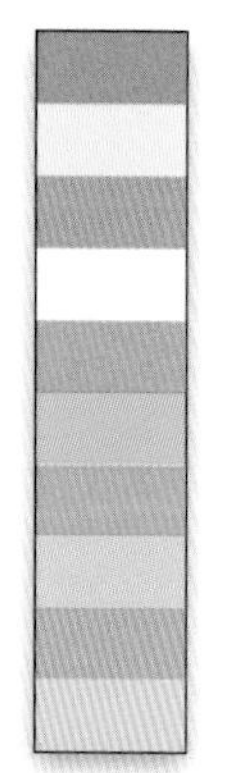

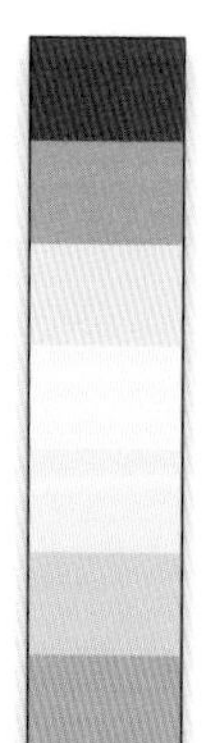

Difference

At the same time, symbols work by being different from other symbols. In addition to resemblance, Ratajski's crop symbols (three pages back) are distinct and different from each other.

You can also create symbols by differentiating a core idea, as with the road symbols below. The same approach is used in the old Latvian bridge and lock symbols to the right. Tilti!

Interstate Highway

State Highway

County Highway

Other Road

Under Construction

Under Construction

Under Construction

Through Road in Populated Place

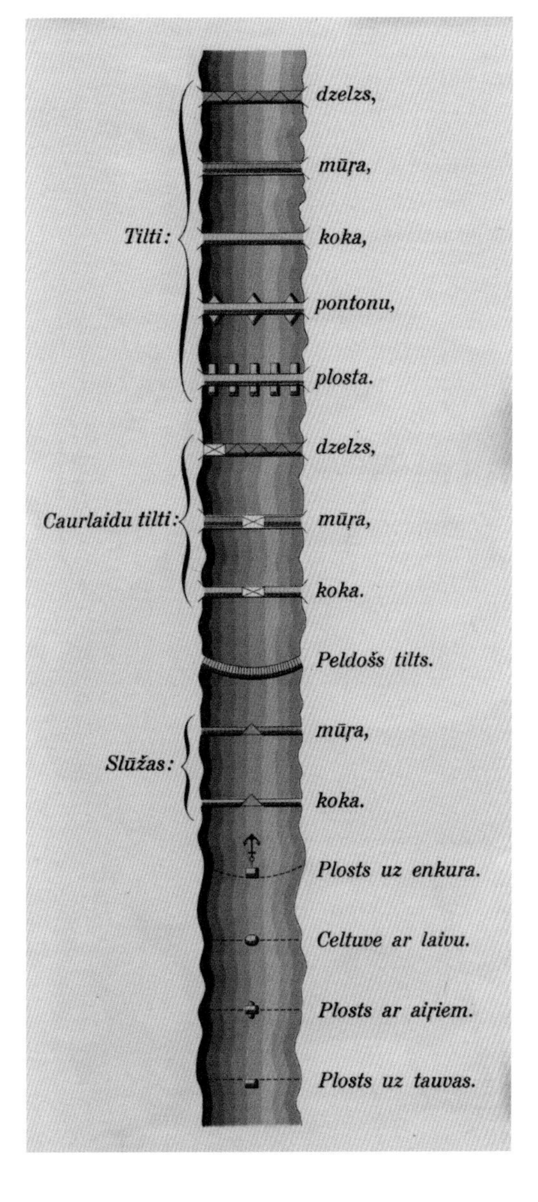

Standardization

Similar to convention, but required! We standardize symbols to clarify and reduce ambiguity. Often this is a property of but a single map – it's internally consistent – but it may officially hold across an entire series of maps. National topographic surveys and geological maps are examples.

Aeronautical charts are created by national agencies, but all belong to the International Civil Aviation Organization (ICAO). The ICAO maintains standards that all aeronautical charts follow. This enables pilots from any nation to use those of another.

On this aeronautical chart around Seattle, Washington, the smooth purple boundary outlines Class E Airspace with a floor 700 ft. above surface. The hashed purple, top, demarcates a Military Operation Area. It's easy to understand why the symbols on aeronautical charts have to be standardized.

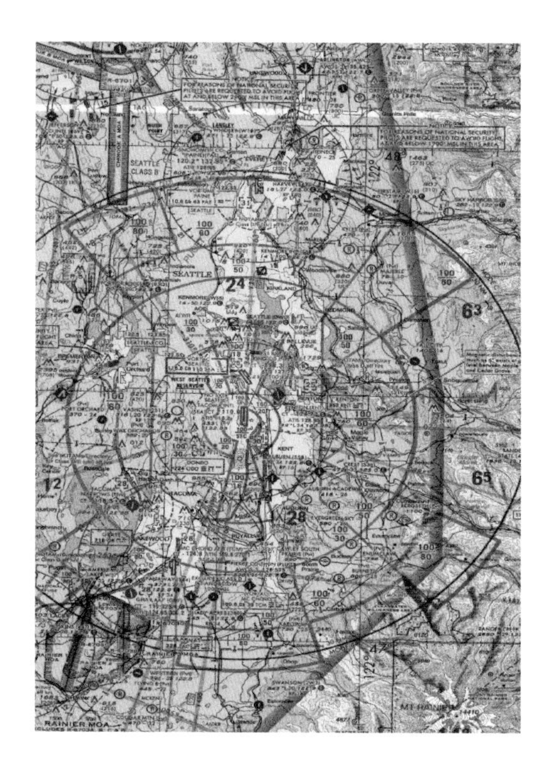

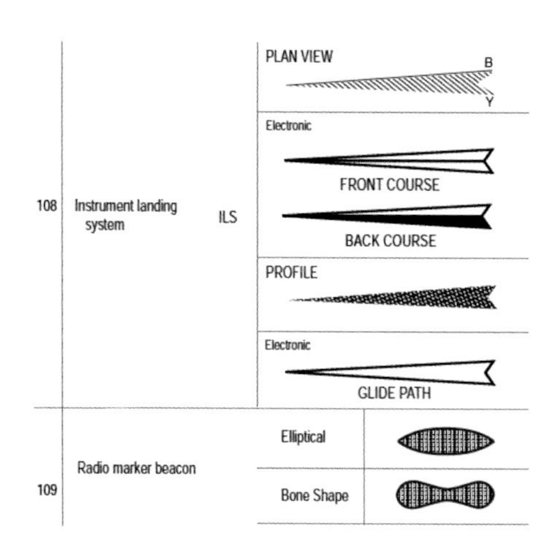

Aeronautical charts display information about the complicated system of flight, from airways to runways. These include symbols for instrument landing systems and radio marker beacons (left). Standardization makes symbol systems like these intelligible, learnable, and flying safer.

Unconvention

There are times when a conventional or standardized symbol would be the wrong choice. A map of the jack-o'-lanterns, one Halloween, on the porches of Boylan Heights homes in Raleigh, North Carolina, locate the porches in the neighborhood and show you each unique pumpkin.

Mapping Terrain

Terrain symbols, which show the vertical dimension of landforms (called relief), are complicated. They simultaneously involve resemblances, relationships, and conventions; they are at once generalized and individualized. Elevations of the surface are points, the symbols are often linear, yet the subject is an area. Terrain symbolization reveals the limitations of every attempt to systematize map making.

The volcano in the upper part of this diagram is depicted head-on in a profile view. For thousands of years this is the only way humans depicted mountains.

A section (below) through North America. Here, instead of looking at the mountains like a bird, we cut through and view relief from the side:

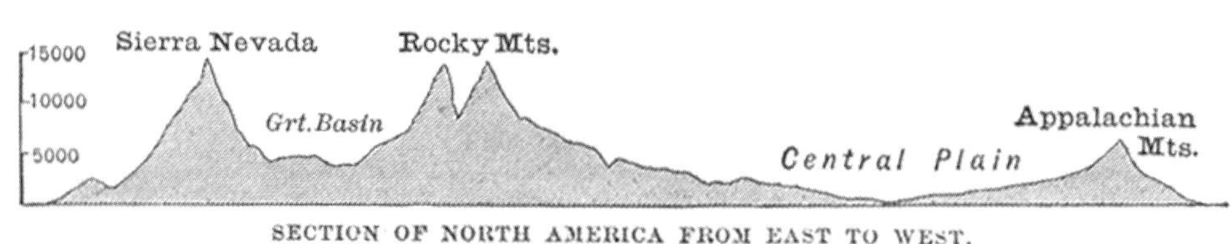

Hachures, developed around 1800, use line orientation and thickness to show relief from above. Eduard Imhof suggested that hachures should:

- Follow the direction of steepest gradient
- Be arranged in horizontal rows
- Have length correspond to the horizontal distance between assumed contours (say, 100′)
- Have a width thicker for steeper slopes
- Maintain density throughout the map

Contours are lines of constant elevation. They are more abstract than hachures, and became a common means of showing relief in the 19th and 20th century.

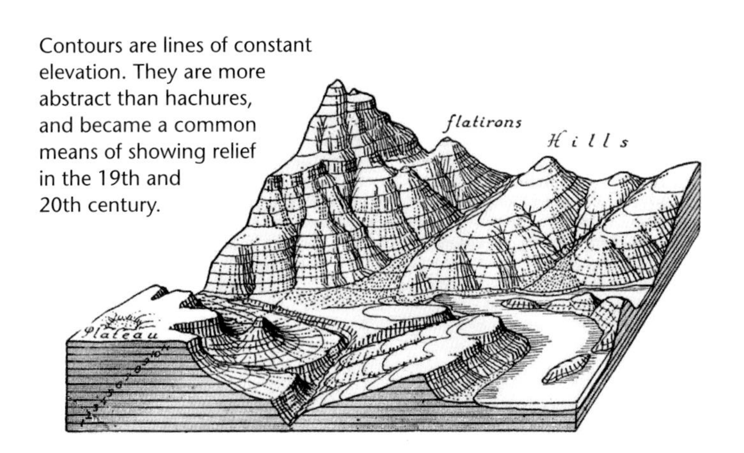

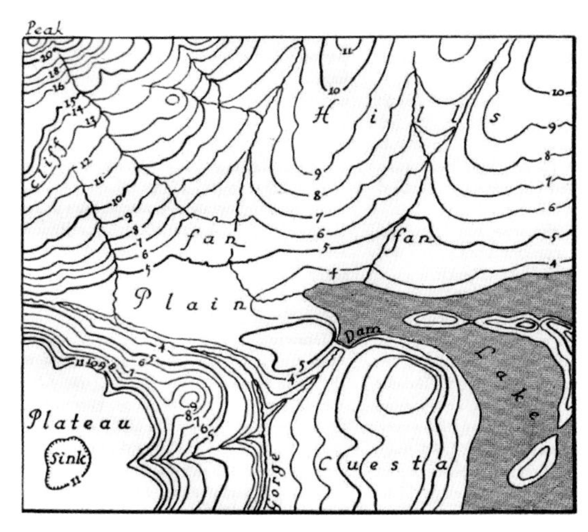

Every point along a contour is the same elevation.
Contours never cross one another.
Moving from one contour to another indicates a change in elevation.
There are always four intermediate (thinner) contours for every index (wider) contour.
The closer contours are to one another, the steeper the slope.
A series of closed contours (making a circle) is a hill.
Contours crossing a stream valley form a "V" pointing uphill and upstream.

People today use all these ways to symbolize hills, as these symbols (at right) drawn by college sophomores reveal. This range of symbols reflects how historical ways of showing terrain have seeped into our collective consciousness.

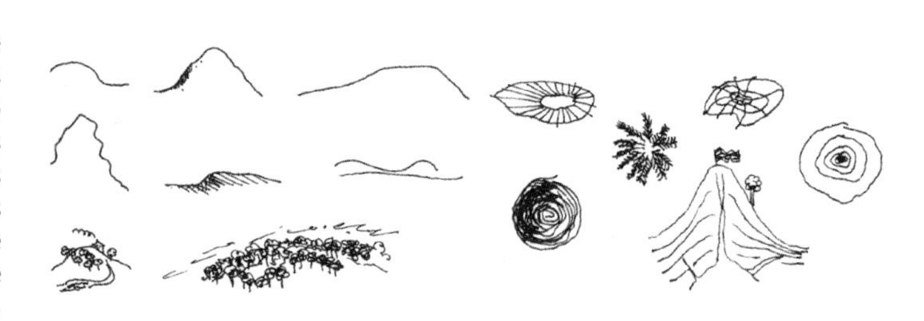

Andre's generic map of nowhere showing everything contains an assortment of map symbols for what Mr. Andre thought were the most common mappable subjects.

Water and water-related features on the map are blue, a **convention** most viewers will understand – even if water is typically not blue in reality.

Farm fields vary in color and texture, suggesting variations on a theme – **differences** – including furrowed crops (striped) versus meadows and different kinds of crops.

The symbols for church and chapel vary in **relationship** to their status: the church with larger type and a bolder shade.

The symbols for trees in the wood and the "clump" are based on **resemblance** to the actual phenomena – trees – they symbolize. The clump!

The **terrain** is shown with hachures, lines of varying width intended to simulate the slope and form of the hills and mountains.

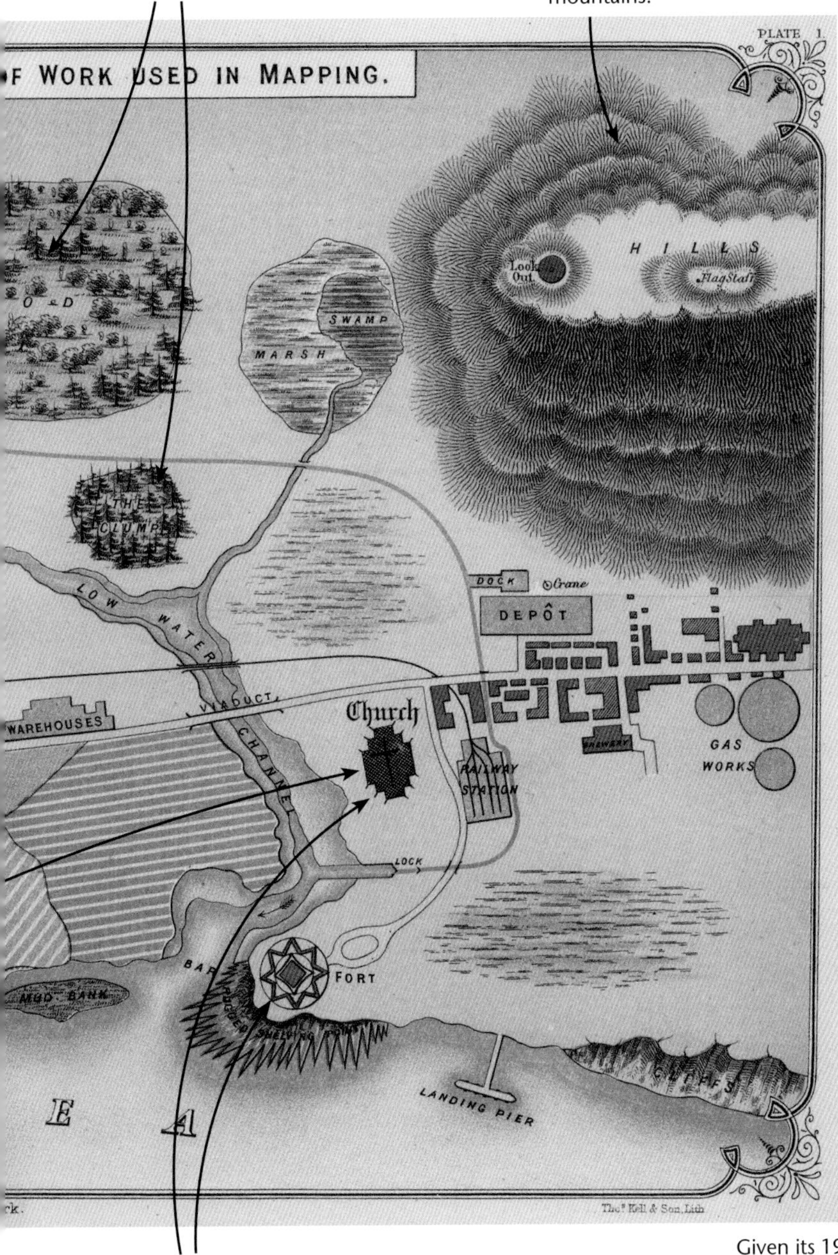

The fort and church are **standardized** – including the cross on the church. The railroad is *not* shown as the standard line with spaced cross tics – common to British maps of the 19th century. Standards are not always standard.

Given its 19th century origins, Andre's map strikes our 21st-century eyes as distinctly **unconventional**. Map symbols and their conventions are always changing.

But to look at the stars always makes me dream, as simply as I dream over the black dots of a map representing towns and villages. Why, I ask myself, should the shining dots of the sky not be as accessible as the black dots on the map of France? If we take the train to get to Tarascon or Rouen, we take death to reach a star. One thing undoubtedly true in this reasoning is this: that while we are alive we cannot get to a star, any more than when we are dead we can take the train.

Vincent van Gogh (1888)

First of all, you will need to choose the correct blue
to indicate water. This should not be too watery.
You must remember: people do not like wet feet.

Emily Hasler, Cartography for Beginners (2012)

...'tis the function of the symbol to glow with the idea symbolized...

Emma Thomas, The New and the Old Geography, *Pennsylvania School Journal* (1901)

The map isn't allowed to get too crazy.

Denis Wood (2014)

More...

Symbolization occupies large sections of any of the previously cited texts. A broader semiological approach to map and graphic symbols can be found in the work of Jacques Bertin, *Graphics and Graphic Information Processing* (1981) and *Semiology of Graphics* (2010). Eduard Imhof's *Cartographic Relief Presentation* (reprinted in 2007) is the classic text on terrain symbolization.

The magisterial volumes in the *History of Cartography* series, published by the University of Chicago Press, provide examples of a diversity of maps from different cultures throughout time.

The blogs for this book, makingmaps.substack.com and makingmaps.net, include an extensive collection of old map symbols, like those covered in this chapter. For the most part, modern cartography books don't devote much space to these kinds of symbols.

Two fun collections of maps by artists – revealing how they have appropriated map symbolization for their own purposes – are Katharine Harmon's *You Are Here: Personal Geographies and Other Maps of the Imagination* (2004) and *The Map as Art* (2010). Also check out *The Art of Cartographics: Designing the Modern Map* by Jasmine Desclaux-Salachas (2018) and *Walking and Mapping: Artists as Cartographers* by Karen O'Rourke (2016).

Sources: Boris Artzybasheff's map of major tropical diseases was published in *Life* (May 1, 1944). "Plan Shewing the Principal Characters of Work Used in Mapping" is reproduced from George Andre's *The Draughtman's Handbook of Plan and Map Drawing* (1891). Lech Ratajski's crop symbols are reproduced from the 1971 *International Yearbook of Cartography* courtesy of the International Cartographic Association. The map of ethnicity in New York City was published in *Harpers Weekly,* January 19, 1895. The Loch Eilt terrain map was published in Vol. 4, Plate 55, sheet 22 of the *Bathymetrical Survey of the Fresh-Water Lochs of Scotland,* 1897-1909. The seven elevation schemes are reproduced from Tom Patterson and Bernhard Jenny, "The Development and Rationale of Cross-blended Hypsometric Tints," *Cartographic Perspectives,* No. 69, 2011, courtesy of the authors. Latvian bridge and lock symbols reproduced from *Apzimejumi Merniecibas un Kulturtechniskiem Planiem* (Ministry of Agriculture, Riga, Latvia, 1928). The aeronautical chart and legend reproduced courtesy of the U.S. Federal Aviation Administration. The jack-o'-lantern map is part of Denis Wood's *Everything Sings* and is reproduced courtesy of Siglio Press. The mountain diagram and geologic profiles are from M.F. Maury's *Physical Geography* (University Publishing, 1898). The hachure map is from Thomas E. French's *A Manual of Engineering Drawing for Students and Draftsmen* (McGraw-Hill, 1911). Hachure rules are from Eduard Imhof's *Cartographic Relief Presentation* (Walter de Gruyter, 1982). The contour and terrain diagram is reproduced from Erwin Raisz's *Principles of Cartography* (McGraw-Hill, 1962) courtesy of McGraw-Hill. Contour line rules are based on Jim Riesterer, *Using a Contour Map* (Geospatial Training and Analysis Cooperative, 2008). Hill sign symbols were collected by students of Denis Wood in 1976.

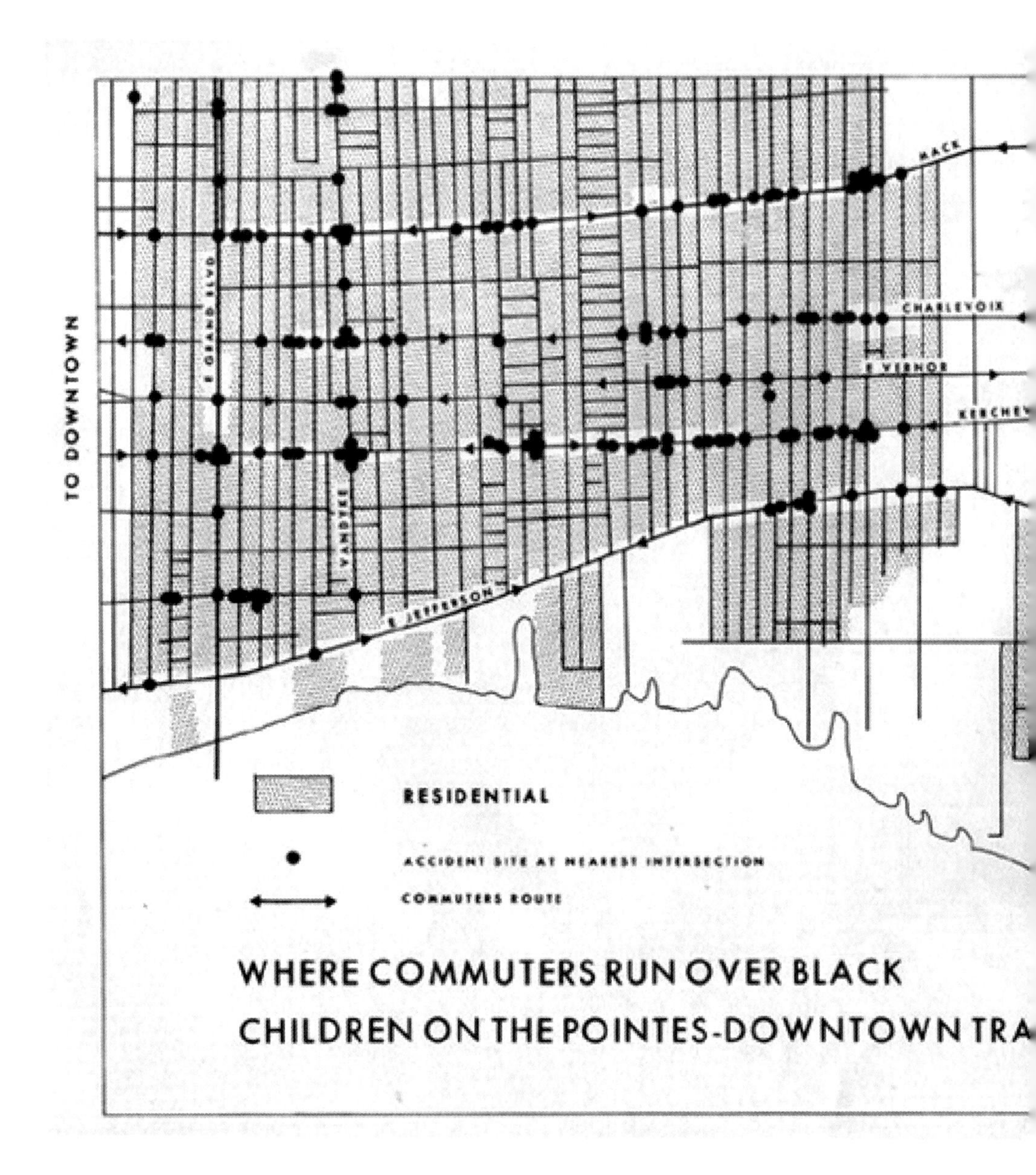
TO DOWNTOWN
E GRAND BLVD
MACK
CHARLEVOIX
E VERNOR
E JEFFERSON
RESIDENTIAL
ACCIDENT SITE AT NEAREST INTERSECTION
COMMUTERS ROUTE
WHERE COMMUTERS RUN OVER BLACK
CHILDREN ON THE POINTES-DOWNTOWN TRA

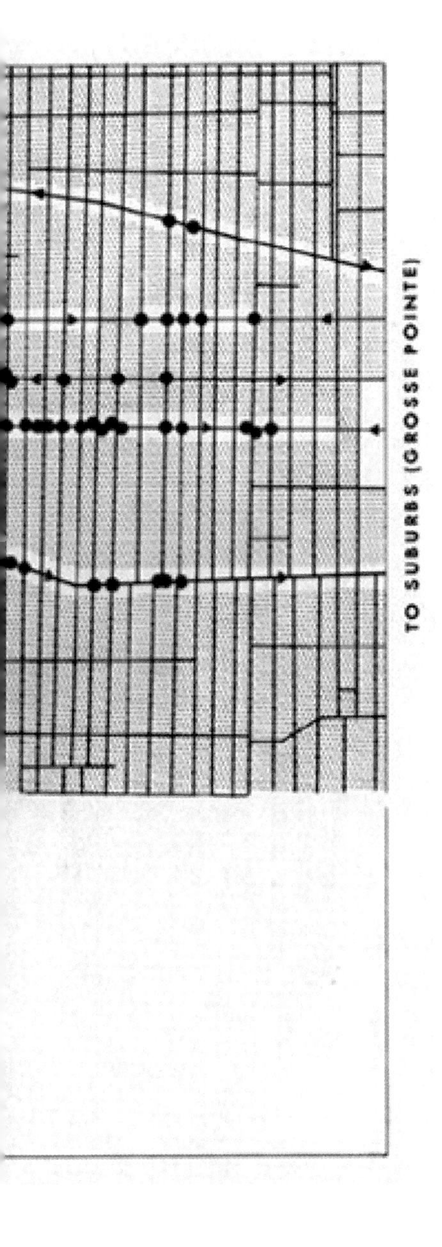

How abstract should your map be?

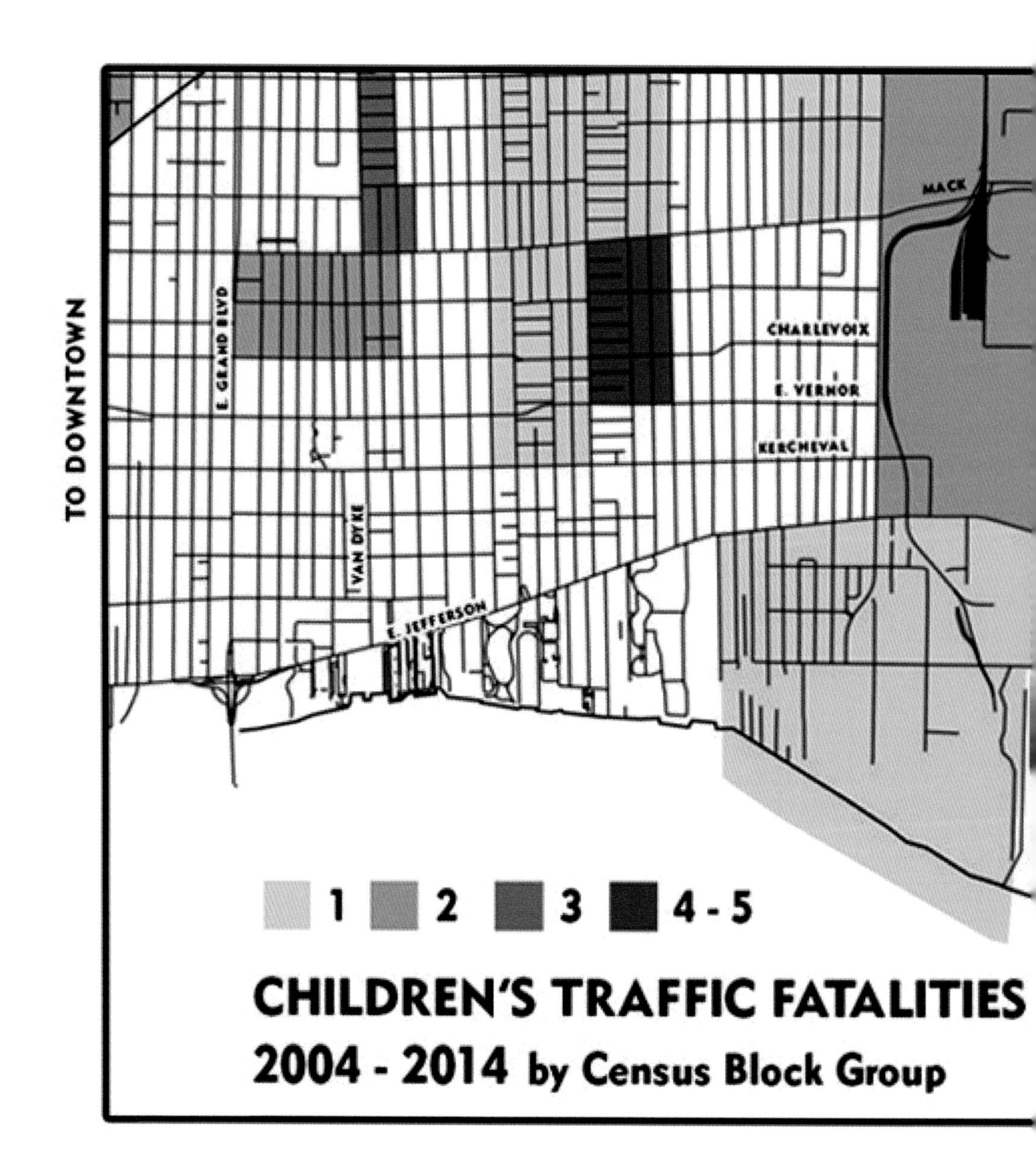
TO DOWNTOWN
MACK
E. GRAND BLVD
CHARLEVOIX
E. VERNOR
KERCHEVAL
VAN DYKE
E. JEFFERSON
1
2
3
4 - 5
CHILDREN'S TRAFFIC FATALITIES
2004 - 2014 by Census Block Group

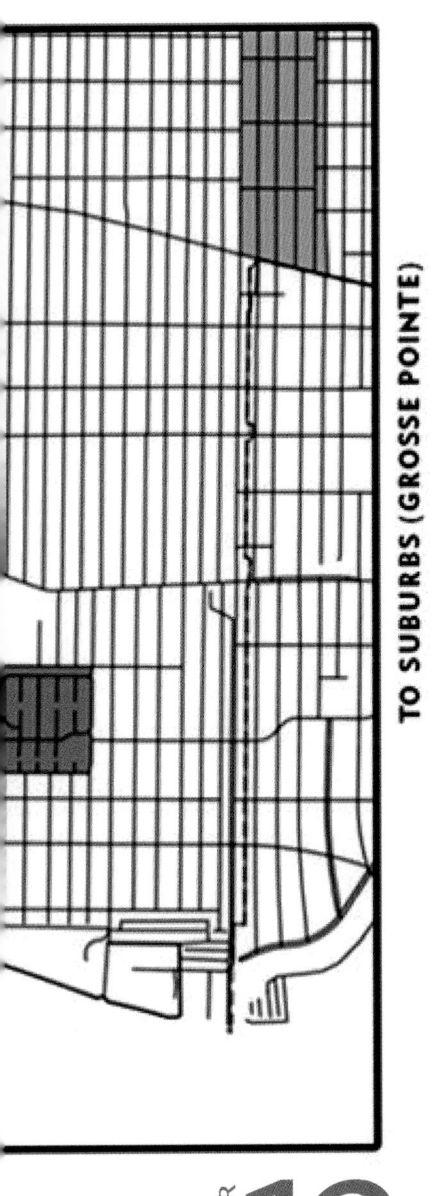

CHAPTER 12 Map Symbol Abstraction

Map symbols are always abstracted from real stuff in the environment. Gwendolyn Warren and Bill Bunge made a map of child pedestrian fatalities in Detroit (opening this chapter). It gets to the point in a visceral manner. More abstract is the map above showing child fatalities in U.S. Census block group areas. The visceral impact of the Warren-Bunge map is weakened. Map abstraction is powerful but desensitizing.

Ways to Think about Map Symbol Abstraction

We often take individual instances (such as traffic deaths) and stick them in areas (census, counties, provinces, etc.). Aggregation and abstraction are convenient for packaging and distributing and mapping data. They can help us think about data. But complexity, substance, and nuance can be masked as a consequence.

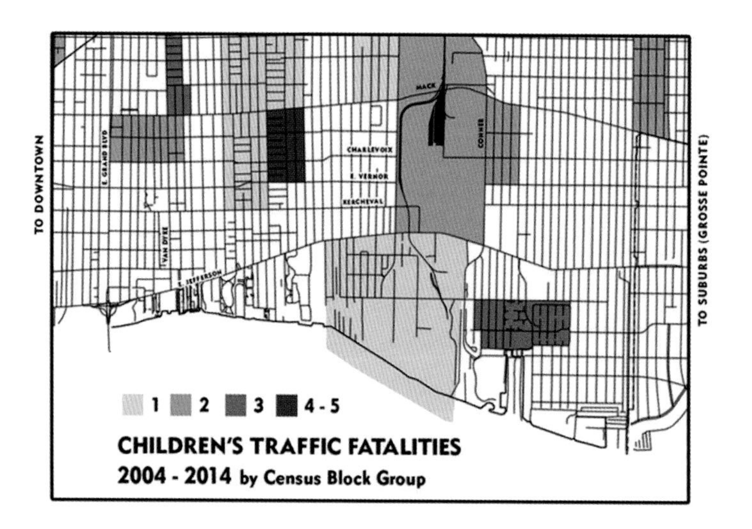

The children's traffic fatalities map by Census Block Group is, according to its author Alex Hill, a stinker. Accidents occur on roads, the dividing lines between block groups. The aggregated data are skewed by the address (side of the street) recorded for the accident.

Dangerous intersections are abstracted away. The title further contributes to this map of the banality of death. Inexcusable rates of pedestrian deaths are subdued. The map lacks the impact of Warren and Bunge's original.

This map of downtown Detroit, created by Ronald Horvath not long after the Warren-Bunge map, isolates the space devoted to roads, parking, and the service and sale of cars. All else is subordinate. It makes clear how much of Detroit is turned over to the car.

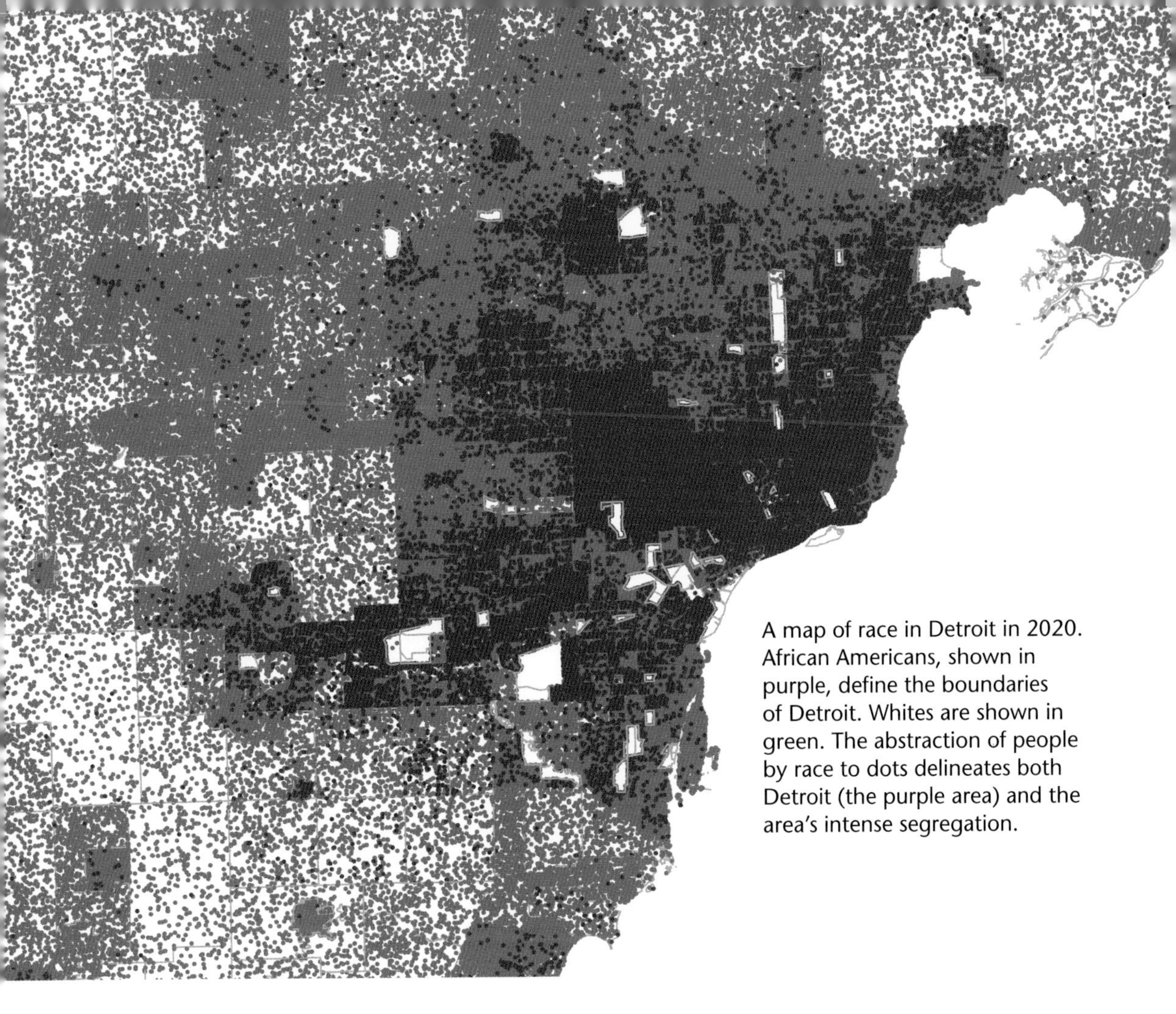

A map of race in Detroit in 2020. African Americans, shown in purple, define the boundaries of Detroit. Whites are shown in green. The abstraction of people by race to dots delineates both Detroit (the purple area) and the area's intense segregation.

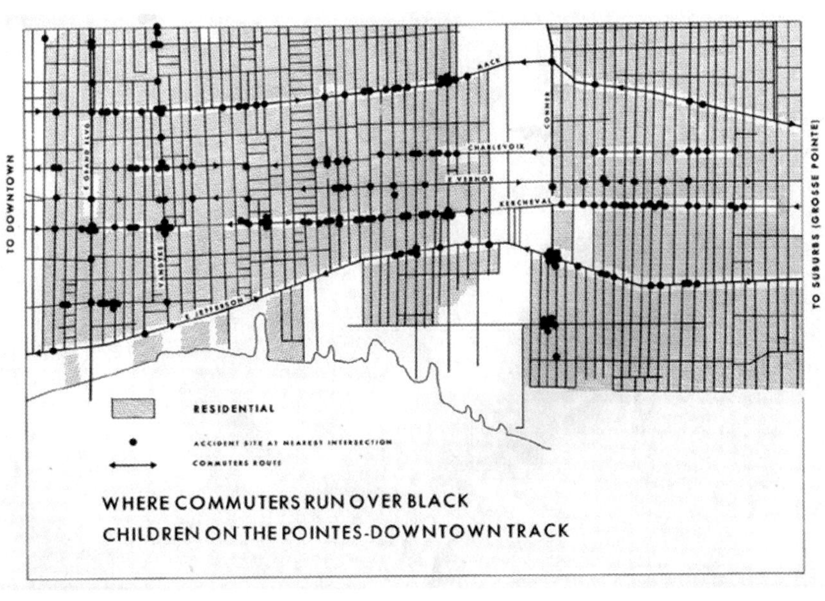

Warren and Bunge were right: commuters run over black kids in Detroit.

Visual Variables

Mappable data vary immensely. One approach to symbolizing your data, the visual variables, guides map symbolization by considering the characteristics of your data. Are your data at points, along lines, or in areas? Are your data qualitative or quantitative? Wedded to a careful consideration of the concepts behind your data, the visual variables serve to guide basic map symbol design.

Points, Lines, or Areas?

Most mappable data are at points (zero dimensions), lines (one dimension), or in areas (two dimensions).

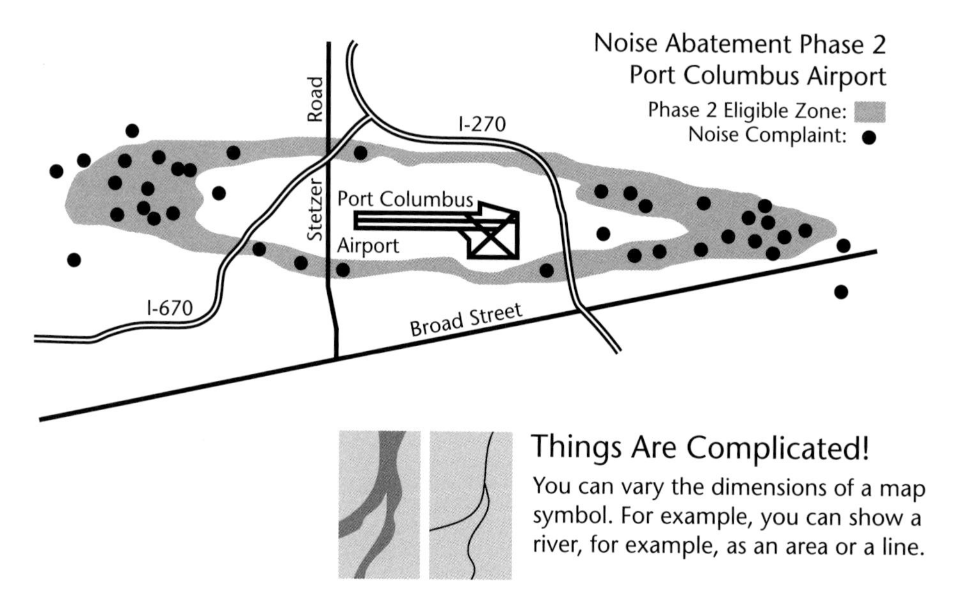

Things Are Complicated!

You can vary the dimensions of a map symbol. For example, you can show a river, for example, as an area or a line.

Qualitative or Quantitative?

Consider next whether your data vary in either quality (differences in kind) or quantity (differences in amount). Some data are not easily qualitative or quantitative.

Qualitative Data

- House location
- Border or boundary
- Land vs. water
- Religious denominations
- Animal species
- Plant types
- Sexual orientation
- Political affiliation

Quantitative Data

Ordinal: distinctions of order with no measurable difference between the ordered data: college rankings

Interval: distinctions of order with measurable differences among the ordered data but no absolute zero: temperature F°

Ratio: distinctions of order with measurable differences between the ordered data and an absolute zero: murder rate per country

Matching Data to Visual Variables

Particular visual variables may suggest important characteristics of your data. If your data are qualitative, choose a visual variable that suggests qualitative differences, such as shape or color hue. If your data are quantitative, choose a visual variable that suggests quantitative differences, such as size or color value. Some visual variables can be manipulated to suggest either qualitative or quantitative differences, such as texture.

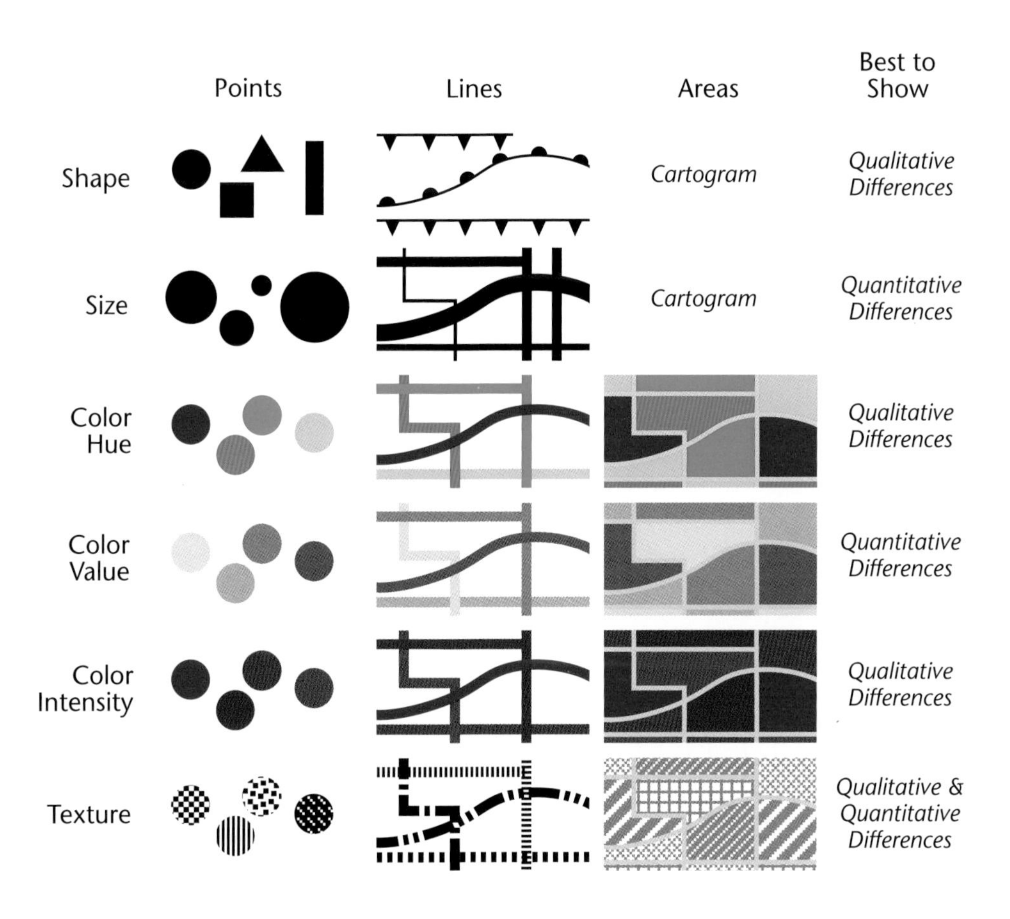

Shape

Map symbols with different shapes imply differences in *quality*. A square is not more or less than a circle, but is different in kind. Map symbol shapes can be pictorial or abstract.

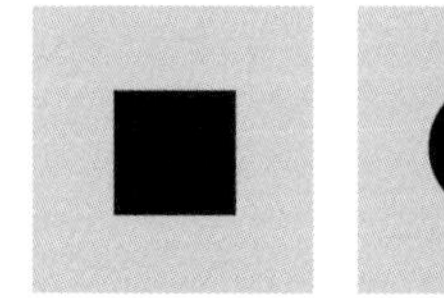

☹ use of shape

★ 53 – 103
♦ 31 – 42
● 21 – 30
■ 11 – 19
▲ 3 – 9

Shape is a poor choice for showing *quantitative* data. Using shape makes it hard to see the patterns on the map, as the symbols do not suggest the order (low to high) in the data.

☺ use of shape

Dominant Hate Group, 2022

Anti-Government / General Hate
Anti-Immigrant
Anti-LGBT
KKK
Neo-Confederate
Neo-Nazi
Racist Skinhead
White Nationalist

Alaska
Hawaii

Shape is a good choice for showing *qualitative* data. Different shapes suggest the qualitatively different groups.

Size

Map symbols with different sizes imply differences in *quantity.* A larger square implies greater quantity than a smaller square.

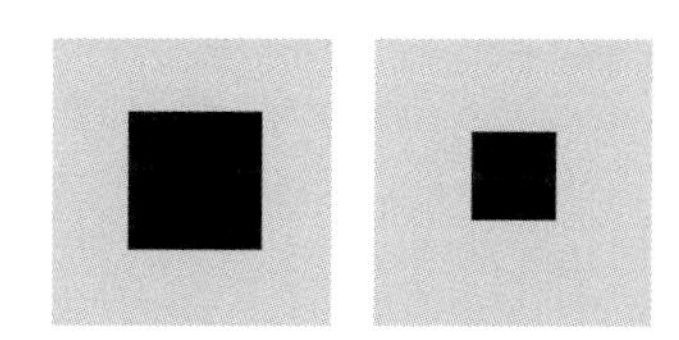

☺ use of size

- 53 – 103
- 31 – 42
- 21 – 30
- 11 – 19
- 3 – 9

Size is a good choice for showing *quantitative* data. The use of one symbol varying in size parallels the order in the data.

☹ use of size

- Anti-Government / General Hate
- Anti-Immigrant
- Anti-LGBT
- KKK
- Neo-Confederate
- Neo-Nazi
- Racist Skinhead
- White Nationalist

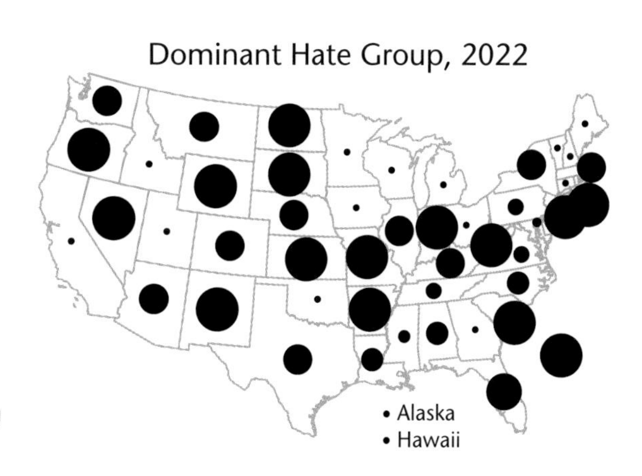

Size is a poor choice for showing *qualitative* data. Different sizes suggest order in the data rather than the qualitatively different groups.

Color Hue

Color hue refers to different colors such as red and green. Symbols with different hues readily imply differences in *kind.* Red is not more or less than green but is different in kind.

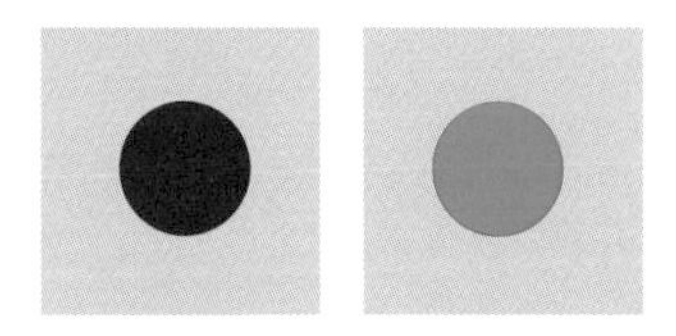

☹ use of color hue

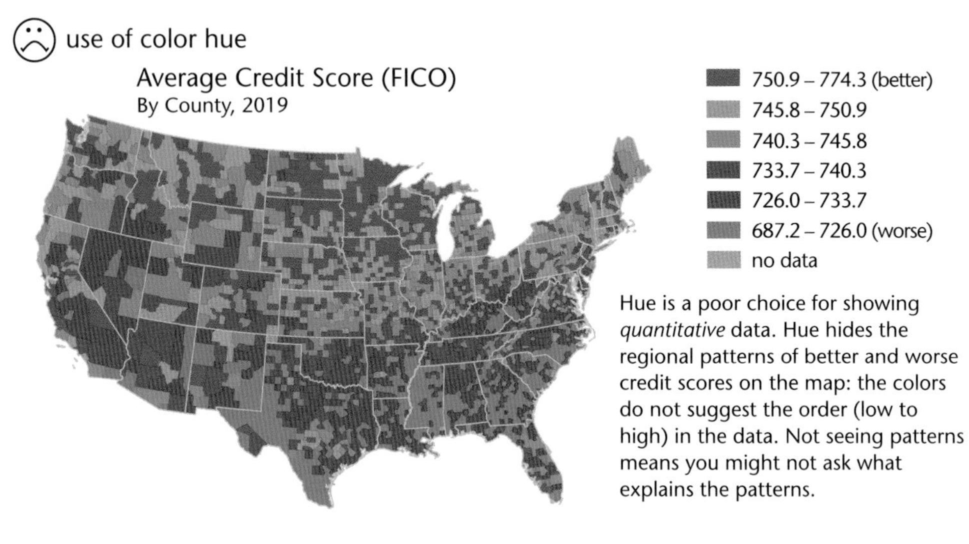

Hue is a poor choice for showing *quantitative* data. Hue hides the regional patterns of better and worse credit scores on the map: the colors do not suggest the order (low to high) in the data. Not seeing patterns means you might not ask what explains the patterns.

☺ use of color hue

Color hue is a good choice for showing *qualitative* data. Qualitatively different hues parallel the qualitatively different data. In this case blue means the state accepted additional federal healthcare funds, and red means the state has not accepted federal healthcare funds.

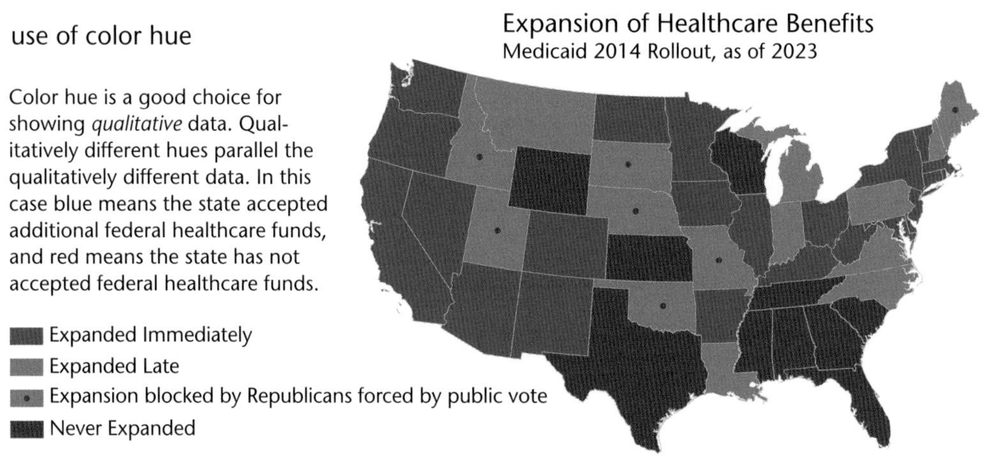

Color Value

Color value refers to different shades of one hue, such as dark and light red. Map symbols with different values readily imply differences in *quantity.* Dark red is more than light red.

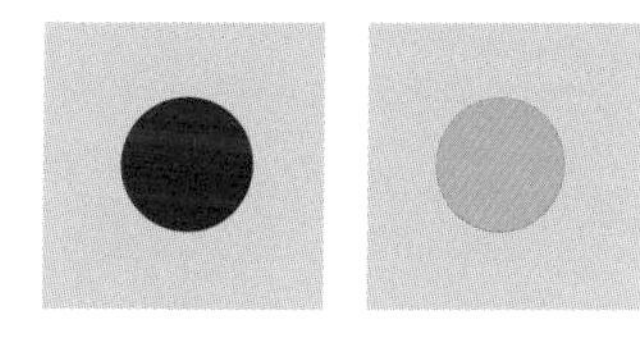

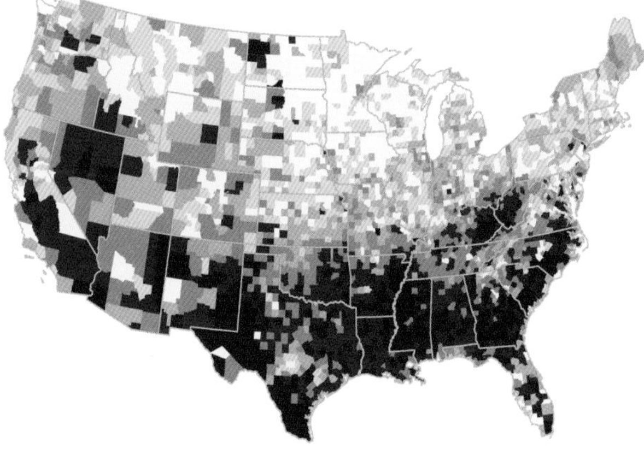

750.9 – 774.3 (better)
745.8 – 750.9
740.3 – 745.8
733.7 – 740.3
726.0 – 733.7
687.2 – 726.0 (worse)
no data

Value is a good choice for showing *quantitative* data. Value change parallels order in the data. It's easy to see poor credit scores, which make people's lives worse, concentrated in the U.S. South. Low scores are driven by medical debt, itself driven by the failure of some states to accept federal healthcare benefits.

use of color value

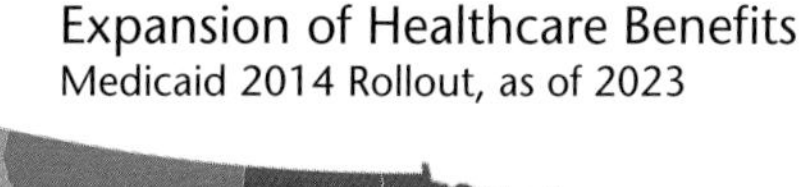

Color value is a meh choice for showing a *qualitative* difference between states that accepted and states that did not accept federal healthcare funds. Value suggests an ordered difference, which masks the distinctive (yes or no) difference in the data.

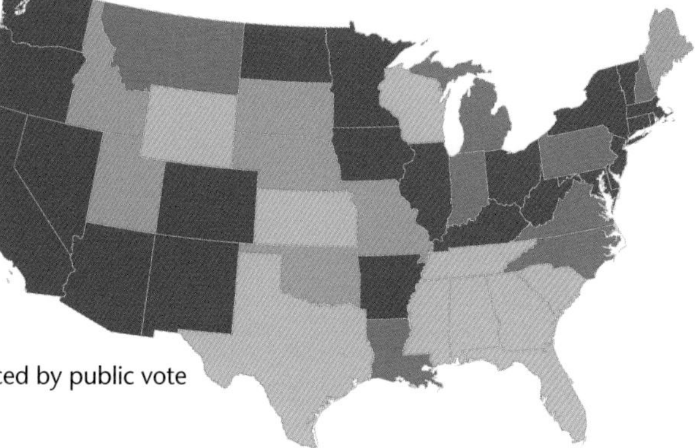

Expanded Immediately
Expanded Late
Expansion blocked by Republicans forced by public vote
Never Expanded

Color Intensity

Color intensity (saturation, chroma) is a subtle visual variable typically used to add nuance to hue or value. It can also be used to show binary (yes or no) data that are not really qualitative or quantitative.

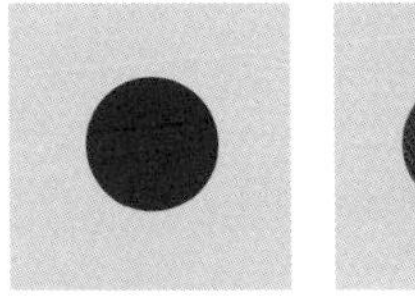

☹ use of color intensity

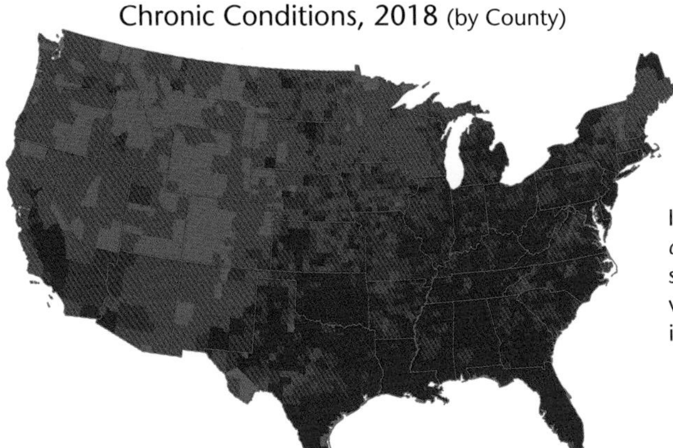

45% – 63.1%
40% – 45%
35% – 40%
25% – 35%
0% – 25%

Intensity is a poor choice for showing *quantitative* data. Intensity may suggest order, but due to the lack of variation in value, the sense of order is weak.

☺ use of color intensity

Percent of Medicare Users with Four or More Chronic Conditions, 2018 (by County)

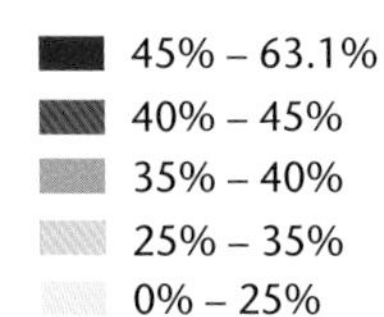

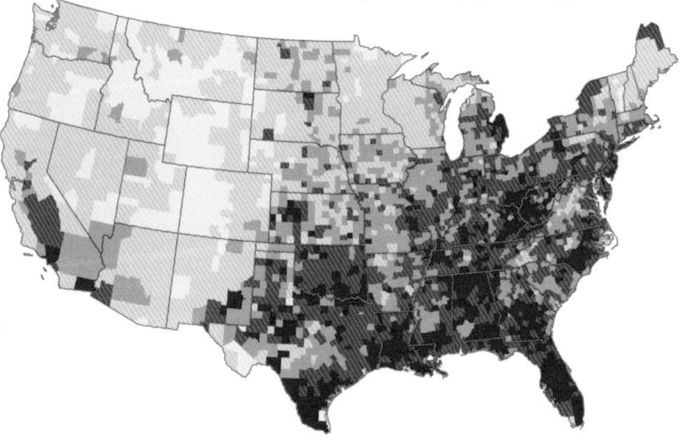

Intensity works well to add nuance to value. The intensity of the highest category is 100%, reduced by 20% for each lower category. This draws attention to, and visually emphasizes, the counties where Medicare users have the worst health.

Texture

Texture (pattern) can imply both qualitative (wetland vs. trees) and quantitative (coarse vs. fine) differences. Select textures so that they suggest the qualitative or quantitative character of your data.

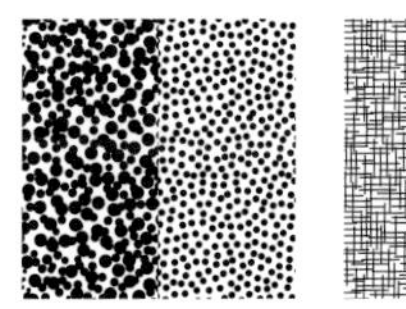

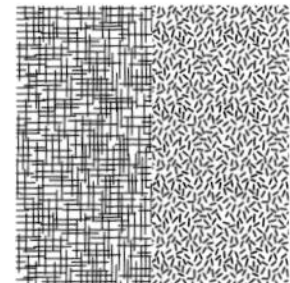

☹ use of texture

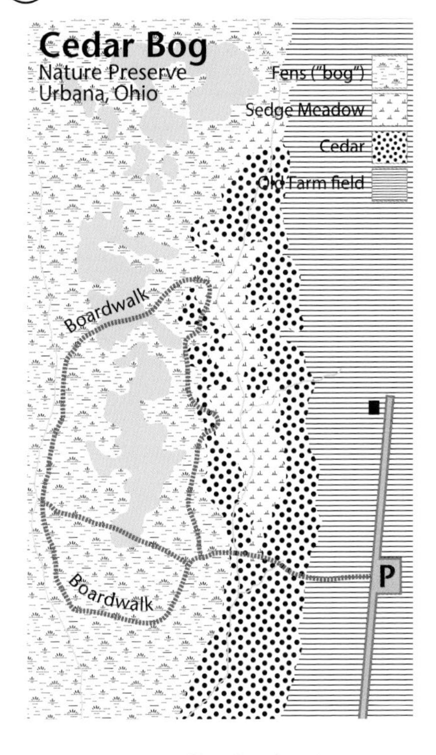

Texture is excellent for showing *qualitative* area data, such as vegetation types in a park. Default textures in software are often visually distracting – so modify them.

☺ use of texture

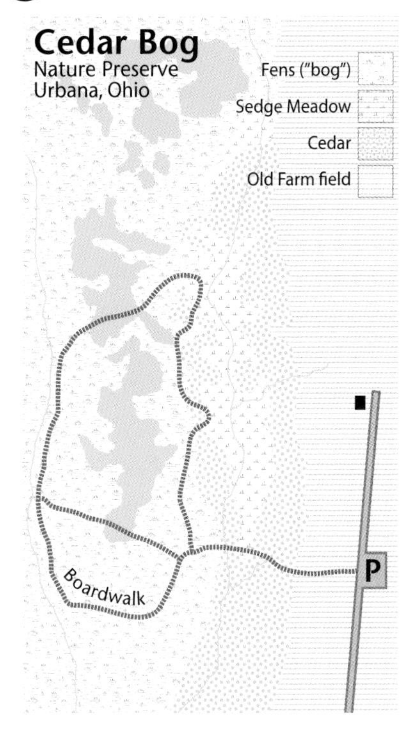

Modifications to texture using hue, value, and intensity distinguish the *qualitatively* different vegetation types and visually balance the textures with the boardwalk, title, and legend.

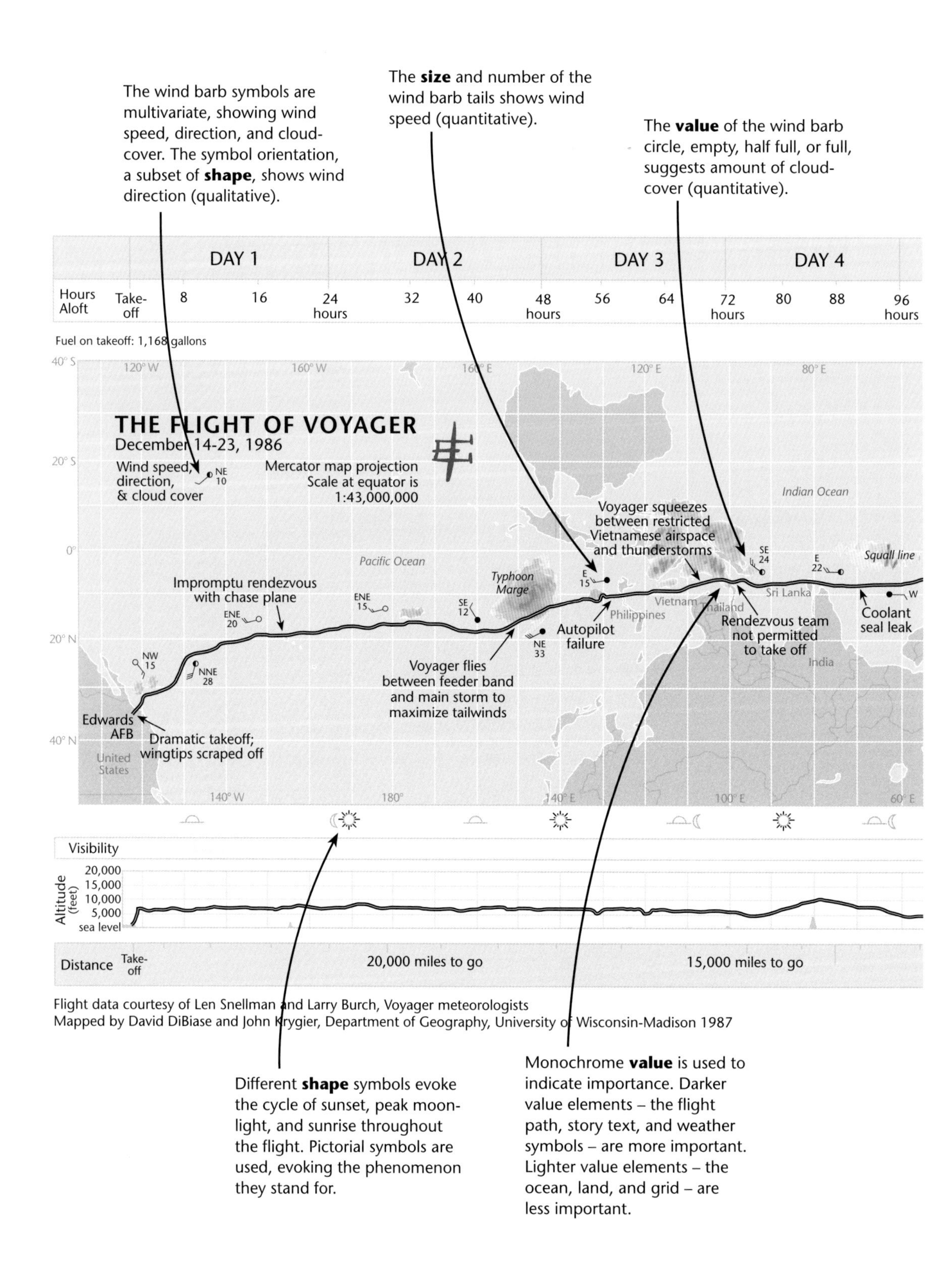
The wind barb symbols are multivariate, showing wind speed, direction, and cloud-cover. The symbol orientation, a subset of **shape**, shows wind direction (qualitative).
The **size** and number of the wind barb tails shows wind speed (quantitative).
The **value** of the wind barb circle, empty, half full, or full, suggests amount of cloud-cover (quantitative).
DAY 1
DAY 2
DAY 3
DAY 4
Hours Aloft
Take-off
8
16
24 hours
32
40
48 hours
56
64
72 hours
80
88
96 hours
Fuel on takeoff: 1,168 gallons
THE FLIGHT OF VOYAGER
December 14-23, 1986
Wind speed, direction, & cloud cover
Mercator map projection
Scale at equator is 1:43,000,000
Pacific Ocean
Indian Ocean
Impromptu rendezvous with chase plane
Typhoon Marge
Voyager squeezes between restricted Vietnamese airspace and thunderstorms
Squall line
Philippines
Vietnam
Thailand
Sri Lanka
India
Autopilot failure
Rendezvous team not permitted to take off
Coolant seal leak
Voyager flies between feeder band and main storm to maximize tailwinds
Edwards AFB
Dramatic takeoff; wingtips scraped off
United States
Visibility
Altitude (feet)
20,000
15,000
10,000
5,000
sea level
Distance
Take-off
20,000 miles to go
15,000 miles to go
Flight data courtesy of Len Snellman and Larry Burch, Voyager meteorologists
Mapped by David DiBiase and John Krygier, Department of Geography, University of Wisconsin-Madison 1987
Different **shape** symbols evoke the cycle of sunset, peak moonlight, and sunrise throughout the flight. Pictorial symbols are used, evoking the phenomenon they stand for.
Monochrome **value** is used to indicate importance. Darker value elements – the flight path, story text, and weather symbols – are more important. Lighter value elements – the ocean, land, and grid – are less important.

The **size** of the type suggests importance (quantitative). Larger-size type labels the more important phenomena, smaller type the less important.

The **shape, size,** and **value** of the flight path – wide with a white core – suggests the symbol is very important. The distinctive symbol shape, size, and value also tie the symbol on the main map to the symbol on the altitude map.

Value is used in the top and bottom data bar to provide overall balance and stability for the map. The gray tones provide a solid base and cap to the overall map.

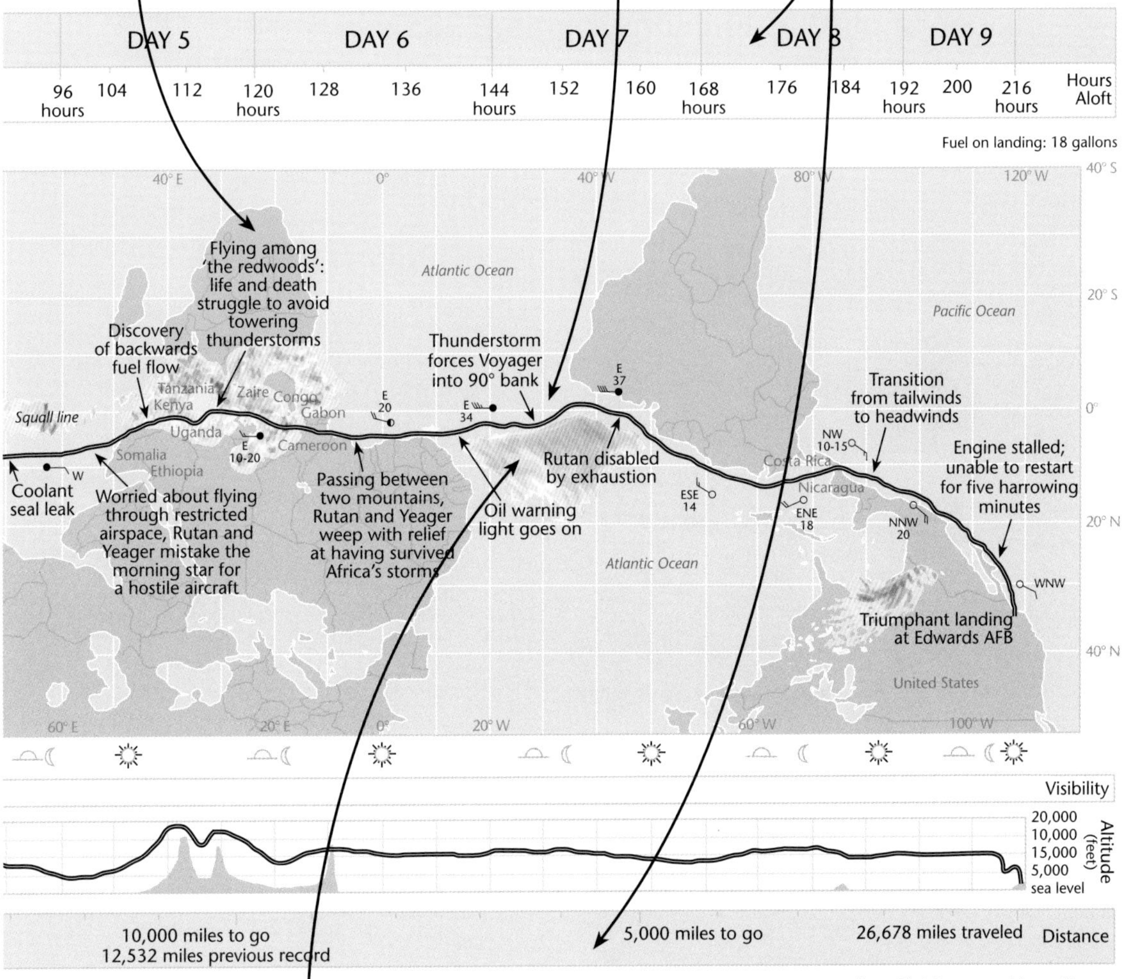

The **texture** of the storm symbols is used to suggest the turbulent character of the phenomenon that provided much of the excitement and terror of the voyage.

On the original printed Voyager map, red **color hue** was used for important map elements, including the route, storms, story type, and weather symbols. Red stands out more than other hues, signifying importance (quantitative).

Symbolizing Aggregate Data

One of the most common types of mappable data is associated with geographic areas: one data value is associated with each geographic area (countries, states, etc.). Symbolizing data grouped (aggregated) into geographic areas is complicated: the same data can be mapped as a choropleth, graduated symbol, cartogram, dot, or isopleth map. Choosing among these methods requires understanding what you are mapping and your goals for the map.

Incidence of AIDS in Pennsylvania, 1985

Geographer Peter Gould was mapping, predicting, and raising awareness of AIDS in 1985. Each county in Pennsylvania has one value: total AIDS cases. The same data are mapped with four methods below. AIDS is a contagious disease unevenly spread through geographic space. How do different map types affect how one thinks about the geography of AIDS?

Choropleth

Darker means more (lighter, fewer) cases.

Graduated Symbol

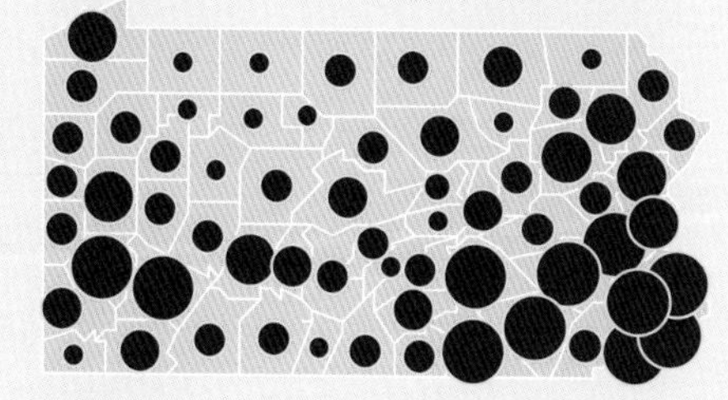

Larger circles mean more (smaller, fewer) cases.

Dot

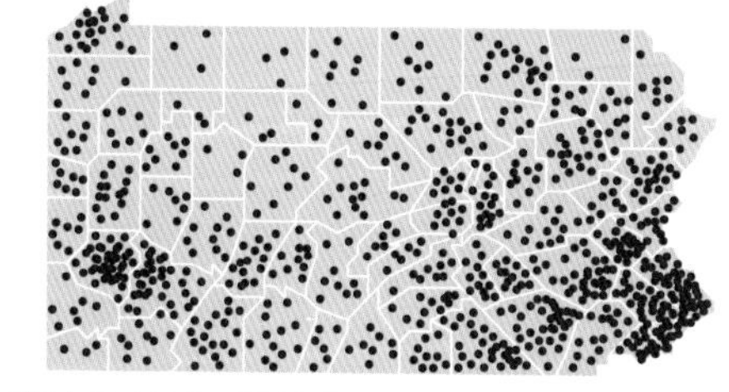

Density of dots reveals density of cases.

Isopleth

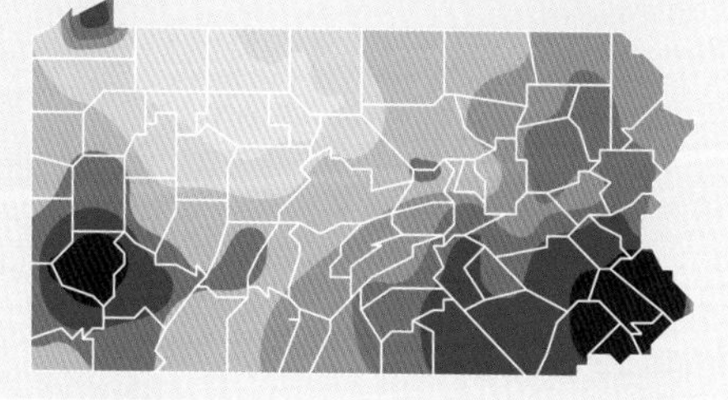

Darker means more (lighter, fewer) cases.

A **choropleth map** varies the shading of each area in tandem with the data value associated with it. This map suggests that the incidence of AIDS is uniform throughout each county, with potentially abrupt changes at county boundaries.

This map is useful for an AIDS educator working at the county level – suggesting that AIDS is everywhere in the county and everyone must take precautions – and to think about the relative count of AIDS in neighboring counties.

A **graduated symbol map** varies the size of a symbol centered on each area in tandem with the data value associated with it. This map suggests a single data value for each county. This map does not suggest a uniform distribution of AIDS cases within each county.

This map is useful for the official yearly health report for the State of Pennsylvania. The map indicates that there are AIDS cases, but the individual symbols subtly suggests that AIDS is contained and under control in most of the counties.

A **dot map** varies the number of dots in each area in tandem with the data value associated with it. On this map, one dot equals 30 AIDS cases. The location of dots does not show the specific location of AIDS cases; rather, the density of dots in each area represents more or fewer AIDS cases in the area.

This map looks like it is showing specific cases of AIDS and that AIDS cases are spread throughout each county in an roughly even manner. Both are false. Dot maps require explanation if they are to be used.

The **isopleth map** creates an abstract surface from the single data value for each area. Imagine a pin stuck in each county, where the height of the pin varies with the number of AIDS cases. Now imagine a surface created by a sheet laid over these pins. This map type is usually reserved for natural phenomena such as temperature.

This map suggests that AIDS is everywhere in the state and highly contagious. This map is useful for a statewide AIDS awareness campaign intended to scare people. Gould liked this one the best, and used it in his campaign.

Choropleth Maps

The choropleth map is one of the most common mapping techniques for data grouped into areas – counties, provinces, states, countries, etc. Choropleth maps vary the shading of each area along with the data value associated with the area.

Appropriate Data: Derived Data (Density, Rates), Sometimes Totals

Mapping total numbers with a choropleth map is usually not recommended, especially when the areas on the map vary in size. A large area may have more people simply because it covers a larger area.

If you map totals (bold, below) and classify the data, an area with 100 people will likely be in a different class than the area with 500 people. The visual difference between the areas on the map is the result of the unequal size of the areas.

100 people
10 sq. miles
10 per sq. mi.

500 people
50 sq. miles
10 per sq. mi.

Mapping totals with a choropleth map can be OK! A marketing company maps the total number of Polish-speaking U.S. citizens by county in the U.S., to assist in a plan to market Polish greeting cards. In this case, what is most important is where the most Polish-speaking folks are. But do consider a graduated symbol map for totals. It may better serve your needs.

Mapping derived data, like population per square mile, takes into account the varying size of areas on the map.

If you map densities (bold, below) and classify the data, both areas have 10 people per square mile and will be in the same class.

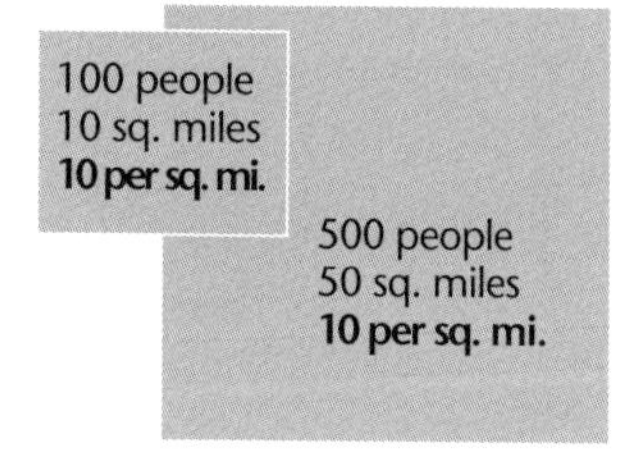

The goals for your map drive your choices! There is no absolute "best" kind of data for a choropleth map independent of your goals for the map. Be aware of the problem of mapping totals with a choropleth map, but if your goals require totals, just do it. Then create a graduated symbol map of the same data, and compare the maps. Please use your brain when making maps.

Choropleth Map Design: Hue, Value, Legend, and Boundaries

☹ hue and boundaries and title

Wisconsin Farm Density

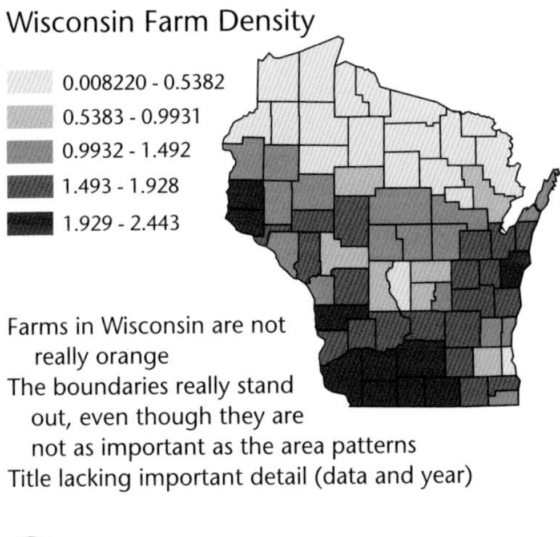

Farms in Wisconsin are not really orange
The boundaries really stand out, even though they are not as important as the area patterns
Title lacking important detail (data and year)

☹ value and boundaries and numbers

Wisconsin Farm Density
Farms per Square Mile
2020

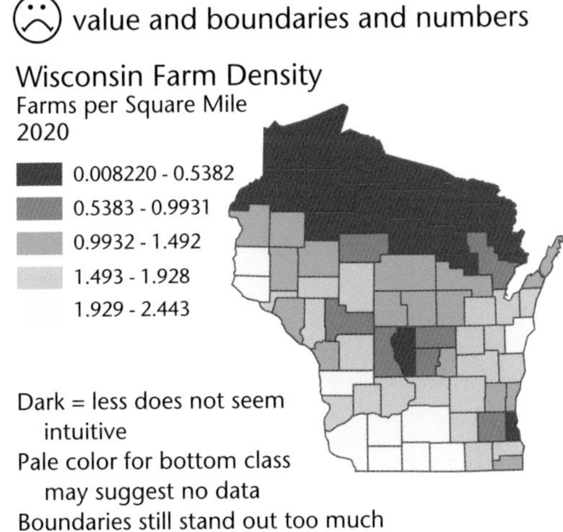

Dark = less does not seem intuitive
Pale color for bottom class may suggest no data
Boundaries still stand out too much
Are the precise numbers difficult to read?

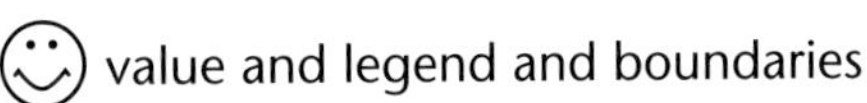

☺ value and legend and boundaries

Wisconsin Farm Density
Farms per Square Mile
2020

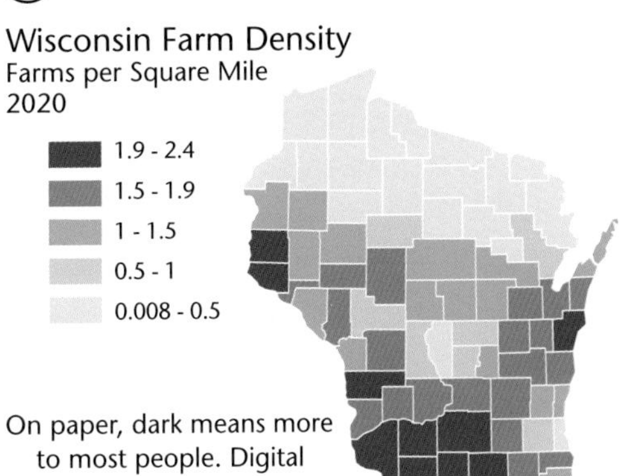

On paper, dark means more to most people. Digital displays: the opposite may be true
"Reversed out" boundaries (white or light gray)
Larger values at the legend top (high = more)
Boundaries less dominant here, but distinct
Reduced the precision of the legend numbers for ease of reading.

Multivariate Choropleth Maps

Multivariate choropleth maps combine two or more kinds of data on a single choropleth map, visually emphasizing the relationship between the different data. Bivariate maps are the most common. These maps follow the basic choropleth rules with a few exceptions.

Choosing the right data to map is vital. Will the two variables correlate in meaningful ways that will appear on the map? What point do you think the map will make about these two variables?

Derived data (density, rates) are best, as with single variable choropleth maps.

Three data classes work well; more classes generate complicated maps that are difficult to interpret, which may be OK.

Data classification schemes can vary: chose one that makes sense for your data.

Color choice makes or breaks a multivariate choropleth map. Look up optimal color schemes for the best options.

Use a qualitatively different hue for the two data variables. Vary value to distinguish low and high

Bivariate color schemes combine the pair of hue and value ranges into a grid

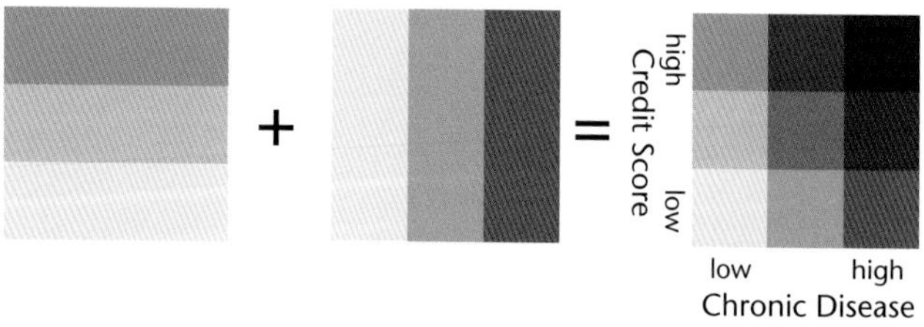

Include a legend that explains how to understand the map and its data, using diagrams like these

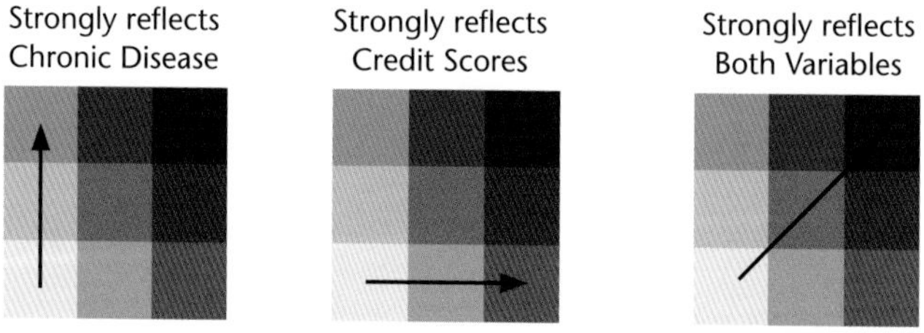

The plan was to use dark for lower credit scores, to highlight counties with low credit scores and high chronic disease. The software default was dark for higher scores. The resulting bivariate map produced an intriguing pattern...

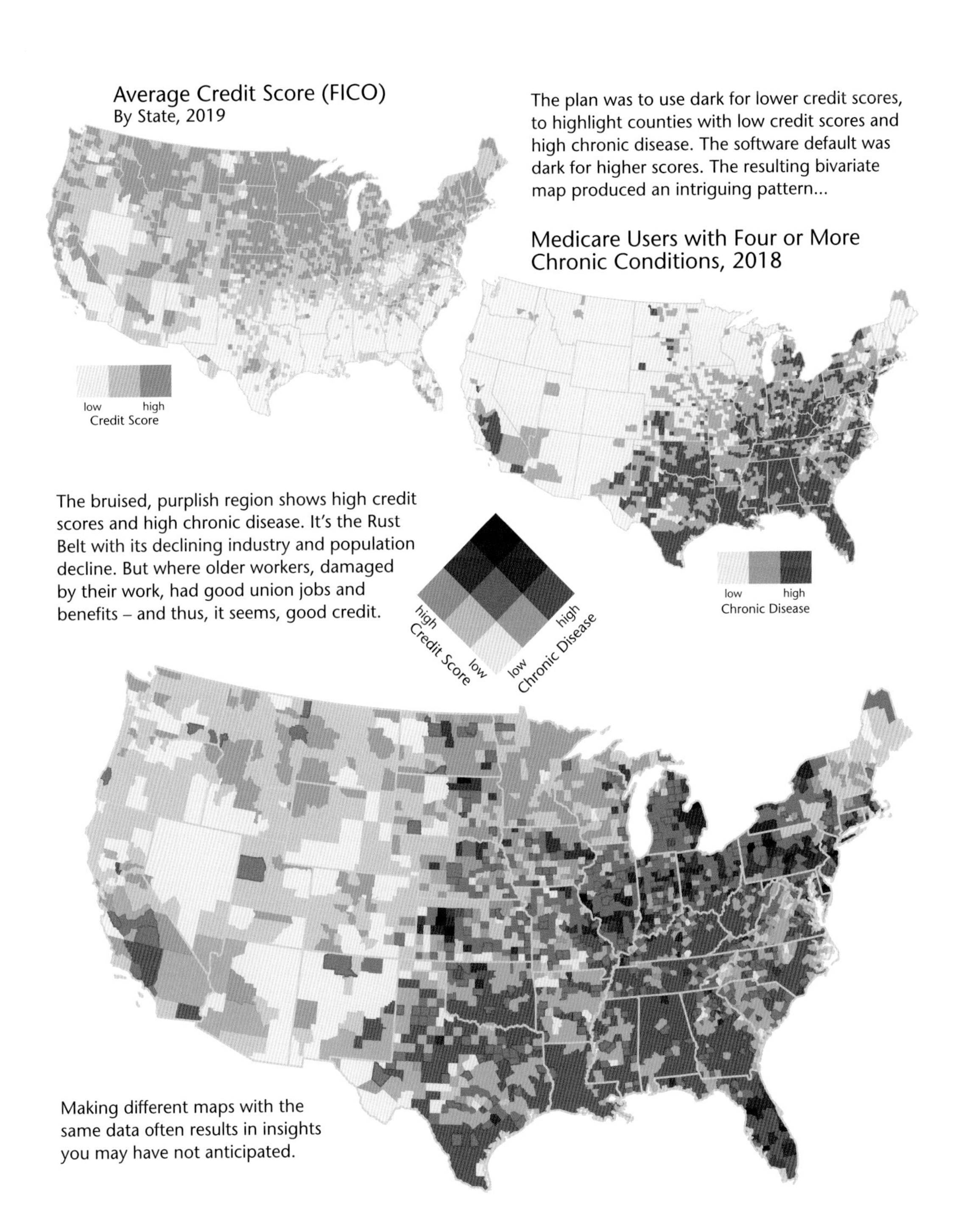

The bruised, purplish region shows high credit scores and high chronic disease. It's the Rust Belt with its declining industry and population decline. But where older workers, damaged by their work, had good union jobs and benefits – and thus, it seems, good credit.

Making different maps with the same data often results in insights you may have not anticipated.

Graduated Symbol Maps

The graduated symbol map varies the size of a single symbol, placed within each geographic area, in tandem with the data value associated with the area.

Appropriate Data: Totals, Sometimes Derived Data (Densities, Rates)

Mapping derived data with graduated symbols is usually not recommended. Graduated symbols readily imply magnitude rather than density or rates.

If you map density (bold, below), the symbols imply no magnitude difference in population between the two areas.

100 people
10 sq. miles
10 per sq. mi.

500 people
50 sq. miles
10 per sq. mi.

Use totals for graduated symbol maps. Mapping totals with graduated symbols suggests the magnitudes inherent in the data.

If you map totals (bold, below), the map implies a difference in total population between the two areas.

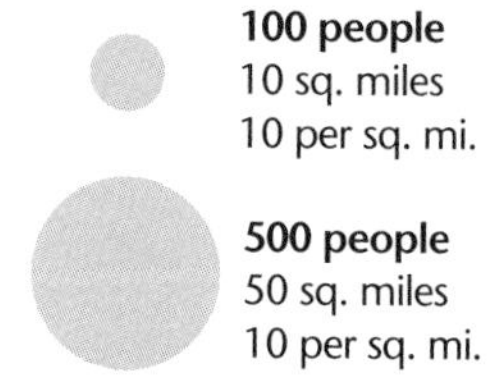

Mapping derived data with a graduated symbol map can be OK! A map of global coffee consumption might, for example, use graduated coffee cups to show the percentage of total coffee consumption in each country then relate the map to a pie chart showing the same data. Coffee and pie, delicious.

Consider a choropleth map for these data; it may serve your needs better. But the graduated coffee cups may be too sweet to pass up.

The goals for your map drive map choices! There is no absolute "best" kind of data for a graduated symbol map – independent of your goals. Be aware of the problem of mapping derived data with a graduated symbol map, but if your goals require it, it's OK to do it.

Graduated Symbol Map: Classification

Classified Legend

 8,001 – 10,000 persons

 5,001 – 8,000 persons

 1,001 – 5,000 persons

Less than 1,000 persons

Classified graduated symbol maps use standard classification schemes. Assign one symbol size for each class.

Less data detail
Easier to match particular symbol on map to legend
Easier to see distinct classes in data

Unclassified Legend

 9,000 persons

 6,500 persons

 2,500 persons

500 persons

Unclassified graduated symbol maps scale each symbol to each value. Legend should include *representative* symbol sizes.

More data detail
Harder to match particular symbol on map to legend
Harder to see distinct classes in data

Graduated Symbol Map: Symbol Design

Squares, Triangles

Less compact symbol
Edgy visual impression good for edgy phenomena

Circles

More compact symbol
Smooth visual impression good for mellow phenomena

Volumetric Shapes

Visually attractive
Suggest volumetric phenomena, avoid these stinkers (unless you are mapping volumes)

Pictographic Shapes

Visually attractive
Easy to understand
Potentially cute and distracting

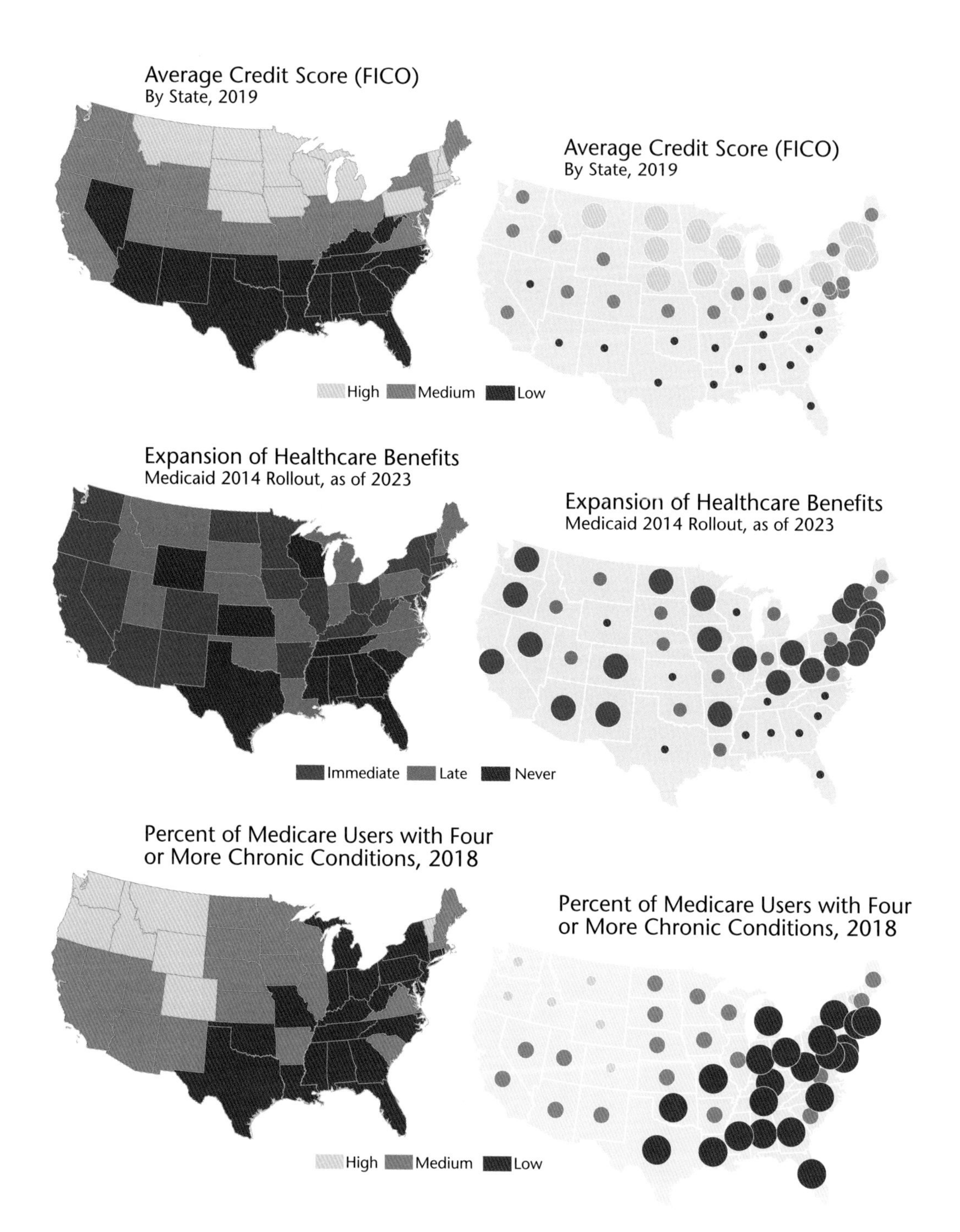
Average Credit Score (FICO)
By State, 2019
High
Medium
Low
Average Credit Score (FICO)
By State, 2019
Expansion of Healthcare Benefits
Medicaid 2014 Rollout, as of 2023
Immediate
Late
Never
Expansion of Healthcare Benefits
Medicaid 2014 Rollout, as of 2023
Percent of Medicare Users with Four or More Chronic Conditions, 2018
High
Medium
Low
Percent of Medicare Users with Four or More Chronic Conditions, 2018

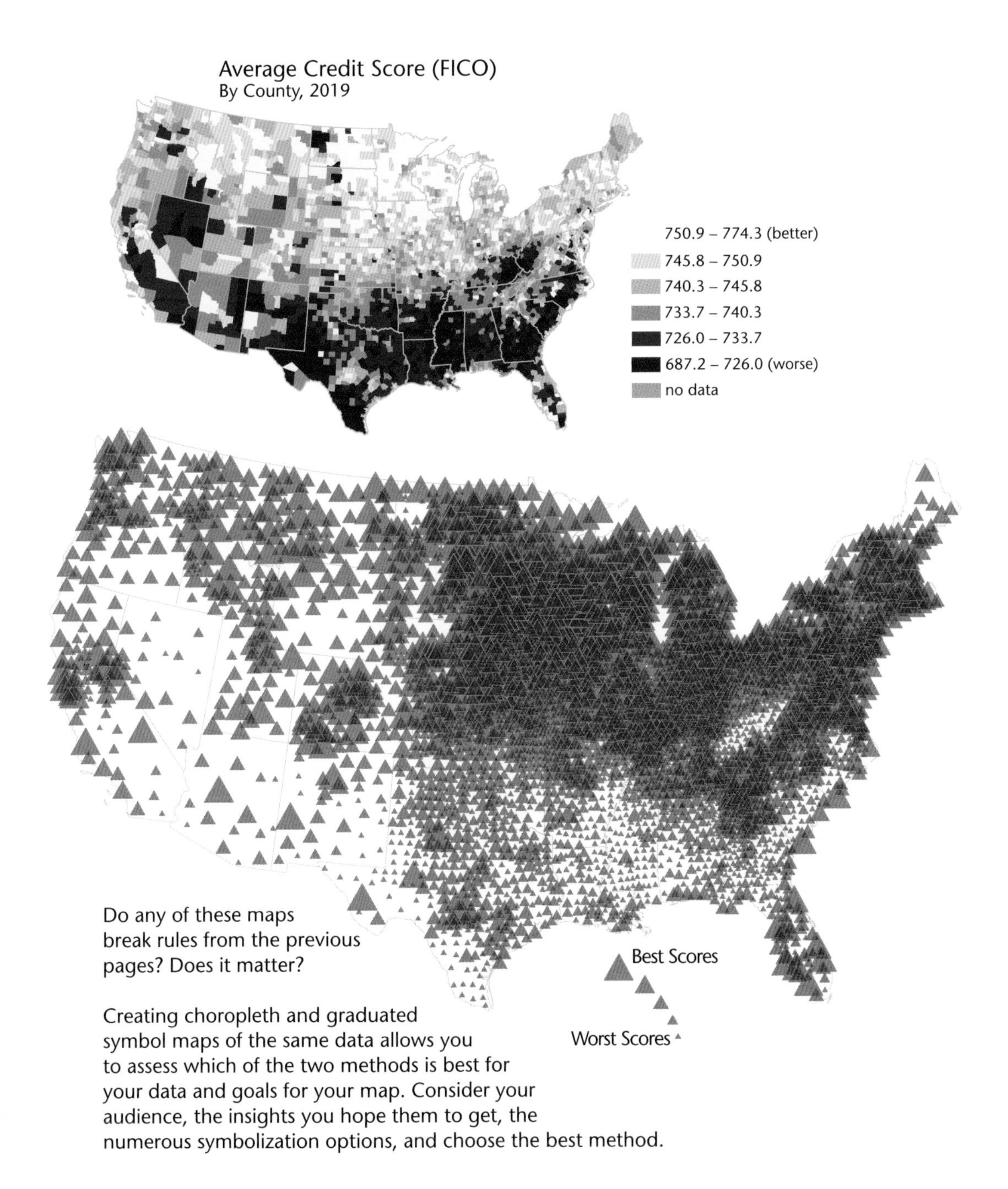

Do any of these maps break rules from the previous pages? Does it matter?

Creating choropleth and graduated symbol maps of the same data allows you to assess which of the two methods is best for your data and goals for your map. Consider your audience, the insights you hope them to get, the numerous symbolization options, and choose the best method.

Cartograms

The cartogram is a variant of the graduated symbol map. Cartograms vary the size of geographic areas (rather than symbols) based on the single data value associated with the area. Cartograms, while difficult to create, are visually striking.

Appropriate Data: Totals and Derived Data (Densities, Rates)

Derived Data: U.S. Suicide Rate

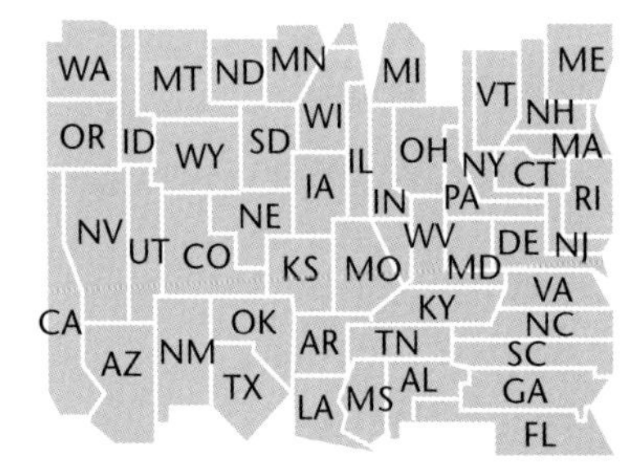

Cartograms of data with minimal variation from area to area can be interesting. Suicide rates don't vary much around the U.S., and thus all the states are about the same size.

Cartograms are not effective when the map reader is not familiar with the geographic areas being varied.

Total Data: 2020 U.S. Election: Elecotral Votes

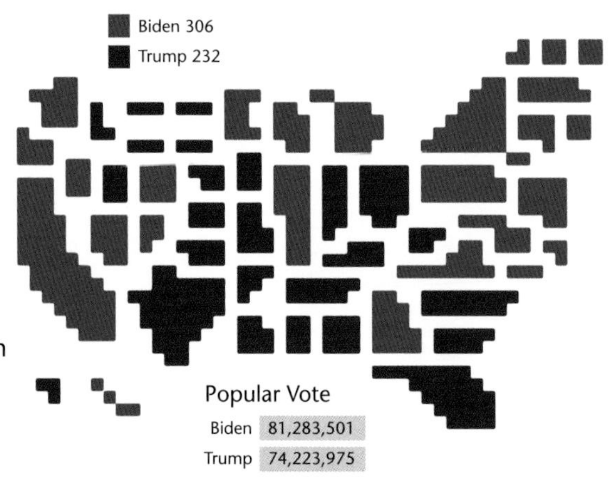

Cartograms are effective when data variation from area to area is significant. A cartogram scaled to electoral votes in each state in the U.S. removes the confusion introduced by state size.

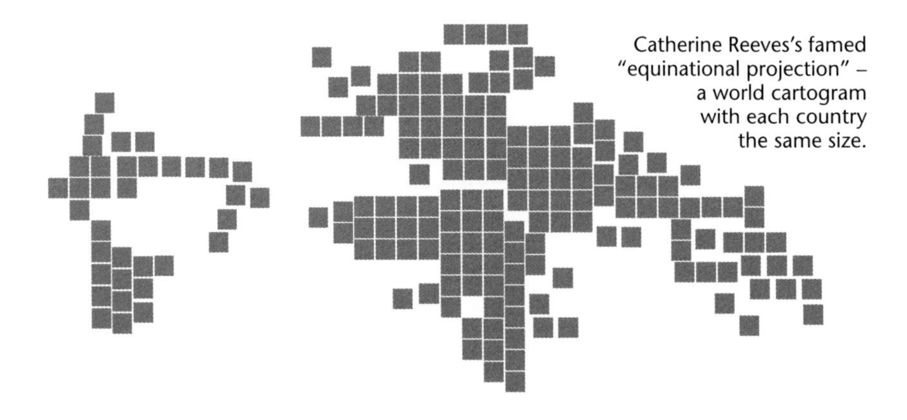

Catherine Reeves's famed "equinational projection" – a world cartogram with each country the same size.

Cartogram: Form

Apportionment Map
William Bailey, 1911

Apportionment means "allotment in proper shares." Each area's size is based on population, not geographic area. Bailey's cartogram is contiguous, roughly retaining adjacency. Compare to noncontiguous cartograms, such as the "equinational projection" (preceding page).

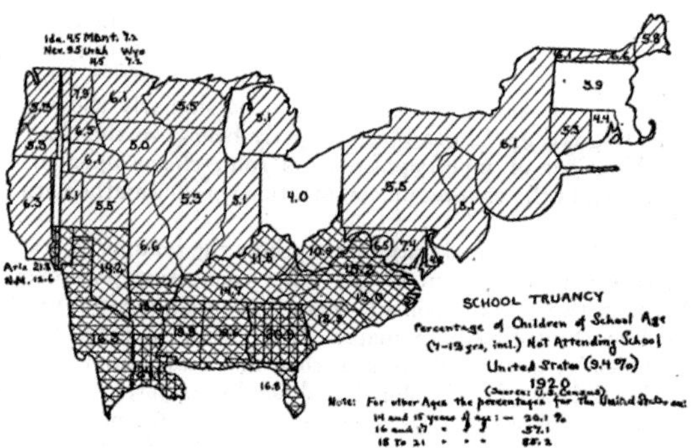

Population Projection
Karl Karsten, 1923

"The corrected areas of the States serve to give an excellent background or evaluation of the importance of the statistics plotted upon the map." Karsten suggested this map be sold, as a blank map, for compiling a second data variable, such as truancy (left), thus allowing the creation of a bivariate cartogram.

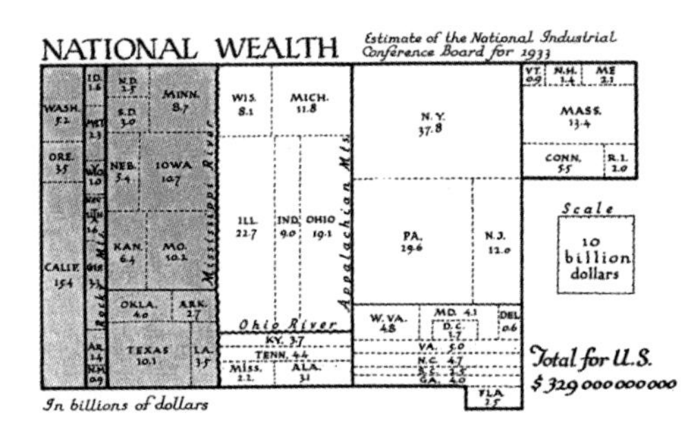

Rectangular Statistical Cartogram
Erwin Raisz, 1934

"It should be emphasized that the statistical cartogram is not a map. The cartogram is purely a geometrical design to visualize certain statistical facts and to work out certain problems of distribution." Raisz's cartogram was similar in form to early European cartograms.

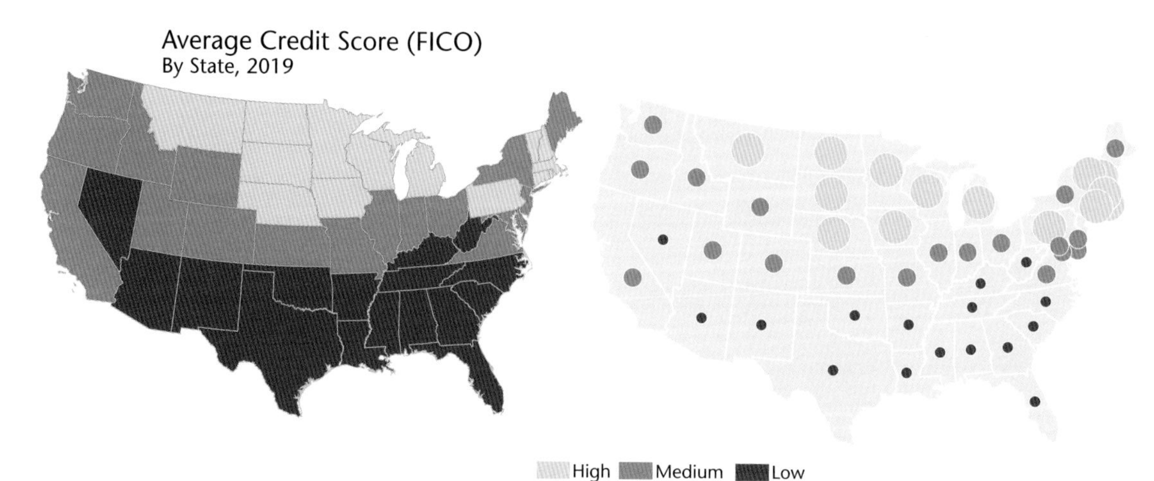

The contiguous cartogram below uses hexagonal tiles to scale each U.S. state to the same size, removing geographic area as a factor.

States are classified into the same three credit score groupings as the choropleth and graduated symbol maps above for comparison.

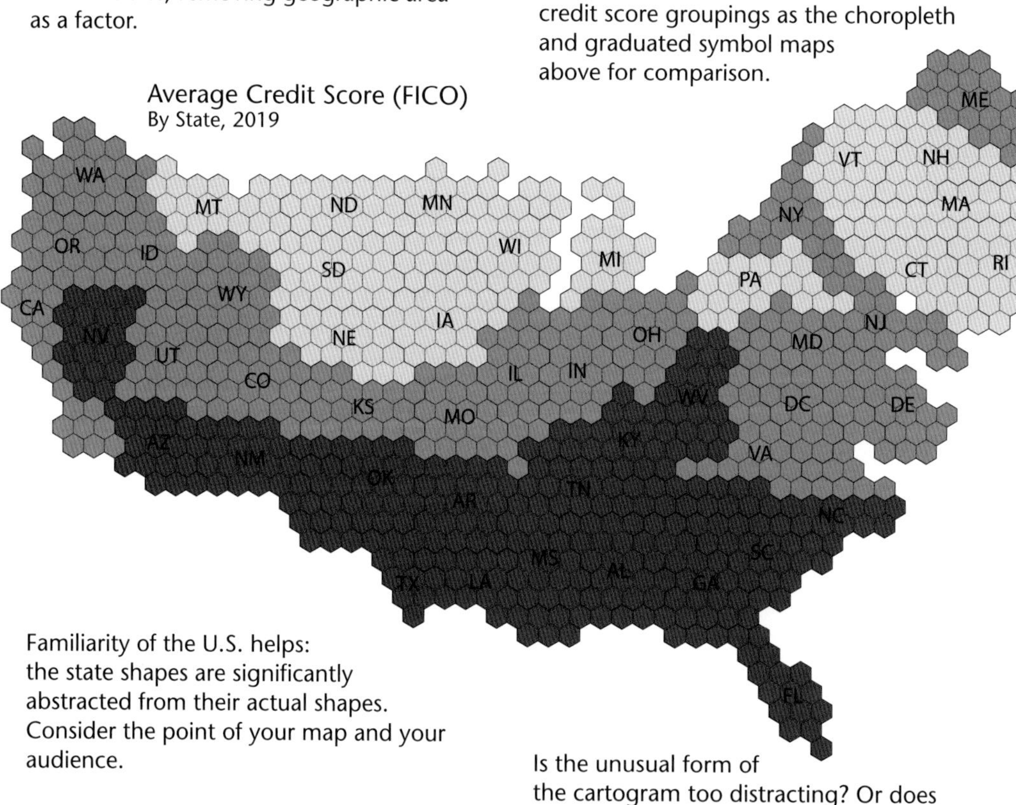

Familiarity of the U.S. helps: the state shapes are significantly abstracted from their actual shapes. Consider the point of your map and your audience.

Is the unusual form of the cartogram too distracting? Or does it more effectively show the distinct patterns of credit scores in the U.S.?

Each state on the cartogram below is scaled to the credit score class (high, medium, low). Colors are redundant, also emphasizing differences between classes.

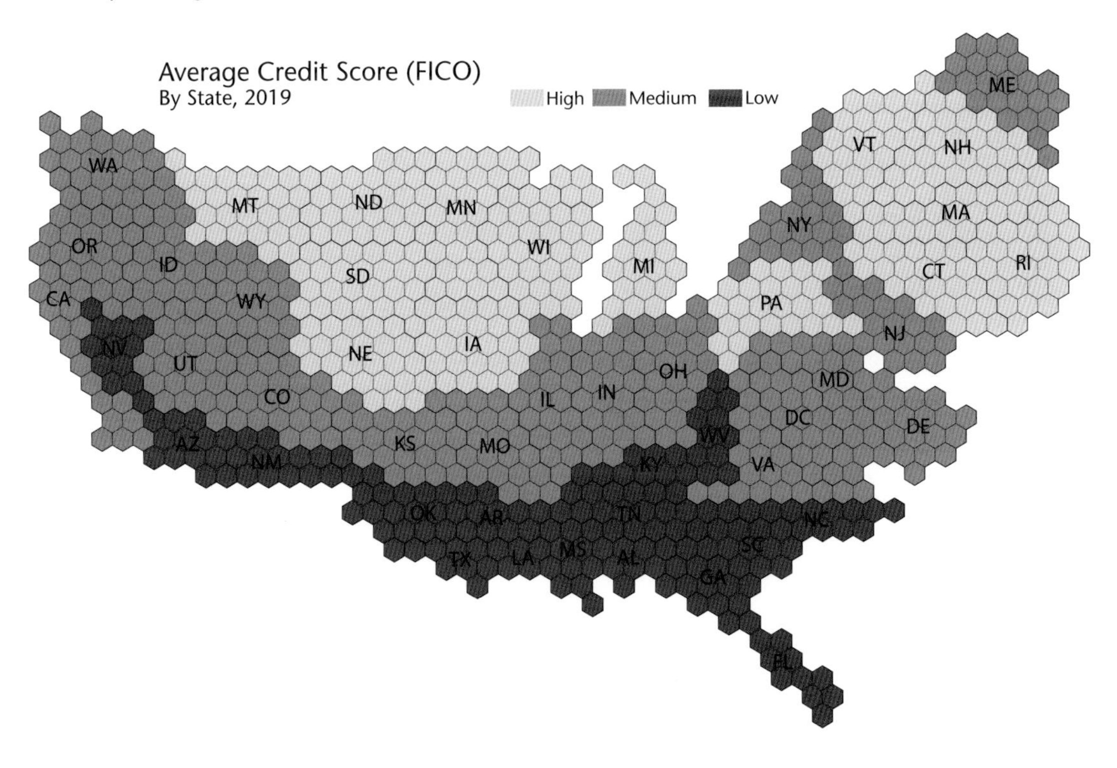

Dot Maps

The dot map varies the number of dots in each geographic area, based on the single data value associated with the area. Dots on a dot map do not represent the specific location of a single instance of some phenomena; rather, the density of dots in a geographic area represents the density of phenomena in that area.

Appropriate Data: Totals, Not Derived Data (Densities, Rates)

Using a dot map for derived data is not recommended. "One dot equals 50 people per square mile" is too weird to think about. Use totals instead. Each dot equals a number of phenomena.

Mapping software typically locates dots randomly in areas on the map. Random concentrations suggest patterns that do not exist. The larger the geographic area, the more likely randomly generated patterns will appear:

☹ Presidential Election, 2020
Data by State

1 dot = 12,000 Biden Voters

Adjust dot value and/or size so that dots begin to coalesce in densest areas and are not too sparse or absent from the least dense areas.

Use different dot values to avoid too little or too much density. Use a round number for dot value (100, not 142).

This is all subtle but important stuff.

Make your dot map using smaller geographic areas and remove the smaller area boundaries:

☺ Presidential Election, 2020
Data by County

Dot Map: Dot Value, and Size

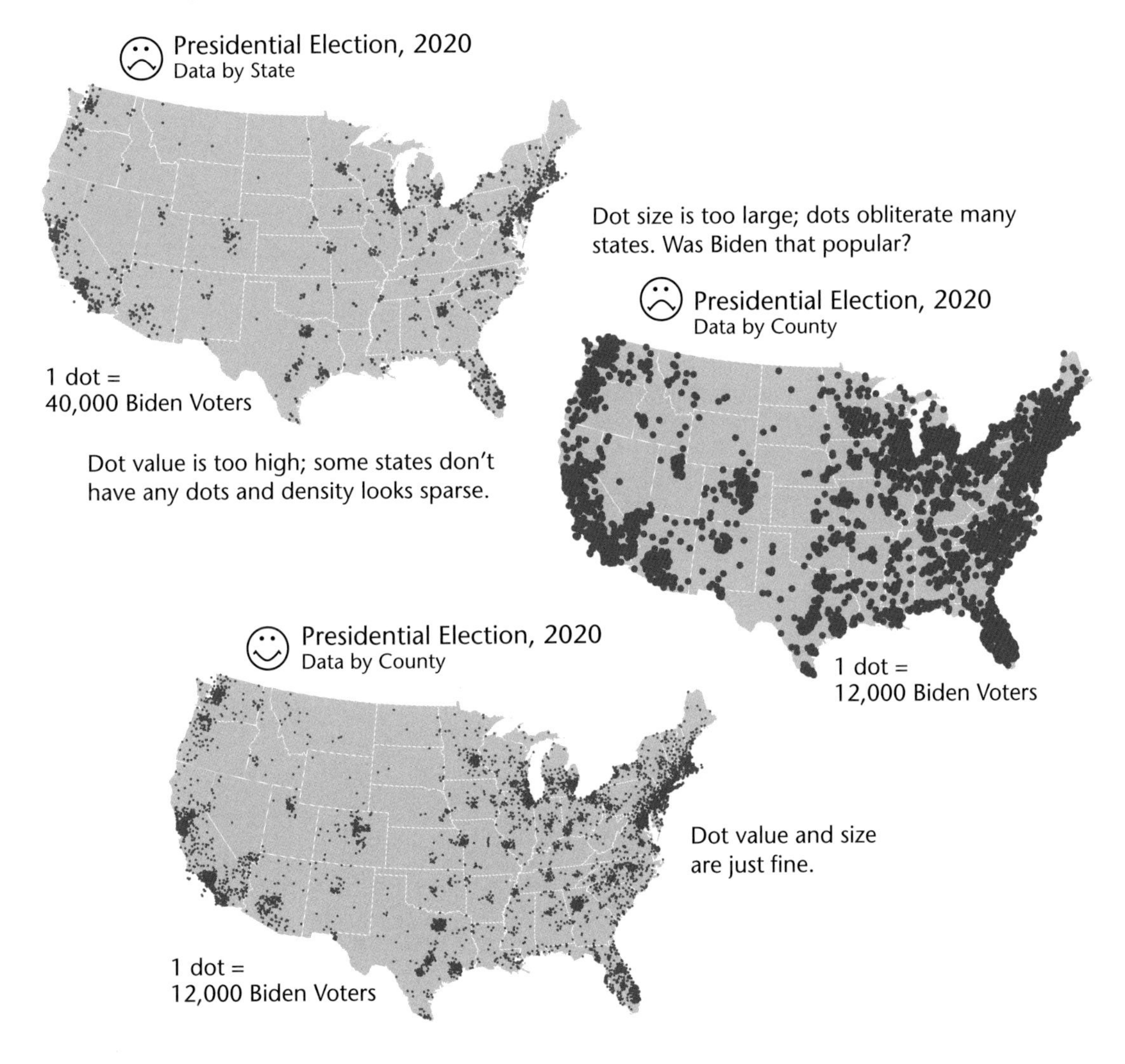

Presidential Election, 2020
Data by State
1 dot =
40,000 Biden Voters
Dot value is too high; some states don't have any dots and density looks sparse.
Dot size is too large; dots obliterate many states. Was Biden that popular?
Presidential Election, 2020
Data by County
1 dot =
12,000 Biden Voters
Presidential Election, 2020
Data by County
1 dot =
12,000 Biden Voters
Dot value and size are just fine.

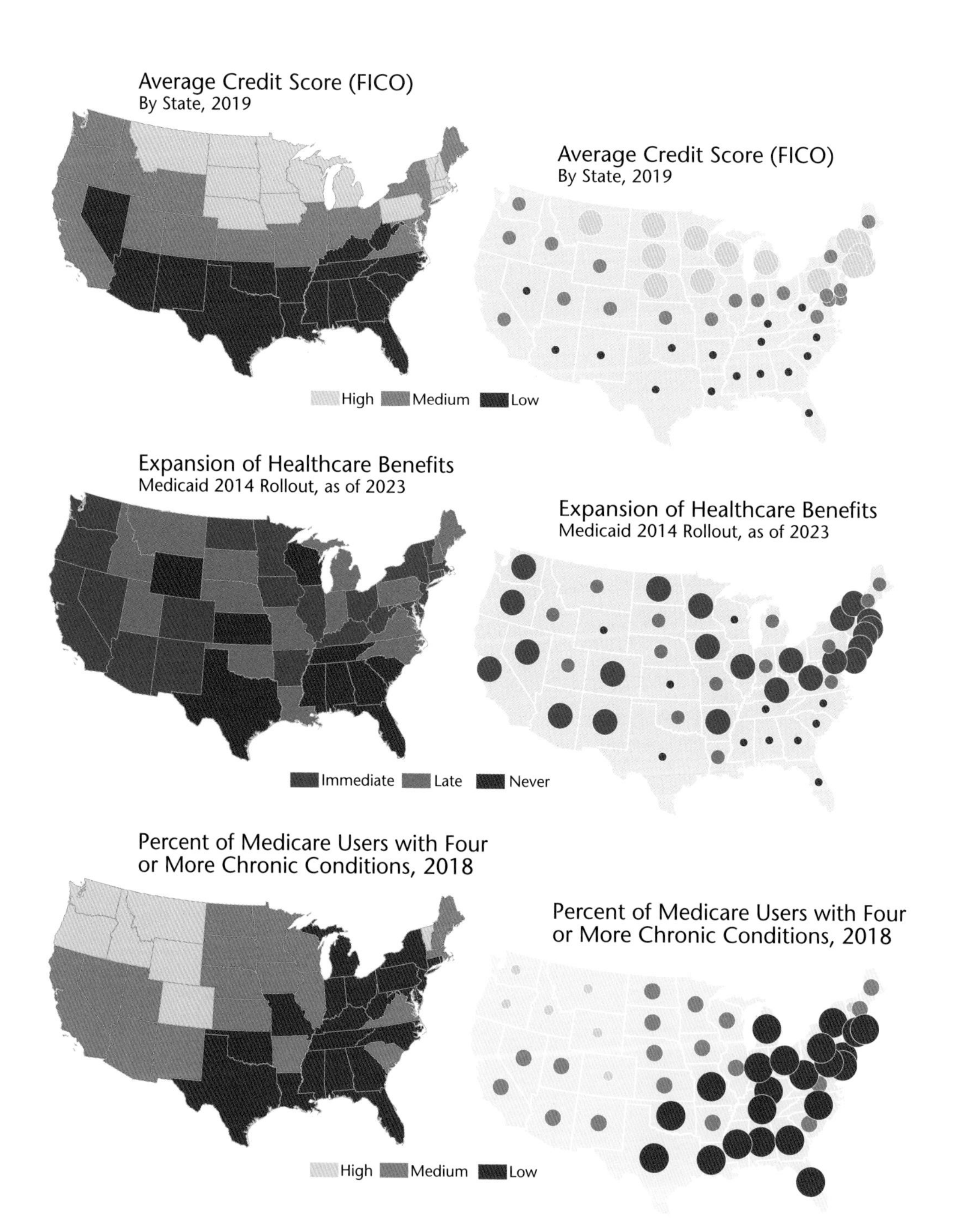
Average Credit Score (FICO)
By State, 2019
High
Medium
Low
Average Credit Score (FICO)
By State, 2019
Expansion of Healthcare Benefits
Medicaid 2014 Rollout, as of 2023
Immediate
Late
Never
Expansion of Healthcare Benefits
Medicaid 2014 Rollout, as of 2023
Percent of Medicare Users with Four or More Chronic Conditions, 2018
High
Medium
Low
Percent of Medicare Users with Four or More Chronic Conditions, 2018

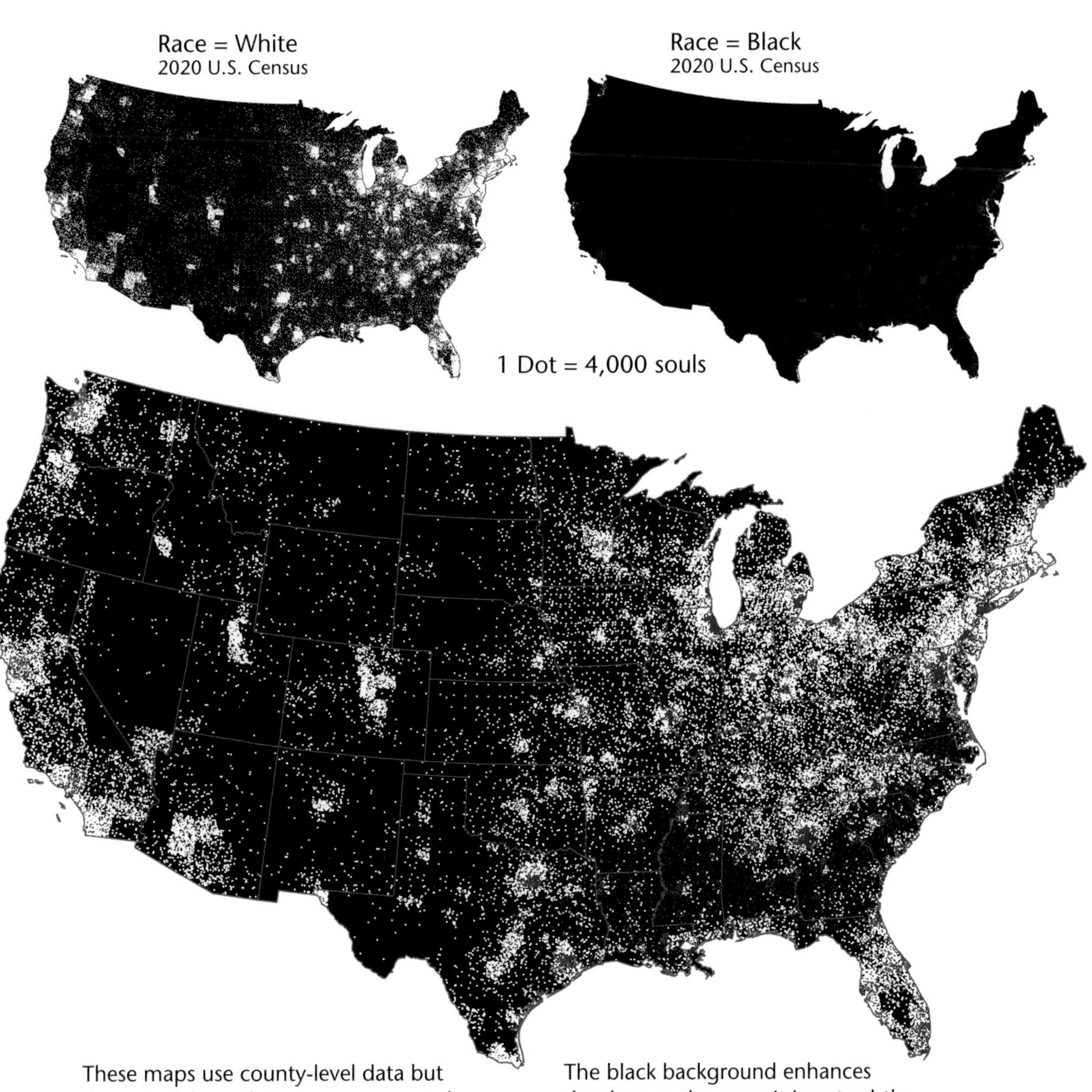

These maps use county-level data but show state boundaries, more accurately revealing the distribution of humans of two races in each state. And vast, empty spaces.

The dot value (1=4,000) and size were chosen to provide an appropriate sense of the greatly varying density of humans in the U.S. Vary dot value and size and use your brain to choose a combination that works for your data and makes your point.

The black background enhances the dots on the map. It is not subtle.

The multivariate dot map (White and Black combined) may or may not work for you

Comparing these maps of White and Black population density in the U.S. with the patterns on the credit score, Medicaid, and chronic health conditions maps to the left may provide additional insights.

Isopleth (Surface, Heat) Maps

Isopleth maps show a continuous, 3D surface created from the one data value associated with each geographic area on your map. A heat map is a broad term that includes isopleth maps and similar surface maps created from data at points.

Appropriate Data: Derived Data (Densities, Rates)

Using an isopleth map for totals is typically not recommended, particularly when the areas on the map vary in size. A large area may have more people simply because it covers a larger area.

If you map totals (below), an area with 100 people will likely be at a different level than an area with 500 people. The visual difference between the areas is the result of the unequal size of the areas.

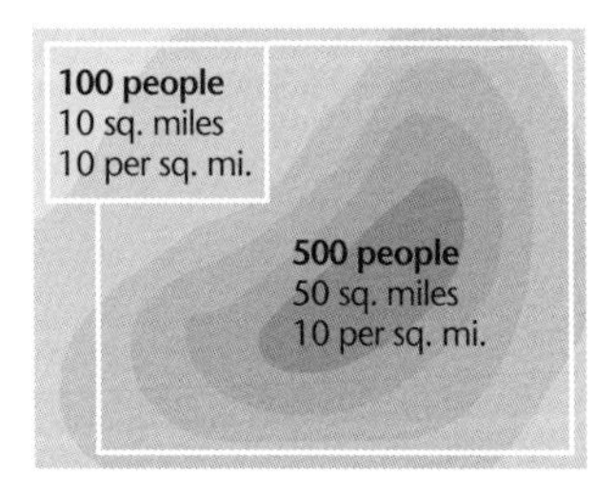

Use derived data such as densities for isopleth maps. Mapping people per square mile takes into account the varying size of areas on the map so the map user can see real differences in the distributions.

If you map densities (below), both areas have 10 people per square mile and will be at the same level. The visual difference (or lack of difference) between the areas is the result of the data.

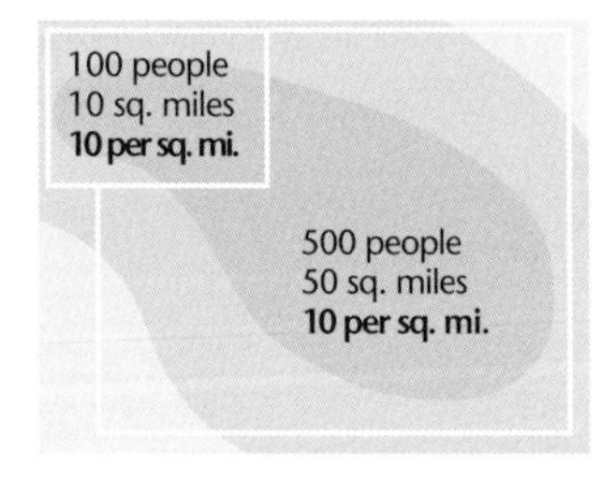

Isopleth maps are tough to make! They require **interpolation:** given the small set of points (one for each of your areas) values are estimated for all other locations. The result (estimated and "real" data) is typically classified with uniform intervals creating contours (lines of constant value). Areas between contours can be filled with shades to suggest high and low areas of the surface.

Isopleth Map: Contours and Shading

Contour lines show constant values slicing through the abstract 3D surface. Contour lines alone may produce a busy map with little room for other data. They may also be misinterpreted as a tangible linear feature.

Filled contours are less busy than contour lines and allow other data to be superimposed. They are more suggestive of a surface. You can keep the contour lines, or leave them off when using filled contours

Percent of Medicare Users with Four or More Chronic Conditions, 2018

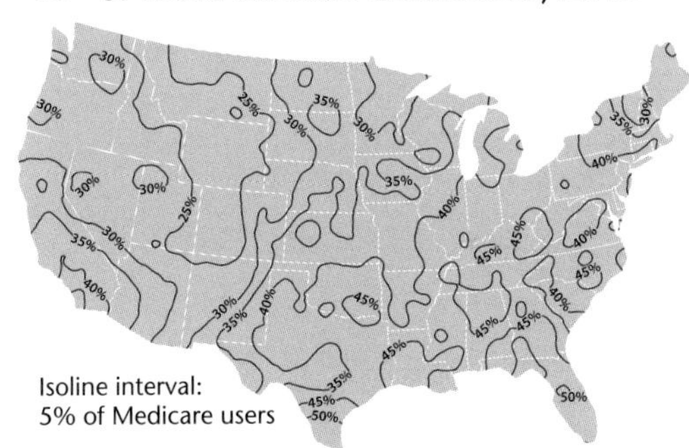

Percent of Medicare Users with Four or More Chronic Conditions, 2018

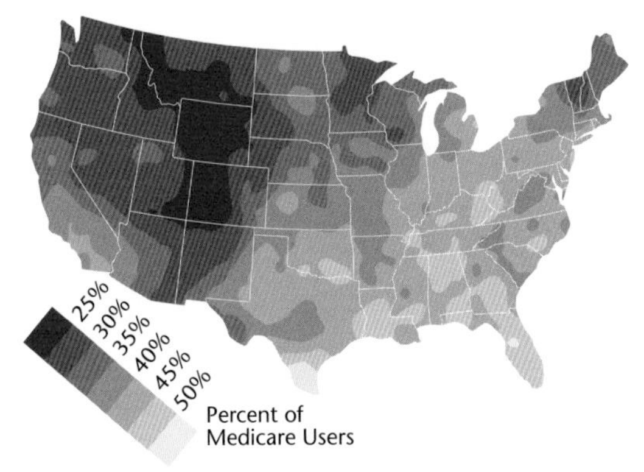

Dark is less? Nah. Unless you want to emphasize low percentages.

Percent of Medicare Users with Four or More Chronic Conditions, 2018

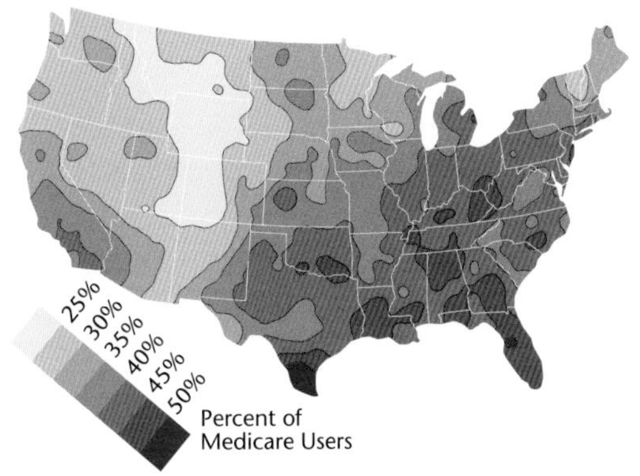

Percent of Medicare Users with Four or More Chronic Conditions, 2018

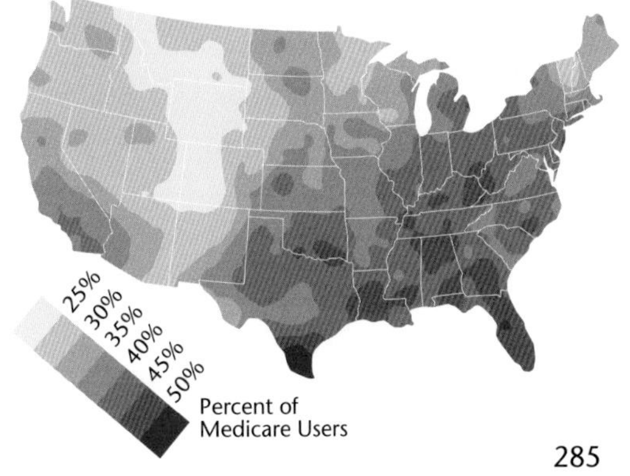

Percent of Medicare Users with Four or More Chronic Conditions, 2018 (by County)

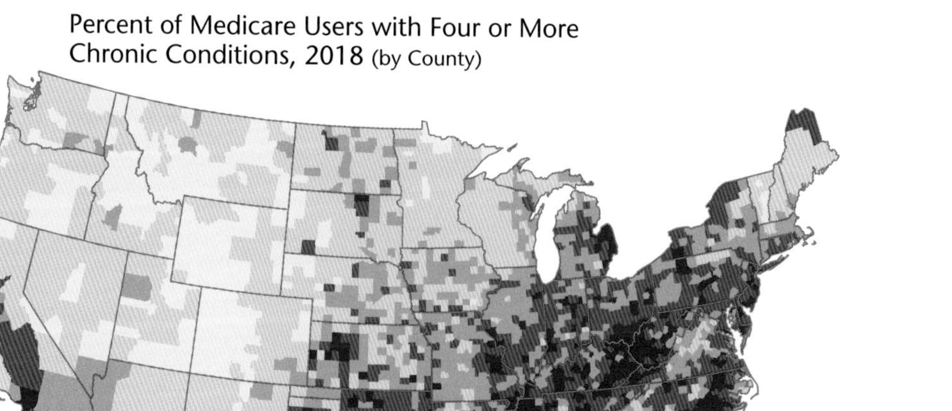

How much generalization
is too much generalization?
Too much abstraction?

A choropleth map (above) shows a single shade for each county corresponding to the data: one number for each county.

Isopleth map interpolation generates data between the one value for each county allowing a surface to be mapped. It can be less generalized (below), looking more like the choropleth map.

Percent of Medicare Users with Four or More Chronic Conditions, 2018 (by County)

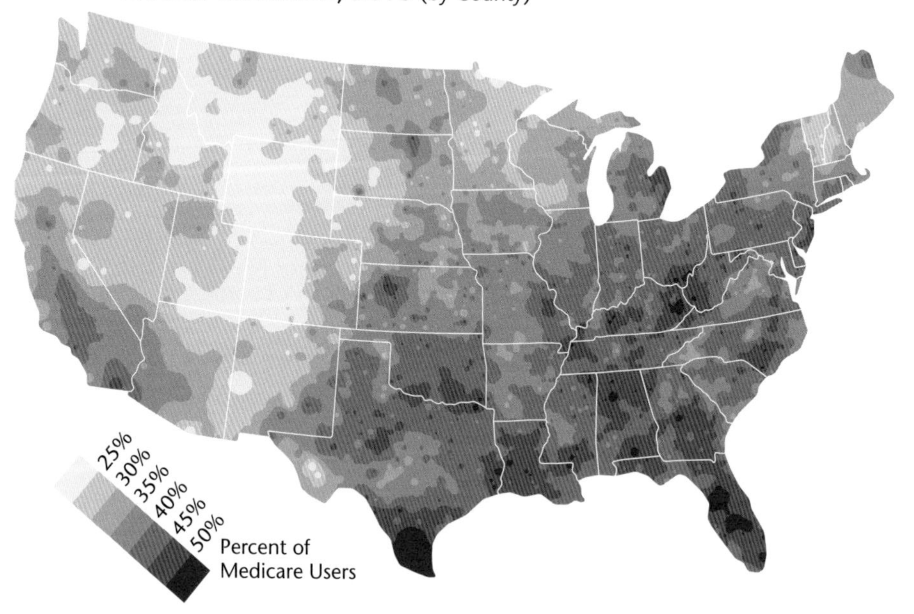

Percent of Medicare Users with Four or More Chronic Conditions, 2018 (by County)

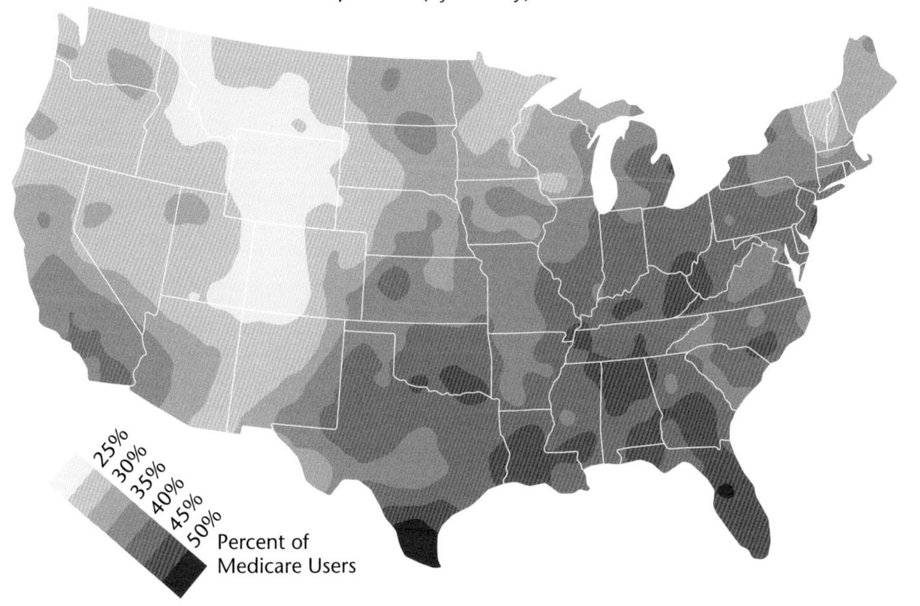

Isopleth map interpolation can be more generalized (above) and even more generalized (below).

Simpler, smoother patterns are easier to grasp; they're also more abstracted from the data and phenomena: really sick people.

Percent of Medicare Users with Four or More Chronic Conditions, 2018 (by County)

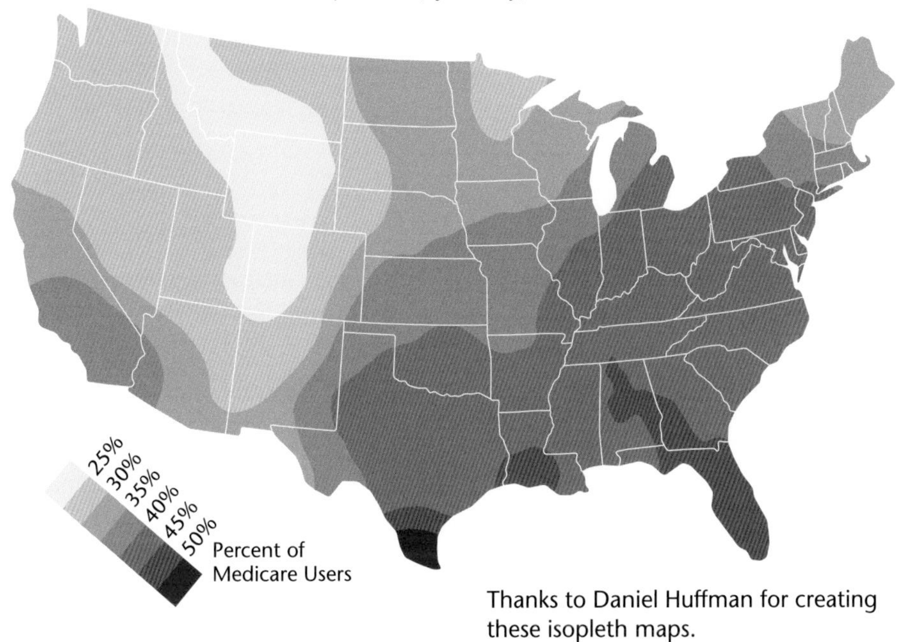

Thanks to Daniel Huffman for creating these isopleth maps.

Multivariate Maps

Visual variables are combined to help you understand your data and make your point. Multivariate map symbols show two or more kinds of data for one location, replacing two or more single variable maps. Data shown on four maps, for example, can be combined on one map using symbols called Chernoff faces, one of many multivariate map techniques. Making effective multivariate maps is really hard.

Hawaii

Average Credit Score (FICO)
By State, 2019

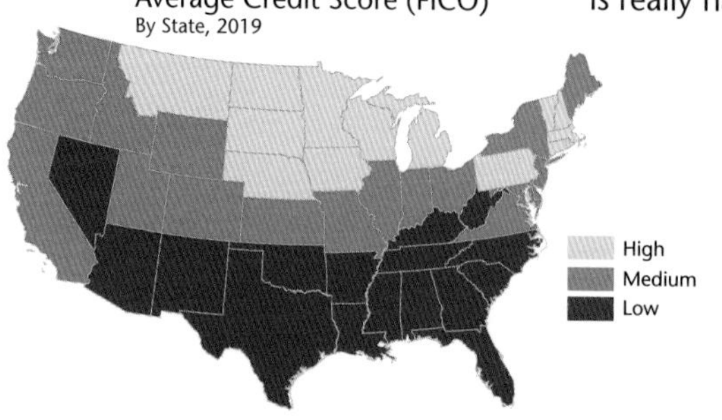

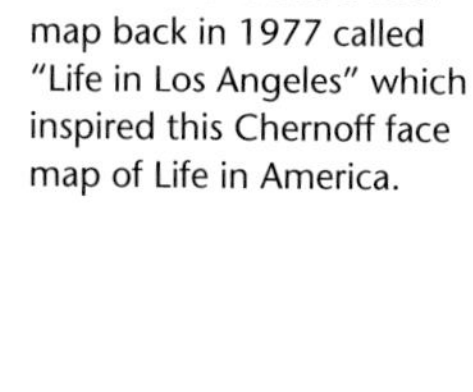

Gene Turner made a cool map back in 1977 called "Life in Los Angeles" which inspired this Chernoff face map of Life in America.

Expansion of Healthcare Benefits
Medicaid 2014 Rollout, as of 2023

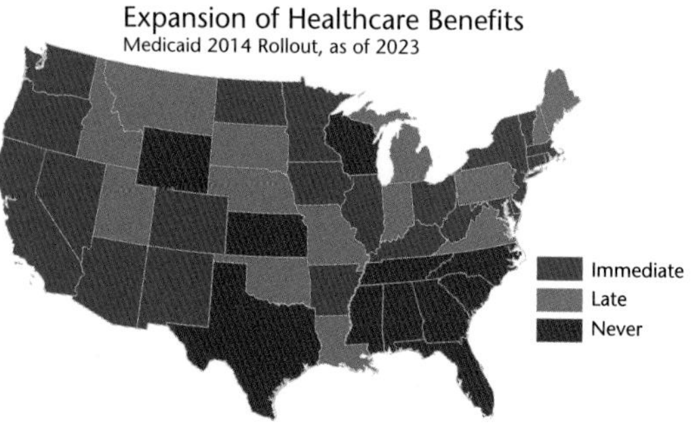

Average Credit Score	Medicare Expansion	Chronic Health Conditions	Support for Trump
High	Yes	Less	Low
Med.	Late	Med.	Med.
Low	No	More	High

Percent of Medicare Users with Four or More Chronic Conditions, 2018

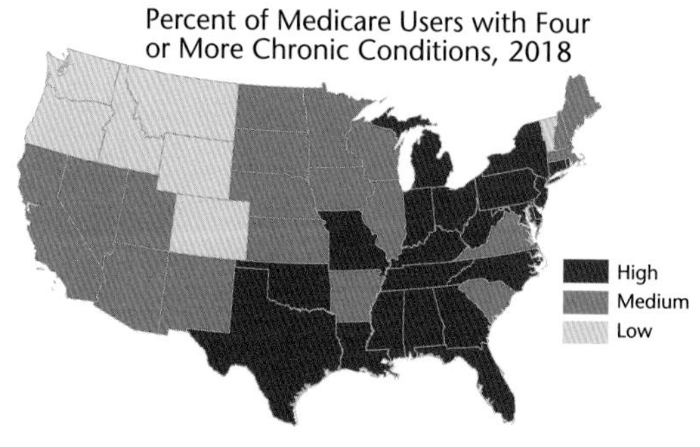

Interpreting multivariate maps is hard. It takes time. More attention. A good legend is essential.

Consider your audience. Will they have the patience for a map like this?

Or would four maps be better?

Support for Trump, 2020 Election

Life in America

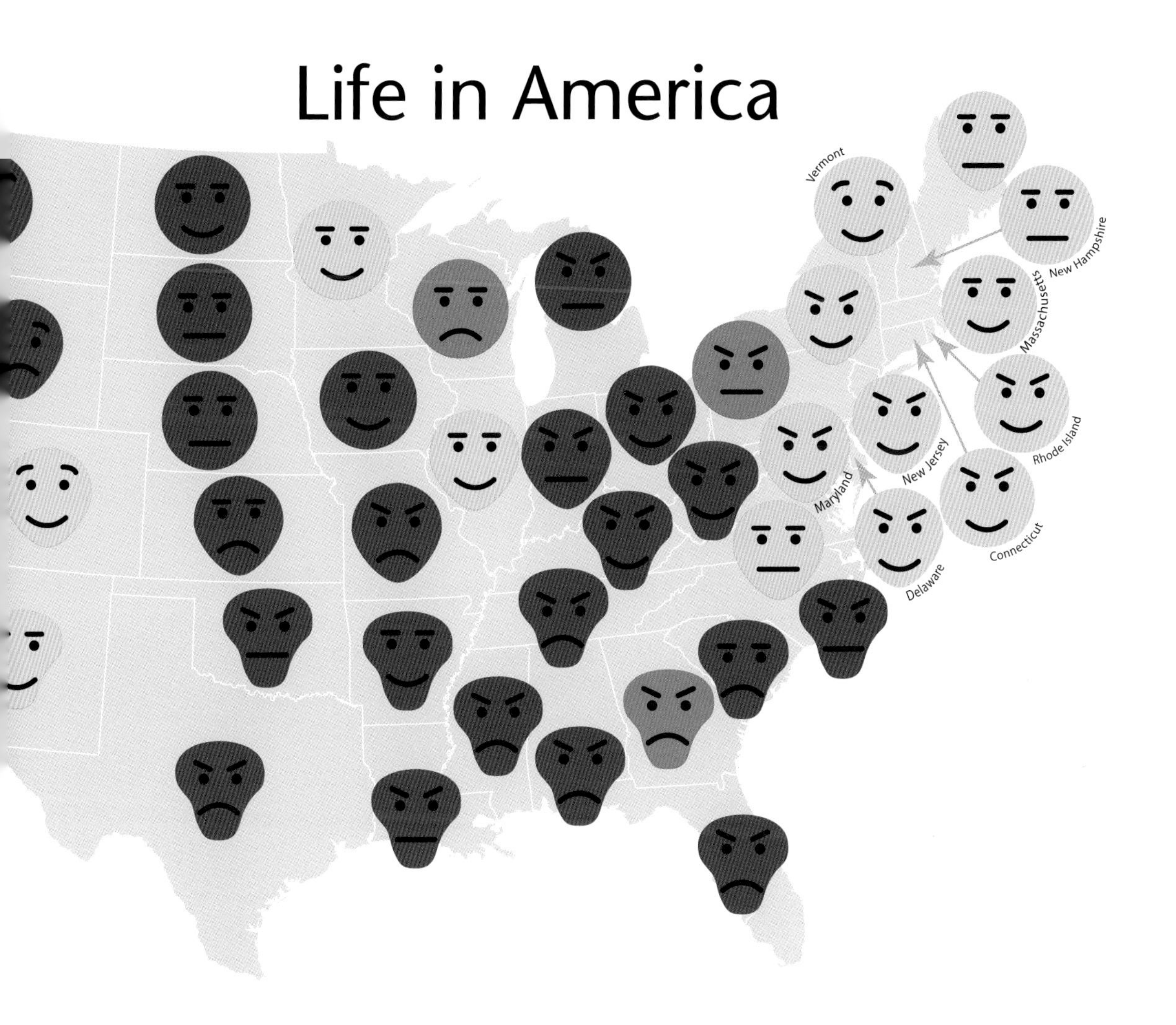

Why faces?

Chernoff faces work to show multiple variables of data because of the human capacity to interpret nuanced facial expressions - unlike numbers or complex colors.

Applying the idea of visual variables, ask if facial components correspond to your data variables. Components of a face seem to be qualitatively different (mouth, eyes) but can suggest both binary and ordered data.

What do the expressions imply?

What is the map asking you to consider?

Chernoff faces work well with up to 4 or 5 variables.

They are better for general impressions of data rather than precise details.

They require thoughtful design to work.

Cultural factors play a role in how we interpret facial expressions.

Design your map to make a point. What's bad? What's good? Faces are never objective. Your choices will strongly shape the impression people get from your map.

What do I care for the colored pins on a General's map? It's not a fair bargain – this exchange of my life for a small part of a colored pin.

Irwin Shaw, *Bury the Dead* (1936)

And then I went to bed, and went to sleep, and slept soundly, and the next morning I sent for the chief engineer of the War Department (our map-maker), and I told him to put the Philippines on the map of the United States (pointing to a large map on the wall of his office), and there they are, and there they will stay while I am President!

President William McKinley (1899)

Map-makers, rather then they will have their maps naked and bald, do periwig them with false hair, and fill up the vacuum with imaginary places...

Thomas Fuller, *History of the Holy Warre* (1639)

More...

Every single traditional cartography text devotes more than ample space to the map methods reviewed in this chapter. The visual variables are central to Jacques Bertin's rubbery-bound *Graphics and Graphic Information-Processing* (1982). Alan MacEachren did a nice job of summarizing map abstraction methods in *Some Truth with Maps: A Primer on Symbolization and Design* (1994). Erwin Raisz includes a panoply of cartograms in his books and atlases, including previously cited texts as well as his *Atlas of Global Geography* (1944), *Atlas de Cuba* (1949), and *Atlas of Florida* (1964). The modern maven of cartograms is Danny Dorling (dannydorling.org).

For early examples of many of the map symbolization methods in this chapter, see Arthur Robinson's *Early Thematic Mapping in the History of Cartography* (1982). Another neat collection is *The History of Topographical Maps: Symbols, Pictures and Surveys* by P.D.A Harvey (1980)

Sources: Gwendolyn Warren and Bill Bunge's map "Where Commuters Run over Black Children" was included in *The Detroit Geographical Expedition and Institute Fieldnotes 3* (mimeographed, 1971). Alex Hill created the maps of Detroit pedestrian fatalities at our request, using data from the Michigan Traffic Crash Facts (MTCF) database. More about Alex's work at detroitography.com. The car versus pedestrian space map is from Ronald Horvath's article "Machine Space" (*Geographical Review,* April 1974) reproduced courtesy of the American Geographical Society. The hate group data are from the Southern Poverty Law Center (splcenter.org). FICO Score data from Sumit Agarwal, Andrea Presbitero, Andre F. Silva, and Carlo Wix. "Who Pays For Your Rewards? Redistribution in the Credit Card Market," *Finance and Economics Discussion Series 2023-007.* Washington: Board of Governors of the Federal Reserve System, 2023. Expansion of healthcare benefits data from www.kff.org. Multiple Chronic Conditions data from Centers for Medicare & Medicaid Services (cms.gov). Cedar Bog map recreated from "Habitat Map: Cedar Bog Nature Preserve" by R.C. Glotzhober found at cedarbognp.org. Maps of Pennsylvania AIDS data are redrawn from maps in Alan MacEachren, *Some Truth with Maps* (Association of American Geographers, 1994). Description of bivariate maps inspired by "Bivariate Choropleth Maps: A How-to Guide" by Joshua Stephens, February 18, 2015 atjoshuastevens.net. The suicide rate cartogram is redrawn from Dent's *Cartography: Thematic Map Design* (2008). The "Equinational" cartogram, designed by Catherine Reeves, is redrawn from *Globehead! The Journal of Extreme Geography* (1994). Bailey's "apportionment map" was published on April 6, 1911, in *The Independent.* Karsten's "population projection" was published as a mimeographed sheet and also reproduced in his *Charts and Graphs* (Prentice Hall, 1925). Erwin Raisz's rectangular statistical cartogram was published in "The Rectangular Statistical Cartogram," *Geographical Review*, 24, no. 2 (April 1934). Credit score cartograms redrawn from maps created by Tilegrams (https://pitchinteractiveinc.github.io/tilegrams/). Isopleth maps created by Daniel Huffman (somethingaboutmaps.com) using QGIS software (qgis.org). The Life in America map was inspired by Gene Turner's map Life in LA (1977).

Crestview Rd.
Proposed Connector
Property Values high med. low
Crestview Rd.
Lincoln Park
MLK High School
African American Community Center
1st African Methodist Episcopal Church
% African American high med. low
Thanks to Brandon's newspaper stories, and presentations made by Jaki, Susan, and a bunch of others to the city council, the council asked the planning department to come up with some alternatives.

What do you think about these alternatives the planning department's come up with?
They're both better than the original connector, but C's going to run into a stiff fight from the homeowners there.
Proposed Pharmaceutical Development
B
A
Downtown Area
C
Proposed Residential Development

Yeah, B may be more expensive than the other two but there's very little environmental impact. I don't know anyone who objects to it.

I've heard that someone's polled the council members. They're all for it. They're really pushing this.

They don't want to lose that plant! They're voting next week.

They're never going to keep A. We need to start planning a victory celebration.

How about a 'we killed the connector' party?

We Killed the
Connector Celebration
Friday September 15
7 pm to ???
African American Community Center
175 Douglas

Amazing what you can do with maps, isn't it?
NITY
CEN
Crestview Rd.
Lincoln Park
Oldest Home in City
Historic City Hall
Historic "Shotgun" Houses
Oberlin Historic District
% Historic Buildings
high
med.
low

A Note to the Users of *Making Maps*

Making Maps is not like other map and cartography texts. It is concise, is graphic (as befits the subject), focuses on the map making process, includes specific guidelines to help you design better maps, asks you to think, and shows maps that matter in the real world, engaged with conflict, human curiosity, politics, discovery, and controversy.

Making Maps was designed for a smart general audience who want to understand and engage in map making. As such, we have provided substantive examples, left out superfluous jargon, and included guidelines that work regardless of the map making tools you are using. Resources included at the end of each chapter and on the book's web spaces (makingmaps.substack.com and makingmaps.net) will lead you to the wonderful abundance of additional information about maps and map making available in books and on the internet.

Rapid development of mapping and GIS software, web map programming and AI is bringing mapping and geographic data analysis and visualization to a rapidly growing number of humans. *Making Maps* necessarily steers clear of software content that will be out of date before the book is printed. The authors of *Making Maps* have sifted through the last 500 years of map making knowledge, winnowing it down to concepts and guidelines we believe to be relevant regardless of the tools you use or the ultimate medium of your maps.

Making Maps was also designed for use in courses on mapping, cartography, and GIS at the introductory or advanced level. It can serve as the sole text in a course or supplement another text. Given the diverse approaches to maps and mapping in courses in and outside of geography programs (where courses on maps are usually taught), we attempt to provide key concepts and good examples relevant to just about any way one might choose to teach about maps. Chapters are substantively organized to mimic the map making process, but they can be used in any order. Course instructors are expected to expand upon the content covered in *Making Maps* in lecture and/or laboratory sessions (which provides a good reason for students to show up for class) and fit the text into their vision of maps and map making. Some course resources are included at makingmaps.substack.com and more will be added. As digital mapping is diverse and constantly evolving, instructors (and students) should stay up to date with blogs, websites, journals books and conferences that document new mapping technologies and applications.

This book, like any book, reflects the personality, quirks, and intellectual interests of its authors. For better or worse, we just didn't think the world needed another boring text on something as interesting as maps.

Materials Reproduced in *Making Maps*

The authors have created almost all of the illustrations in this book. Reproduced illustrations are indicated in the Sources section at the end of each chapter. Every attempt has been made to secure the reproduction rights to non-original material in this book.

Acknowledgments

Denis Wood: I need once again to thank Christine Baukus and Irv Coats for their continuing support and John Krygier for inviting me to join him on what has turned out to be a really cool trip. Chandler has been a lot of fun to work with too!

John Krygier: Feedback from readers and reviewers, along with Denis questioning everything, has substantially contributed to this 4th edition of *Making Maps.* Ohio Wesleyan University has helped to defray costs associated with the book. A big thanks to my family: Patti, John Riley, and Annabelle.

Thanks especially to this edition's reviewers for their constructive critiques. Major changes in this edition grew from your thorough comments. Nathan Burtch (George Mason University), Melinda Shimizu (SUNY Cortland), Barbara Trapido-Luria (Arizona State University), Michael Trust (Northeastern University), and Blake Walker (Simon Fraser University). Nat Case (INCase, LLC) shared his map design and production expertise as well as philosophical thoughts and both are embedded throughout this new edition. The reviewers made this a much better book.

Book design and production by John Krygier using ArcGIS Pro, Adobe Illustrator, Photoshop, and InDesign. And coffee. The majority of this book was made at Cup O Joe Coffee in Clintonville (Columbus, OH).

About the Authors

John Krygier teaches in the Department of Environment and Sustainability at Ohio Wesleyan University, with teaching and research specializations in cartography, geographic information systems (GIS), and environmental and human geography. He has made lots of maps and published on map design, educational technology, cultural geography, multimedia in cartography, planning, the history of cartography, and participatory GIS. He has a master's degree from the University of Wisconsin, where he worked with David Woodward, and a PhD from The Pennsylvania State University, where he worked with Alan MacEachren. See krygier.owu.edu for more information.

Denis Wood holds a PhD in geography from Clark University, where he studied map making under George McCleary. He curated the award-winning *Power of Maps* exhibition for the Smithsonian and writes widely about maps. His books include *Rethinking the Power of Maps* (Guilford Press, 2010), *Everything Sings: Maps for a Narrative Atlas* (Siglio, 2013), and *Weaponizing Maps,* with Joe Bryan (Guilford Press, 2015). A former professor of design at North Carolina State University, Wood is currently an independent scholar living in Raleigh, North Carolina. See deniswood.net for more information.

Chandler Wood is an illustrator and story-board artist residing in Los Angeles, California.

Index

D

E

F

G

N

O

P

Q

R

S

T